REGIONAL DEVELOPMENT AND PLANNING FOR THE 21ST CENTURY

Regional Development and Planning for the 21st Century

New priorities, new philosophies

Edited by
ALLEN G. NOBLE
FRANK J. COSTA
ASHOK K. DUTT
ROBERT B. KENT
The University of Akron

Routledge
Taylor & Francis Group

LONDON AND NEW YORK

First published 1998 by Ashgate Publishing

Reissued 2018 by Routledge
2 Park Square, Milton Park, Abingdon, Oxon OX14 4RN
711 Third Avenue, New York, NY 10017, USA

Routledge is an imprint of the Taylor & Francis Group, an informa business

Publisher's Note
The publisher has gone to great lengths to ensure the quality of this reprint but points out that some imperfections in the original copies may be apparent.

Disclaimer
The publisher has made every effort to trace copyright holders and welcomes correspondence from those they have been unable to contact.

A Library of Congress record exists under LC control number : 98072805

ISBN 13: 978-1-138-32560-9 (hbk)
ISBN 13: 978-1-138-32563-0 (pbk)
ISBN 13: 978-0-429-45032-7 (ebk)

Contents

List of Figures

List of Tables

List of Contributors

Rais Akhtar is Professor and Head of the Department of Geography, University of Kashmir, Srinigar, India. He is a recognized authority on medical geography.

Brian Coffey is Associate Professor and Chair of Geography and Director of the Urban Studies Program at the State University of New York, Geneseo, N.Y., U.S.A.

Frank J. Costa is Professor of Geography and Planning at the University of Akron, Akron, Ohio, U.S.A. He is a former director of the University's Center for Urban Studies.

Christopher Cusack is a doctoral student in Planning at the University of Akron, Akron, Ohio, U.S.A.

Ashok K. Dutt is Professor of Geography and Planning at the University of Akron, Akron, Ohio, U.S.A. A former planner with the Calcutta Metropolitan Planning Organization, he is an Asian specialist.

Elisha Efrat is Professor of Geography at Tel Aviv University, Tel Aviv, Israel. He has written several books on Israeli planning topics.

Edgar Guardia is a Bolivian economist. A 1987-1988 Humphrey Fellow at Cornell University, he is the executive director of DESEC, a Bolivian NGO involved in rural development projects.

Animesh Halder is Deputy Director, Calcutta Metropolitan Development Authority. He has published on the urban poor of Calcutta.

Eric J. Heikkila is a member of the faculty of the School of Urban and Regional Planning, University of Southern California, Los Angeles, CA, U.S.A. He is Executive Secretary and Co-Founder of the Pacific Rim Council on Urban Development.

Shantha K. Hennayake is on the faculty of the Department of Geography, University of Peredeniya, Sri Lanka.

Bronwyn Irwin is a graduate in International Relations from the State University of New York, Geneseo, N.Y., U.S.A. She is currently pursuing graduate work in international development.

David J. Keeling is Associate Professor in the Department of Geography and Geology, Western Kentucky University, Bowling Green, KY, U.S.A. He is also Director of the Latin American Studies Program.

Robert B. Kent is Professor of Geography and Planning at the University of Akron, Akron, Ohio, U.S.A. He has been an urban and regional planner in Huancayo and Cajamarca, Peru and a consultant on urban and regional planning in Bolivia.

Thomas Klak is an Associate Professor of Geography, Miami University, Oxford, Ohio, U.S.A. He is also Director of the Miami University Program in Latin American Studies.

Chandreyee Mittra is a doctoral student in Geography at the Ohio State University, Columbus, Ohio, U.S.A.

Garth Myers is Assistant Professor of Geography, University of Kansas, Lawrence, Kansas, U.S.A.

Allen G. Noble is Distinguished Professor of Geography and Planning at the University of Akron, Akron, Ohio, U.S.A. He is a former U.S. Foreign Service Officer.

Mauri Palomaki is retired Rektor of Vaasa University, Vaasa, Finland. He has written widely on development topics.

Bernard L. Panditharatne is retired Vice-Chancellor of the University of Peredeniya, Sri Lanka. He is a specialist on urban geography.

George M. Pomeroy is a doctoral student in Planning at the University of Akron, Akron, Ohio, U.S.A. He is chairman of the Regional Development and Planning Specialty Group of the Association of American Geographers.

Victor R. Savage is Deputy Head, Department of Geography, National University of Singapore.

Joseph L. Scarpaci is a Professor in the Department of Urban Affairs and Planning, Virginia Tech, Blacksburg, VA, U.S.A. He has authored books on privatization, health care policy and delivery in Latin America.

Olav K. Sibille, a geographer and urban planner, is a member of the Center for Urban Studies, Lima Geographical Society. He worked in Cuzco, Peru, on the Project for Strengthening Local Institutions (COINCIDE), funded by the Inter-American Foundation.

Betty E. Smith is Assistant Professor of Geography at Eastern Illinois University, Charleston, Illinois, U.S.A. She has conducted research in Ecuador, Bolivia and Argentina on urban topics and regional land use.

Bruce Taylor is Quality Assurance Coordinator at The Open University of Hong Kong. An urban planner, he has taught planning courses at the University of Macau and the Chinese University of Hong Kong.

Theresa L. Urban is Internet Developer for INDUS Corporation. She was a Fulbright Scholar at the Universidad Pedagógica Nacional de Francisco Morazan, Tegucigalpa, Honduras in 1995.

Coenrad van der Wal is a former planner with the IJsselmeerpolders Development Authority. Among other tasks he worked on the planning of the Almere settlement.

Vandana Wadhwa is a doctoral student in Planning at the University of Akron, Akron, Ohio, U.S.A.

Patrick H. Wirtz is a doctoral candidate in Urban Planning and Development at the University of Southern California, Los Angeles.

Acknowledgements

Editing a volume such as this one, with authors from ten different countries, is a daunting task, made at all possible only by the cooperation of the chapter authors and by the friendly cooperation, timely assistance and unflagging devotion of numerous staff members. All have made valuable contributions to this volume. Traci Baker, Christina Dalrymple, Julie Perez, Melissa Rambler, Sharon Mauer, David Pickens, Monica Idley and Dorothy Tudanca typed numerous revisions of chapter texts as the book evolved. Joseph Stoll and Claudia James of the University of Akron, Department of Geography and Planning provided technical and cartographic assistance and Claudia supervised the final production of the camera-ready copy. Deborah Sauder undertook the task of producing the manuscript in its final form. Her willing cooperation, excellent judgement and unfailing good humor have lightened the editors' burden. Without the good efforts of all these people, and especially the contributions of Ms. James and Ms. Sauder, the work could not have been completed. We also thank Professor Charles B. Monroe, who as Head of the Department of Geography and Planning, University of Akron supported the research and facilitated the production of the volume. All have the editors' gratitude.

1 Currents of Change: Urban Planning and Regional Development

FRANK J. COSTA, ALLEN G. NOBLE, ASHOK K. DUTT, ROBERT B. KENT

Large scale, highly centralized governments are under assault throughout the world. As President Clinton declared in his 1996 State of the Union address, "the era of the big government is over". The impact of this shift away from centralization of the various functions of government is substantial. At the same time, growing population numbers place enormous pressure on existing facilities and the environment. Both these issues are addressed by chapters in this volume.

In this book we are concerned with the changing nature of regional development and planning activities, along with the problems of government generated by these activities. We also are concerned with how these are affected by decentralization pressures in various parts of the world. Pressures are reflected in a series of political, economic, environmental, and network adjustments, as well as the growth of problems emanating from rapid urban growth, housing shortages, and diseases.

Centralization and Recent Planning

For much of the post World War II period, the organizational structure for regional development and planning demonstrated six characteristics. First, large centralized bureaucracies controlled most, or all, political and economic resources. Marxist-Leninist ideology invested the state with supreme power. Other levels of government were purely instrumental and possessed no intrinsic authority. Their function was to carry out centrally-determined policy. The entire Socialist Bloc and much of the developing world adhered to this perspective.

Second, a pervasive element of centralization was the development and operation of centralized planning institutions, whose functions included the determination of the economic factors of production, including the allocation of raw materials and labor to the production of goods and services. A further characteristic of these planning institutions was their central role in the development of multi-year plans covering all aspects of national life.

Third, since economic centralization and political centralization are often parallel, many nations practicing centralized planning were, and are, authoritarian. Little or no role exists for public participation and ratification in the decision-making process. The specific concerns of the localities or regions adversely affected by central decision-making usually went unheeded. A classic example of centralized force overwhelming local dissent is the decision to proceed with the construction of the Three Gorges Dam on the Yangtze River in China (Spence, 1997).

Fourth, the primary theoretical foundation for centralized planning is an approach which can be termed *comprehensive rationality*. The premise of this approach is that a central unit is capable of obtaining comprehensive knowledge about the conditions of national life including economic conditions, social conditions, and all other significant elements making up the fabric of society. Once obtained this knowledge becomes the basis for rational decision-making by the central unit which now knows best, because of its comprehensive knowledge base, about how to proceed along the path of national development.

Fifth, an obvious outcome of political and economic centralization and its consequent political authoritarianism, is the frequently forced replacement of local or regional administrative and planning practices by those which are nationally-imposed. Thus, for instance, in many Latin American nations and elsewhere guidelines for town plans, municipal administration, and municipal accounting are mandated from central government agencies.

Sixth, in many nations, localized or regional ethnic and religious identity has longstanding historic roots. The multi-ethnic character of Switzerland has been enshrined in its political institutions. Centralized states have frequently tried to eliminate local or regional ethnic and religious identities. These have been viewed as "relics" of a less-developed, even feudal, past which would have to give way to the historically more progressive national state with its rationally-based cooperative institutions and lifestyle focusing on the nation.

Acculturation, to a single set of national norms or standards including, at times, language, has been a feature of 20th century life in much of the world. The banning of Catalan in Spain during the Franco era or initiatives to use English as the official language of the United States are examples of imposing centralized or national cultural norms.

Thus in the era of centralized planning, local and regional planning initiatives were imposed from the top. They were devised and implemented within the context of national goals and were, essentially, instruments for the realization of national planning initiatives. These patterns of central control were usually somewhat less pervasive in Western democracies.

Changes in the Centralized Structure

The strong edifice of centralized power has, in recent years, come under unremitting assault. What were considered accepted dogma and historically sanctioned institutions have been vilified as historical anachronisms subscribing to outdated theories of political and economic organization. Several factors have contributed to this "reversal of fortune" for centralized bureaucracies.

One of the most significant has been the collapse of the Socialist Bloc. Bipolar political and ideological division characterized most of the post-war period. With the collapse of the Socialist Bloc, centralized political and economic power and ideology fell into disrepute.

Capitalism flourishes most where government controls are least. Thus the end of bi-polar ideological conflict has strengthened the "historical correctness" of reduced authority for centralized government and economic planning. Coupled with the collapse of the Socialist Bloc has been a resurgence of conservative political power in the Western democracies, which took the form of reducing the power of central government and of central guarantees for social welfare. The European welfare state has undergone significant "down-sizing" during the 1980s and 1990s, as has the entire social welfare safety net in the United States.

The resurgence of what had been considered feudal and nearly extinct allegiances to local or regional ethnic and religious identities undermines the assumptions about the eventual triumph of national over regional norms and identity. The maps of central and eastern Europe, as well as central Asia, have been redrawn several times in recent years to take account of the large number of new states created out of the old multi-ethnic and centralized states including the former Soviet Union, Yugoslavia, and Czechoslovakia.

Even in long-established nations, regional identities are being reasserted and, in some cases, threaten national unity. Spain has had to contend with violent Basque and Catalan separatist movements, as well as demands for increasing regional autonomy from Galicia and Valencia. Italy currently faces the prospect of a rising secessionist movement in its wealthy northern provinces. Belgium has in recent years recreated itself as a loose federation of

Dutch and French speaking regions which display little enthusiasm for continued association. Even the United Kingdom has experienced strong devolution pressures in Scotland, and more recently, in Wales. In North America, a strong separatist movement remains active in Canada, where French speakers in Quebec have been agitating for greater political autonomy.

This erosion of national identity often assumes a violent aspect. Witness the turmoil and violence spawned in Northern Ireland or the states of the former Yugoslavia. Potential challenges to national authority have surfaced elsewhere as evidenced by the growth of the local white militia movement in the United States. In this book, analyses of the impact of ethnic nationalism are given for Israel, where the Palestinian minority is agitating for nationhood, and Sri Lanka, where the Tamils in the northeast have been fighting to establish their own state.

Political decentralization is now an established pattern throughout the world. National coordination of economic policy has been reduced as a result of political decentralization and the retreat from the welfare provisions associated with the most advanced industrial democracies. Even the United States, where government has not had as strong a presence in social welfare as in other industrial democracies, is witnessing the diminution of the federal role in almost all aspects of social policy. This is what President Clinton meant when he described our era as one of limited and diminishing government.

In this new social and economic environment, governments are seeking to reduce the provision of public services by the increasing imposition of user fees. Hence, public services or amenities once provided under the umbrella of general tax revenues, no longer are provided or subsidized in this fashion. Thus, access to National Forest lands in the United States, once free, now requires payment of day use fees and parking fees. Toll roads, essentially unknown in Latin America 30 years ago, are an increasingly common feature of the landscape in Argentina, Peru, Bolivia and Mexico.

However, while these strong nationally-based movements toward devolution move forward, an internationally-based movement toward economic centralization is underway. A form of capitalistic globalization is occurring which owes its growing strength to a number of developments. Included are the growth of multi-national corporations, which function most effectively in an era of eroding national power, and increasing local or regional autonomy since these conditions favor freer trade and fewer political barriers to trade. A parallel and equally significant development is the growth of multi-national trading blocs or common markets. The earliest and most developed of these is the European Common Market, originally established in 1957 and now known as the European Union. The name change is indicative of an expansion in the authority for the multi-national unit beyond purely economic into political and social concerns. This in turn, results in a further erosion of national sovereignty. Within such a developing setting, cities and regions are able to seek out international opportunities for trade and commerce with fewer restraints from national authorities. A situation is developing in which entrepreneurial cities and regions will be able to function in a semi-autonomous manner similar in many respects to the medieval "city-state" (Jacobs 1984).

Emerging Organizational and Theoretical Structures

The single most ubiquitous manifestation of change in the organizational structure for planning is the political devolution process underway in one form or another throughout the world. Pressures for local decision-making are being exerted upon central authorities in both the developing and developed world. "Out of 75 developing and transitional countries with a population greater than five million, all but twelve claim to be embarking on some form of transfer of political power to units of local government" (Kingsley 1996, 419). Decentralization pressures have even reached Nepal, an isolated country with a long history of autocratic rule, where formal delegation of planning authority from the central government to the district and sub-district levels recently occurred (Bienin, et al 1990, 66).

In the developed world equally strong pressures for decentralization decision-making are being felt. Several of the highly centralized nation-states of Western Europe are experiencing a power shift from national capitals to state and provincial capitals. "Regionalism, whether within or across national borders, is Europe's current and future dynamic" and much of this dynamic is being promoted by regional centers as a protest, in part, "against the authority of national capitals" (Newhouse 1997, 68). The instances are many and include Milan in Italy and Barcelona in Spain, both of which see themselves as the true centers of power in their respective regions of Lombardy and Catalonia, and which are also the wealthiest regions in each country. Even France, which Newhouse characterizes as the world's "consummate nation-state", decentralization pressures are emerging, especially in the large and economically vital region of Rhone-Alpes where Lyon, the capital, seeks to substitute its authority for that of Paris (Newhouse 1997, 78).

What are the underlying causes and the intellectual arguments undergirding decentralization? Two factors seem to stand out from the rest. These are the increasing globalization of capital mentioned earlier and an apparently universal dissatisfaction with unresponsive central bureaucracies. Loss of control by central authorities over economic policy within the context of the nation-state is one outcome of the globalization of capital; "the powerlessness of politicians has contributed to a profound malaise" within national governments and the "competitive demands of the global market place" have reduced national leaders to reactive roles. "Supranational forces" tell them "that they no longer control their destiny" (Cohen 1996, Section 4: 1). Thus, nation-states are witnessing the erosion of their primacy in economic policy. Within such a setting, regions or decentralized units of government are seizing opportunities to create their own economic destiny. "Many and probably most of the wealthiest provinces of Western Europe are interacting with one another and together creating super-regions-large economic zones that transcend national boundaries" (Newhouse 1997, 69).

Yet some caveats can be noted. Some central government agencies command great respect in the new global environment. When the U.S. Federal Reserve Bank or the German Bundesbank speak, markets listen and respond throughout the world. Indeed, some might also argue that the preference for supposedly responsive regional governments by multi-national companies is driven largely by the capacity of such firms to bully or control smaller government entities with greater ease than with national governments.

Thus in summary, two seemingly divergent yet related economic forces are reducing the importance of national borders in economic issues. One of these is the globalization of capital and the emergence of a global market which acts to reduce the ability of nation-states to maintain economic policy in conflict with global economic realities. Many nations have been forced to scale back on welfare provisions for their citizens, and labor unions have been unable to protect wages and benefits for workers. The second economic force is a resurgence of regional, and, in some cases, local autonomy in economic matters including foreign trade. Decentralized units are freer to pursue cooperative economic arrangements beyond national borders. Successful regions and cities, have always sought to establish economic and political ties to cities and regions beyond their immediate geographic confines (Gottman 1991, 277). Today's economically successful cities are part of a global network which reaches beyond the borders of their nations.

Alongside this economic transition is a parallel and related political one that also fosters decentralization. Kingsley (1996, 419) refers to the growing strength of political decentralization in the developing world as a reaction in large part to the "rigid, unresponsive, inefficient, and often corrupt" performance of most central governments in these countries. What is required are locally-derived solutions to meet local problems. Policies imposed by central authorities often fail. For example, the Kampong Improvement Programme of Indonesia was in jeopardy until it was reorganized to involve local residents in planning decision-making (Choguill 1994, 943-944).

An idea has taken hold which affirms that local solutions to local problems are far more effective than solutions imposed from outside the local area. This idea can be seen in the resurgent regionalism of Western Europe, as well as in the power shift from the national to the state governments in the United States. At the more local level, individual cities are affirming their right to political autonomy and economic outreach in what some call a new era of the "city-state". Even within cities, forces for devolution are emerging. Kingsley (1996, 424) describes a proposal for devolution of some political authority to community groups as a means to revitalize poor neighborhoods in American cities. These groups should be "grass roots" organizations which would adopt a comprehensive approach to the redevelopment of their areas through a highly participatory form of neighborhood government.

This approach reflects a substantial shift in the theoretical paradigms underlying planning practice. Whereas earlier the theoretical foundation was rational comprehensiveness which supported the centralized structure for planning, today the developing theoretical bases for planning stress small-scale political communities, local solutions to local problems, participatory decision-making and democratic governance. However, while national level governments have often asserted a willingness to devolve power to local and regional governments, the chasm between rhetoric and reality can be vast. Latin American governments, for instance, have proclaimed support for political and administrative decentralization, but in practice progress has been very limited.

These concepts are not new. They have motivated political thinking since the time of Plato and Aristotle, both of whom preferred the small-scale political community or "polis" to larger and, they reasoned, more autocratic kingdoms or empires. In the 19th century Peter Kropotkin believed that new technological breakthroughs, especially the electric power grid, allowed the deconcentration of people from industrial cities into small, semi-autonomous, but technologically connected communities (Kropotkin 1975).

In the 20th century, Schumacher (1973) has written about the need to free local economies, especially those in the developing world, from the centralizing tendencies of national bureaucracies which favored policies geared toward increases in manufacturing output over the need to protect and enhance local employment. The more radical thinkers of today advocate a far-reaching devolution to loose networks or confederations of independent regions and cities. Prominent among them is Kirkpatrick Sale, an advocate for "bioregionalism" which is a highly anti-hierarchical movement that stresses the importance of local autonomy and decentralization. Sale and other proponents of this approach call for the creation of a network of autonomous, but linked "bioregions whose rough boundaries are determined by natural characteristics rather than human dictates" (Sale 1985, 55). Bookchin (1995) advocates political autonomy for cities and the creation of confederations of independent city-states which will eventually replace the nation-state through a gradual process of transfer of authority from central governments to local governments.

Friedmann (1985, 160) sums up the current argument for small scale political communities by asserting that these satisfy a deeply felt need for self-determination and that they can accommodate with greater facility the need for "fine-grained planning of ecological balances and the built environment" which, in turn, will help to sustain the social life of a community "that requires a balanced natural environment". Finally, according to Friedmann, the small scale environment can respond more quickly and effectively to natural and cultural change.

The Growing Importance of Urbanization

The city had its origins in the dim recesses of eastern Mediterranean history. Over the years the urban place has continued to evolve until today it represents the dominant environment for a significant portion of humanity. The future promises to see urbanization become even more important. Compared to 29.3 percent in 1950 and 45.3 percent in 1995, 61.1 percent of the world's population will be living in urban areas in the year 2025, 82.5 percent of people will live in cities in the developed countries, and 53.8 percent in the less developed countries (U.N. 1995, 78-79). In terms of numbers, 2.48 billion persons will be added to the 1995 urban population by 2025. As the world in general continues to urbanize, the urbanization rate in less developed countries continues to grow faster because: a) the developed countries have already reached a saturation point in urban development, and b) the less developed countries, with the majority of their population living in rural areas, have undergone a population explosion after World War II, which also bolstered urban population. Such a differential pace of urbanization will continue at least through 2025 (Dutt et al, 1994; Dutt and Noble, 1996).

Only relatively recently has mankind recognized that careful and far sighted planning is required to protect advances which have already occurred and to ensure that progress, defined in terms of maximum well being and benefits, can be extended to the greatest number.

At the same time that social and economic returns have been achieved through urbanization, as witnessed by higher standards of living, better education, and material comforts, a significant range of dilemmas, problems and challenges to well-being have been created. Social disorganization, environmental degradation, and economic disparities have all grown in most urbanized settings of the developed world, spawning crimes of violence, chemical substance abuse, family breakdown, individual stress and frustrations, society tensions, environmental pollution, waste of natural resources, and over-enthusiastic commercialization.

Let us examine two universal urban problems: traffic congestion and air pollution. As the number of registered vehicles in the world increased from 46 million in 1946 to 600 million in 1991, urban traffic congestion increased to the extent that new and expanded road networks failed to keep pace with that increase. In Bangkok alone the annual cost of such congestion has been estimated at about $400 million. The dust and lead pollution in Bangkok, Jakarta and Kuala Lumpur costs an annual $5 billion, i.e., 10 percent of city income (World Bank 1997, 118-120). Suspended particulate matter and sulfur dioxide pollution resulting from industrialization and burning of fossil fuel plague urban areas in ever increasing amounts. The World Bank, (1997) named four cities of the world (Beijing, Hong Kong, Teheran, and New York), where *both* suspended particulate matter and sulfur dioxide exceeded World Health Organization guidelines. Of the 24 cities examined, Calcutta recorded the highest suspended particulate matter (393 annual mean micrograms per cubic meter), while Teheran had the highest sulfur dioxide (165 annual mean micrograms per cubic meter) (World Bank, 1997, 121). Chronic respiratory disease, heart problems and damage to plant life result from such pollution.

In developing societies, although the fact of urbanization is being achieved, the promises anticipated have not been. In many countries, standards of living have actually fallen, housing quality has declined, basic urban services have broken down, unemployment has remained high and the social stability incorporated in village/rural societies has been lost. Urbanization has indeed been a mixed blessing for large parts of the world.

Political, Economic and Social Change

Political and economic settings and social change are intertwined. Their combined effects impact regional development, as well as planning. Regardless of the stage of local development, significant adjustments are being required in order for government entities at all levels to successfully cope with changing political, economic and social conditions. These challenges and the responses to them are dealt with in the first part of this volume, illustrating that different country settings have required different adjustments to cope with their unique circumstances.

Palomaki, Costa and Noble in Chapter 2 examine the differing administrative structures for planning in Finland and the United States. Finland's highly centralized system is undergoing a decentralization of administrative and planning authority to a new set of regions. Here, again, is evidence of the general trend toward decentralized decision-making current throughout most of the world.

In Chapter 3, Elisha Efrat addresses the effects which the Israeli-Palestinian negotiations have had on regional planning and development in Israel. In large part, the planning challenges of interior areas in Israel have been subordinated to the more politically imperative demands for development in areas along the West Bank and Jordanian borders.

Somewhat different in scope and orientation, has been the question of the impact of Tamil insurgency on Sri Lankan planning and development efforts. Shantha K. Hennayake and Bernard L. Panditharatne, in Chapter 4 conclude that development and planning initiatives are doomed to failure unless political stability is achieved. For successful planning results one needs political stability. In any event, the solutions can only be reached by compromise and will also involve considerable expenditure.

Chapters 5 and 6 in this volume are more focused to economic development and social issues and policies in Third World countries. Each author selects a different issue and investigates the problem in a particular setting. Together they illustrate the broad range of the critical questions which face planners and other leaders in the Third World.

George M. Pomeroy and Ashok Dutt in Chapter 5 provide a comparative analysis of the impact of development on population change in China and India. They discuss how the quite distinctive development path chosen by each country has contributed to diverse approaches to policy formulation and, subsequently, to population control measures. It is clear that the non-democratic, socialist administration of China is able to implement population control policies more effectively than the democratic government of India.

David Keeling's concern in Chapter 6 is with transportation infrastructure and its significance for promoting economic development. His analysis is based on transportation improvements in Mexico and their potential for improving or enhancing regional accessibility to markets both within and outside Mexico.

The Problem of Urban Housing

Another theme that binds urban places together is the problem of providing adequate housing as numbers of urban residents grow rapidly and steadily. Particularly vexing is the difficulty of providing shelter to families of low income. A United Nations report (1996, 337) concluded that, a) lower income urban housing in the developing countries has not improved in recent years, when affordability, tenure, standards of quality, and access to services are taken into consideration, b) the number of people living in unacceptable housing has grown, c) but housing conditions have improved in many countries, and d) in many cities of both developed and developing areas of the world inequality in incomes and in housing conditions have increased, together with the attendant dangers of social and political conflict. The policies utilized by planners in developed countries have limited relevance for developing areas. The success which planners in Bangkok have experienced (Dowall 1989) has not been replicated elsewhere for a variety of reasons, one of the most important of which is that most developing countries do not have the advantage of the rapidly growing economy of Thailand.

Adenrele Awotona (1996, 56) has identified seven specific problem areas concerning mass housing in west Africa which need more careful study and analysis. They are (1) employment and housing; (2) the economics of housing; (3) the squatter problem; (4) housing and health; (5) the adaptation of modern building techniques to low-cost housing; (6) housing standards and regulations; and (7) the role of public, private and informal sectors in the provision of housing. Awotona's research agenda is appropriate for other developing areas as well as west Africa.

In the current volume, three chapters investigate elements identified by Awotona, but in countries far removed from west Africa. In Chapter 7, Dutt, Mittra, and Halder investigate the relationship of employment and housing for slum areas in Calcutta. Calcutta, India is home to almost 11 million people, one-third of whom live in slums consisting of sub-standard housing. The already high pressure of population and the inability of slum dwellers to pay for better housing contributes to increasing size and density of the slums. They suggest that employment patterns are intimately connected to migration patterns. Taken together these regulate the location of below-standard housing slums and their persistence.

Housing and health is considered by Akhtar, Dutt and Wadhwa in Chapter 8. Malaria eradication has been touted as a medical success story, but its effectiveness is actually quite uneven. The resurgence of malaria in India is shown to be closely related to the expansion of urban housing, especially in northwestern India.

Finally, Betty Smith in Chapter 9 examines two problem areas—housing standards and regulations and the role of public, private and informal sections in the provision of middle class housing. The most important trends for medium-sized cities in highland Ecuador are declining residential property size, increasing cost of development, and increasing demand for lots. The roles of various sectors are explored for local land development.

The Environment, Development and Sustainability

Among the most significant currents of change in the world today is the increasing concern for the environment in which the world's people must exist. As world population grows and as peoples in developing countries struggle to achieve levels of development which will sustain reasonable standards of living, an awareness is growing of the necessity to reach that development through processes which will be sustainable in future. The idea of sustainable development must be understood in its proper context. It is related to the management of human activities and has two basic concerns. First, development ought to be encouraged bridging the gap between "haves" and "have nots". Second, one must ensure "that development does not damage the planet's support systems or in other ways jeopardize the interest of future generations" (United Nations 1996, 421). The four chapters of part III address these concerns.

The contemporary economic liberalization of the Peoples Republic of China has brought to the forefront several tensions in Chinese society, particularly the relationship between modernism and tradition. Patrick Wirtz and Eric Heikkila examine this situation as it applies to sustainable development. This concern is an important theme in the recent literature of cities and regions but it has been proposed much more frequently than it has been adopted. Sustainable use of resources in a developing setting is even less frequently encountered.

One of the most promising areas of prospective economic growth in which sustainability can be maintained is that of tourism. Coffey, Irwin and Urban in Chapter 11 describe efforts in Costa Rica to employ international tourism as a basis for economic development. At the same time, a concern for environmental protection and conservation motivates Costa Rican tourism planning as evidenced by the decision to implement a coastal zone management program to protect coastal areas from overdevelopment.

Joseph Scarpaci examines the growth of tourism in Cuba. Government policy in tourism development is based on the need to find new hard currency sources for the Cuban economy after the collapse of traditional markets with the demise of communism in the former Soviet Union and in eastern Europe. Since most tourism investment is from overseas sources, the major change that will ultimately result will be the growth of a private sector which, eventually, will provide a challenge to state control of the economy.

Still another aspect of the dilemma facing Third World countries attempting to achieve sustainability is explored by Klak and Myers in Chapter 13. These countries are being pressured by multi-lateral organizations and industrialized countries to structure their economies so as to best accommodate foreign investment. The authors conclude that short-term payoffs are likely to jeopardize the long-term economic growth in many of the Third World countries. Klak and Myers discuss the tactic of industrialization promotion used by many developing countries in their effort to attract foreign investment. The authors draw upon the promotional materials devised by these countries as part of their attempt to attract outside capital. Government sponsored foreign aid is now being supplemented and, in some cases, substituted by private sector investment.

The Situation of Mega-Cities

Urban growth is occurring most rapidly in cities which are already among the largest urban places. These mega-cities are gigantic urban agglomerations facing all the urban problems of smaller centers, but in addition, a great range of challenges which are the product of their very size. The United Nations (1995) has reported that while only seven mega-cities of over 5,000,000 people, with a total population of 42 million, existed in 1950, in 1995 the number had risen to 23 containing 167 million people. Furthermore, it projected that by the year 2015 the number would be 44 and they would total 282 million people. The problem is aggravated by the fact that 36 of the 44 will be in the developing countries where financial resources are meagre to cope with resolving the problems of providing adequate housing and infrastructure.

Another most startling, and at the same time, one of the most depressing statistics is the rate at which urban population is growing, most of it in the developing countries. Even at this late stage in the developed world, the connection between continued population growth and virtually all problems of environmental pollution is yet to be made effectively in the popular mind. It is increasingly fashionable to talk of a sustainable development, in which resources are not consumed so as to be unavailable for future use. However, the ingredient which is usually overlooked is the continually growing number of people which places ever larger pressures on resources. In the developing world, even less thought is given to these considerations. This, in spite of the fact that such growth is overwhelmingly concentrated in developing countries. Charles Setchell (1995, 1) reminds us that over six million additional urban dwellers are added to developing world cities *every month*. Urban areas in developing countries are faced with rapidly increasing populations, as well as challenges of providing both adequate services and the required and ever-enlarging infrastructure necessary to provide adequately for that population. Environmental quality often suffers as urban efforts fall behind growth rates. What is needed is innovative strategies to analyze future trends of both population growth and planning efforts.

Christopher Cusack suggests, in Chapter 14, that Calcutta serves as a global example of the problem, as well as the potential, of mega-cities in the developing world. Since Indian independence in 1947, Calcutta has largely continued to follow Western planning techniques, but these have proven unsuccessful for the most part. The chapter identifies and explores likely future planning directions.

Cities, especially mega-cities, are in a state of constant flux. In the case of Hong Kong the condition is augmented by proximity of fundamental political change. Political restructuring will involve an ultimately radical, although perhaps gradually introduced, shift from private sector initiatives to state investment. Chapter 15, by Bruce Taylor, investigates likely modifications in Hong Kong, and also in Macau, which reverts to China in 1999. The discussion is framed in the late colonial orientation of planning.

Finally, Victor Savage examines how planning is utilized as a tool of nation building in what can be viewed as the best example of a modern city state. Singapore may well set the parameters for mega-city development in the future. The situation in Singapore is especially significant because it may be affected by the Chinese assumption of power in Hong Kong.

Urban Networks

The final section of this volume examines the relationship of cities to one another as part of a system or network of urban places. Urban places are not discreet locations. They are intimately connected to their surroundings which, of course, provided the very rationale for their existence, at least in the beginning. Cities are linked to other urban places in a network of inter-relationships. Thus, it is hazardous to discuss any city without reference to its spatial situation. Two chapters of this book address considerations of spatial context.

Chapter 17 by Coenrad van der Wal is concerned with the evolution of efforts to plan urban settlements on the IJsselmeer Polder in the Netherlands. The long history of successful reclamation of land from the North Sea has provided an opportunity for Dutch planners to create a network of interdependent small urban places. At the same time, high population pressures in the rest of the country have emphasized the urgency of effectively establishing additional urban settlements.

In Chapter 18, Kent, Guardia, and Sibille examine legislative reforms from the mid-1990s in Bolivia which are designed to promote citizen participation in the process of planning and governance at the local level and to decentralize the structure of regional and local government. They detail the new legal environment that exists between the central government and local governments with respect to revenue sharing, local taxation, decentralization, public participation, and how these changes have transformed the process of urban and regional planning. The new Bolivian legal context portends significant changes in the relationships between not only the central government and local governments, but also among cities at different levels of the nation's urban hierarchy.

The Past as Prologue to the Future

As that great critic of urbanization, Lewis Mumford (1938, 4), observed, "cities are a product of time". Urban settlements are organisms that grow (and sometimes decline). Their current form, functions, growth directions, and even personalities are determined in large part by what has gone on before. To understand the present and to hope to predict the future one must know the past.

Throughout much of history, humans have attempted to design more effective settlements, or less often, more attractive settlements. During the Industrial Revolution, the rapidly growing cities of Europe faced a range of problems generated by that change in economy. Technical solutions to urban population growth problems were slowly achieved. Drinking water supplies were safe-guarded, although not before some catastrophic events had taken place. The connection between drinking water and epidemic disease was only gradually established (Stamp 1964, 34). Sewage disposal represents another area in which long effort was needed to achieve important results. In many cities even in the developed world, the problem of solid waste disposal has still not been effectively solved. Urban transportation was not significantly improved until well past the middle of the 19th century. Even street lighting was not provided widely in cities until near the end of the 19th century.

Much more intransigent were the difficulties of a broad social and economic nature. Such problems arose in large part as a result of the uncontrolled growth and spread of cities which was in turn a consequence of increasing urban dependence on industrialization.

Toward the end of the 19th century, the New Towns movement, which originated in Great Britain, addressed these broad problems through creative planning solutions involving the planting of aesthetically pleasing, economically self-sufficient and politically separated communities.

The need for continuing innovation to address urbanization problems has not decreased today. These concerns are addressed by Christopher Cusack in the final chapter. The use of non-governmental organizations (NGO's) is documented as is the continually growing significance of technology. He suggests that the new global regionalism is grounded in what he describes as "knowledge - based development". The future promises to be especially challenging for the centers of the developing world. Third World cities must find their own solutions to many of the problems which have already been tackled in developed countries, whose solutions may no longer be appropriate or effective. Futhermore, many difficult urban planning questions, even in the developed world, remain to be addressed. Chapters in this volume, explore the situations which must be faced by planners in a world moving toward fuller development with greater emphasis on regional economies.

The "new" trends of devolution, decentralization, and sustainability present planners and government officials at all levels with new challenges. Although the scale, geographical focus, and orientation frequently is changed, the basic problems remain awaiting new and innovative approaches.

References

Awotona, Adenrele, 1996, "A review of housing research in West Africa", *Third World Planning Review*, 18:1:45-57.

Bienen, Henry, Devesh Kapur, James Park and Jeffrey Riedinger, 1990, "Decentralization in Nepal", *World Development*, 18:1.

Bookchin, Murray, 1995, *The Philosophy of Social Ecology: Essays on Dialectical Naturalism*, New York: Black Rose Books

Choguill, C. L. 1994, "Crisis, Chaos, Crunch? Planning for Urban Growth in the Developing World", *Urban Studies*, 31:6: 935-945.

Cohen, Roger, 1996, "Global forces batter politics", *New York Times*, (November 17), Section 4, page 1.

Dowall, D.E. 1989, "Bangkok: a profile of an efficiently performing housing market", *Urban Studies*, 26:327-339.

Dutt, Ashok K., Frank J. Costa, Allen G. Noble and Surinder Aggarwal, 1994, "An Introduction to the Asian City", pp. 1-12 in Ashok K. Dutt et al. (eds), *The Asian City*, Dordrecht: Kluwer Academic Publishers.

Dutt, Ashok K. and Allen G. Noble, 1996, "Urbanization Trends in Asia", pp. 1-14 in Lan-Hung Nora Chiang et al. (eds), *Fourth Asian Urbanization Conference Proceedings*, East Lansing: Michigan State University, Asian Studies Center.

Friedmann, J. 1985, "Political and technical movements in development: agropolitan development revisited", *Environment and Planning D*, 3:155-167.

Gottman, Jean, 1991, "The dynamics of city networks in an expanding world", *Ekistics*, 350/351:277-281.

Jacobs, Jane, 1984, *Cities and the Wealth of Nations: Principles of Economic Life*, New York: Random House.

Kingsley, G. Thomas, 1996, "Perspectives on Devolution", *Journal of the American Planning Association*, 46:4:419-426.

Kropotkin, Peter, 1975, *Fields, Factories, and Workshops Tomorrow*, New York: Harper Torchbooks.

Mumford, Lewis, 1938, *The Culture of Cities*, New York: Harcourt, Brace.

Newhouse, James, 1997, "Europe's rising regionalism", *Foreign Affairs*, 76:1:67-84.

Sale, Kirkpatrick, 1985, *Dwellers in the Land: the Bioregional Vision*, San Francisco: The Sierra Club.

Schumacher, E.F. 1973, *Small is Beautiful: Economics as if People Mattered*, New York: Harper & Row.

Setchell, Charles A. 1995, "The growing environmental crisis in the world's mega cities: the case of Bangkok", *Third World Planning Review*, 17:1:1-18.

Spence, Jonathan, 1997, "A Flood of Troubles", *New York Times Magazine*, (January 5).

Stamp, L. Dudley, 1964, *The Geography of Life and Death*, Ithaca, N.Y.: Cornell University Press.

United Nations, 1995, *World Urbanization Prospects*, New York: United Nations.

United Nations, Center for Human Settlements, 1996, *An Urbanizing World*, New York: Oxford University Press.

World Bank, 1997, *1997 World Development Indicators*, Washington, D.C.: The World Bank.

2 Finnish and American Planning: A Comparative Analysis

MAURI PALOMAKI, FRANK J. COSTA, ALLEN G. NOBLE

The content of public planning activity and the expected outcomes for planning vary from one national setting to another. Many factors account for this, but the most significant relates to prevailing national values about the importance of collective action to achieve common goals for economic and spatial development. In this chapter, we compare the current planning systems of the United States and Finland and examine recent changes in planning structures, philosophies and approaches. To do this, we need to understand the global context for public planning.

Characteristics of nations with strongly developed public planning programs, or in a broader sense, collective action goals, include a highly centralized political system with limited local or regional autonomy, an authoritative decision making process, an ideologically-rooted emphasis on social equality, the dominance of public rights in property to the exclusion of private rights, and a centrally planned economic system. These conditions characterized the former Soviet Union and the communist states of eastern Europe. Today, such settings are relatively rare, the most extreme example probably being North Korea.

The converse of these characteristics is found in national settings where public planning is less important, and restricted primarily to the control of land subdivision and the provision of transportation. Business interests, which dominate in these settings, normally, seek to remove or, at least greatly reduce, public controls over the use of private property. Land is a market commodity. Once again there are few, if any, nations that conform completely to this extreme, but the United States seems to come close to fitting this paradigm.

Finland, with its traditionally centralized planning structure, contrasts significantly with the United States. Additionally, Finland faces challenges to its planning approaches by its accession to the European Union and the necessity to adapt both new structures and methods. Both of these considerations warrant a comparative examination of the planning philosophies and structures of the two nations.

19

Finland

Finland was dominated, both culturally and politically, by Sweden for several centuries. This domination ended in the early 19th century, when Finland became an autonomous Grand Dutchy within the Russian Empire. Finnish nationalists welcomed the break with Sweden and began to lay the foundations for an independent Finnish society. The national capital was moved from Turku to Helsinki in 1812, in part to create more geographic distance between the center of Finnish national life and Sweden (Blomstrom and Meller 1991, 184-189). This period of nascent Finnish nationalism coincided with similar national movements then underway throughout Europe.

Until the 1890s, Russia respected Finland's domestic autonomy in most essential matters. For example, Finland was able to establish its own monetary system in 1865. In the early years of this century, a "Russification" campaign was initiated, but this met with strong resistance by the Finns. A step toward eventual independence took place in 1906, when Finland established its own Parliament and adopted a policy of universal suffrage. When the old Russian Empire collapsed in 1917, Finland achieved complete independence after a struggle, which developed into a brief, but bitter, civil war (Blomstrom and Meller 1991, 184-189).

The Evolution of Government in Finland

Finland is a democratic republic with a constitutional form of government. Legislative power in Finland is vested in a unicameral Parliament - the Eduskunta. Executive power is vested in a State Council of Ministers led by a strong president elected by popular vote for a renewable six-year term of office (Mead 1968, 99-148).

Through the 1980s, Finland had a highly centralized political system with limited authority devolving to the regional level. This was a consequence not only of the nation's political history, but also its population size, which is not much larger than an average-sized European province. Finland is also very unevenly settled. The core area is close to the southern coast and most activities have concentrated there. Representative government existed at the national level and at the local level, but many local and regional functions, such as highway planning and construction, environmental planning and control, and labor policy were exercised through delegated authority directly from the central government and through the regional jurisdictions of central government ministries.

Political power in Finland has traditionally been highly centralized. The county and regional levels of government have been directly controlled by the central government. The national territory consists of twelve counties each headed by a governor appointed by the president. Eleven of the twelve do not have elected provincial legislatures, the only exception being the heavily Swedish-speaking island province of Aland. Counties were divided into communes, which elected their own communal councils. These, in turn, chose their own executive board (which could include members who are not elected communal representatives). The executive board was responsible for local health, education and social services. Law enforcement was centrally controlled and not part of the responsibility of communal executive boards (Mead 1968).

The structure just described existed in Finland up to the current decade, but recently proposed realignments are altering the situation profoundly. Although much of the earlier government apparatus continues in place, government is clearly in a state of transition in Finland with greater local autonomy the aim.

Planning in Finland

The structure of public planning in Finland also traditionally has been highly centralized. A multilevel planning system prevails (local, regional and national), augmented by a spectrum of special purpose agencies. The Ministry of the Interior was the national agency charged with final responsibility for planning and building decisions at the national, regional and local levels. The Ministry has been advised by a National Planning Office established in 1956 (Strong 1971, 79). This Office, now dismantled, recommended national policy guidelines for planning to the Ministry, which makes final policy decisions. An example of a recommended policy was the decision by the Ministry in 1959 to require regional planning. Once prepared, all regional plans were submitted to the Ministry for approval.

At the communal level, local master plans as well as building regulations, must also be approved by the Ministry. There is no intervening approval required by regional authorities for these plans or regulations, although local planners are to take account of regional plans (Strong 1971, 79). Once fully approved, local plans confer a decision-making monopoly on communal government.

Several changes in recent years have strengthened the regional planning function. Development area policies in the 1960s "focused on the promotion of manufacturing employment in underdeveloped areas". By the late 1970s the scope for regional policy expanded beyond purely economic questions to encompass issues such as the expansion of public services and containing the increasing concentration of the national population in southern Finland. By the 1980s, counties were given the responsibility of preparing employment and population targets as a basis for regional planning. Most recently, Finnish regional development policy has been characterized as "knowledge-based economic development"(Sarrinen 1993,11-13). In this evolving process of enhancing the role of regional government in planning, universities have performed a central role. One outcome of this has been the successful regionalization of higher education. Saarinen (1993,11-13) refers to it as the most successful effort in regionalization of services. An Organization for Economic Cooperation and Development (OECD) study reached the same conclusion as Saarinen about area development policy in Finland (OECD 1983,5-6).

Restructuring the Finnish Administrative System

The early 1990s were a time of radical change in the administrative and planning system in Finland. The responsibilities for planning have been transferred to the Ministry of Environment. Procedures similar to those utilized by the Interior Ministry have been adopted by the Environment Ministry. The most significant trend is the reduction of central government control on planning and decision-making. The basis for this is the belief, that with a steadily higher level of education in Finland, local decision-makers can reach better and more feasible decisions than was likely in the past.

The most important change is taking place at the regional level. Nineteen newly-created provinces are assuming a more democratic organization than that which characterized the county administration over the past several hundred years. This new direction has been created by amalgamating the former "Provincial Associations" and regional cultural and economic interests and organizations (both ethnic and functional) and regional planning associations.

Democratic provincial government has been a long time goal of Finnish authorities, dating from the last decades of Swedish rule in the late 18th century. This goal became even stronger during the period of Russian rule (Sipponen 1985, 27-52). However, it was not until 1992 that the State Council decided to implement provincial rule, and it is not yet functioning completely as a fully autonomous provincial administration, because the members of the boards are not elected by the people, but nominated by municipal councils. Provinces are also not autonomous economically, because they have no right of taxation. Rather, they receive funding from grants by municipalities and the state. This situation is likely to change, as the more general European system overtakes local practice in many fields, including public administration.

Regional planning associations also have been incorporated into the new provinces. Earlier they had functioned as associations of communes, with planning duties sanctioned by the construction law. Many of them began as voluntary associations, because the central government was not rapid enough in organizing public planning agencies. Regional plans were made compulsory for the whole country in 1968, although planning had begun in the late 1950s on a voluntary basis.

The other regional administrative body, the county, will remain a state government agency in regions with general competence in the administration system. Counties have lost some of their former activities to provincial bodies. All their former tasks cannot, however, be delegated, (e.g. judicial oversight of a commune's decisions). Currently there is discussion on the need to reduce the number of counties from 12 to less than half that number. Consequently, regions would be much greater both in area and population. If this occurs, the general regional administration system will very much resemble that operating in France today.

A third change will occur at the lower end of the administrative hierarchy. Many local government agencies will be combined into a single quite versatile agency called a "new circuit". These, numbering approximately 100, will operate in larger areas than now. At this time, the new circuits will have police operations, local court, and minor duties.

Alongside general administration, there is in Finland a strong tradition of single purpose planning and administration, which also may have three levels. Nearly every Ministry has, in addition to central planning and control, other levels, as well. A central agency at the national level (School Development Agency, Environmental Centre) will also have a regional level (Agricultural District, Road District, Labour Force district, Courts of Appeal) with their areas of jurisdiction. In the formation of Districts, Ministries have followed their own needs. Therefore, the number of special districts varies, which causes some incoherence in boundary delineation. State administrative authorities also have a local presence (Police Districts, Local Courts, Road Maintenance Districts). Their number is around 200. Their jurisdictional areas vary, with the same kind of consequences for boundary delineation.

The Impact of Change on Planning Activities

For planning, these changes have created a new orientation, too. Regional economic policy planning has been moved from counties to the new provincial governments. The county environmental planning and control function has been discontinued. These tasks have been allocated to a special agency and thus, from general administration to a specialized one under the Ministry of Environmental Affairs.

Finland's decision to join the European Union (EU) has also meant radical changes, resulting in the choice of new planning goals and in the internationalizing of decision-making by allocating a part of national sovereignty to all-European bodies and agencies. The borders have also become more open and cross-national cooperation more common than before.

This internationalization process has added a new European level to Finnish administration and planning. Earlier, closed borders resulted in economic peripheries (as in Eastern Finland), where the standard of living was low and economic growth slow. Now there is more cross-border cooperation which encourages economic growth. Although not totally new, the movement seems to be intensifying and finding new directions (Palomaki 1994, 238-246). The Schengen agreement, if implemented as now proposed, will in effect eliminate country borders in the European Union. Its ultimate effects, however, are as yet unknown.

The direct influence of the European Union on Finnish regional policy and planning is twofold. The Union is capable of influencing the operation of planning organizations by offering partial financing to programs that are in accordance with the Union's own programs, e.g. the European highway development program. The Union's impact can also be negative by withdrawing or withholding support for other programs. These actions can have an influence on national legislation concerning planning policies. Although the European Union budget for development planning and activities is small, it has already had its impact.

The European Union has identified five principles to be used in determining whether a project presented for funding is approvable or not. The first principle, *Concentration*, means that only projects falling within five specific object areas-agriculture, sparsely populated areas, transportation, declining industrial areas and economically peripheral areas-can be considered. The EU is thus focusing development efforts on selected issues and regions. The second principle, *Programming*, demands that single projects must be connected with each other to achieve a coherent planning and regional development strategy. Especially the strengths and weaknesses of regions must be stressed.

This has increased the effectiveness of planning. The third principle, *Partnership*, requires a close consultation between Commissions, member states, local authorities and the private sector. This aims at a more balanced view of tasks. The fourth principle, *Additionability*, stresses that while the EU partially finances projects, all participants have to share in financing, too. The fifth principle, *Subsidiarity*, demands that decisions must be made at the level where their effects take place (Mäkinen 1995). This fifth principle has been well received in Finland, because it promises to reduce the geographic handicaps incurred by a country located at the northern edge of Europe. "Subsidiarity" aims at encouraging decision making as close as possible to the people concerned.

So far, Finland has been well treated in the Union's organizations and policies. To benefit most from Union membership, it seems important to adopt an activist stance and to come forth with positive proposals. The size and location of a country seems to be of secondary importance.

One sign of weakening of the national state is the tendency to create networks of close cooperation between large cities in different countries, although such networks tend to omit the national capitals. In the northern countries, however, the corresponding network seems to be between capital cities and Hamburg, because the size of other cities may be too small.

One important new requirement which will become a central theme in all regional planning is the determination of the environmental impact of all plans. Although not new, this approach is useful in its attempt to orient planning in a feasible and comprehensive way. The process is difficult, because there are many alternatives, even down to the special function of the field assessor. This is not so surprising, when environmental impact assessment is considered to belong to the same family of concepts as sustainable development. Impact assessment in Finland is viewed as a means to move toward the sustainable development in lifestyle prevailing now in the Western World.

There also have been changes in planning needs at all levels. The regional Master Plans (*seutukaavat*), for instance, are now more or less operational. Only plan monitoring and feedback needs remain. But the changes are much more far reaching. Lauri Hautamäki (1996), a prominent Finnish geographer and planning theorist, has elucidated the changes likely to come in future.

The Future of Regional Policy and Planning in Finland

The fundamental idea behind regional planning earlier was the construction of the welfare state, with its operational environment centered only on Finland.

The power to make decisions was entrusted to the national government. Now, the objective is the securing of sustainable development. Now, regional planning is globalized, with a league of states involved, but still the local dimension remains important.

The interaction formerly was vertical, from the top downwards and from the bottom upwards. Now, it is also horizontal. Before, it was essentially hierarchical. Earlier planning was in the hands of public authorities; now private firms and other types of organizations may participate. Earlier, change was thought to be predictable and the planning environment stable. Now, the direction of change is uncertain, and the environment complex and stormy. Politics was rationalistic and planning was based on rational decisions. This has changed to a more random style. Planning organization, which was formal and hierarchical, has evolved into a more network-like and flexible structure.

Earlier the theories behind procedures were centralized and rationalistic, striving to totalities, incrementalistic, and of a "mixed scanning" type. They are now concentrated on more local areal development using strategic thinking and planning, and involving negotiative and synergistic aspects. Classic location theory, central place systems and growth pole theories were the theoretical bases in the past. The change has been towards creative thinking, visionary scenarios, operations analysis and network economy. Earlier, the plans were implemented by public agencies, now many others are involved, such as associations of communes and local or regional enterprises.

Hautamäki (1996) also argues that the concrete objects of planning are changing, because responsibilities are becoming more complex. The process will resemble the specialization process operating in geographical studies in general. The all-embracing geography of welfare was the starting point. Planning directions included, among others, planning of regional structure, rural planning, city planning, urban systems and research on developing countries. At an even more basic level, many specialized sub-fields grew up in regional planning studies on Europe, to include rural and urban politics, industrial politics, regional development and sustainable development, and environmental politics.

In such a fluid state the role of the planner has also changed. In the past the planner was a many-facetted professional, involved in decision-making as a public official or bureaucrat, in research as a theoretician and methods constructor. In problem solving the planner acted as researcher, collector of data, developer, organizer and professional expert. Now planners are, in addition, educators, designers and conductors of research, working on programming as strategists, evaluators, or project planners. Planners also have diplomatic tasks as experts, negotiators, consultants and arbitrators. They have to be leaders with new ideas, innovations, and experiments. The planner in Finland and elsewhere must be a visionary, utopian and an agent for change.

The Background of American Government

The United States is a federal republic with a constitutional form of government. The U.S. Constitution establishes a national government of limited and expressly enumerated powers. These include taxation, monetary and currency, war, and postal powers, plus the responsibility to conduct diplomacy and to regulate interstate commerce. All other powers are given to the individual states, which together constitute the second level of government. Thus, the U.S. Constitution, without using the term, created a federal system in which sovereignty is divided between two levels or spheres of government within the same nation.

A second feature of the American governmental system is the separation of powers clause of the Constitution. National power is distributed among three branches—the legislative, the executive and the judicial. Each can work to check the power of the others. For example, Congress has impeachment powers over the executive, the executive may veto bills, and the courts can invalidate congressional legislation or Presidential action. The system of separated power is also a system of "checks and balances" which works to reduce concentration of power in any one person or branch of government. State constitutions contain similar provisions for separation of powers within state governments.

The American judicial system is based primarily upon the traditions of English common law. Common law differs from codified law in that it is based upon legal precedent. Under this system, earlier court decisions are recorded "as precedents to be respected and adhered to in adjudicating subsequent disputes"(Lai 1988, 15). Codified law, on the other hand, is based upon the *a priori* determination of appropriate outcomes and penalties for criminal and civil matters by a group of legal scholars.

In common law, *a priori* judgement is replaced by *a posteriori* or an after the event determination (innocent until proven guilty). The long-term impact of common law has been to accord individuals more protection from the laws and regulations of the government. In American jurisprudence this feature of the common law has established the quasi-sacred concept of "due process" which, in theory, protects each person from unreasonable governmental action by guaranteeing him a "fair hearing" and protection from the public actions deemed by courts to be unreasonable, arbitrary and capricious" (Rose, 1989).

Thus, in summary, the American governmental system is a federal one with two sovereign spheres or levels—the nation and the individual states. Within each sphere, power is separated into three distinct branches to avoid concentration and the threat of authoritarian rule. Finally, the judicial branch is governed by the English common law tradition designed to accord citizens greater protection from an intrusive government.

The American Planning System

One of the powers not specifically enumerated in the U.S. Constitution is the general police power, which allows governments the right to enact and enforce laws guaranteeing the "health, safety and general welfare" of citizens. This power is thus a power reserved to the states. Within the broad scope of the police power resides the power to enact land use and planning laws. In the American system the national government cannot adopt laws or policies affecting land use or planning.

Within states the power to plan for and regulate development of land is usually delegated to local governments. Generally this delegation can create serious difficulties if no overarching or coordinative state policy exists. Although, constitutionally, there are only two sovereign levels in American government, practice, over time, has created lower or local levels. Since most state governments are unable to cope with the details of local administration, they have delegated administrative powers to locally established governments. A tradition of strong local autonomy has evolved over time. In theory, local powers are, in effect, state powers. In practice, local governments have become a third sovereign level of American government.

In the area of land use planning and regulation, this devolvement of state power to a multiplicity of local jurisdictions has in most cases created an uncoordinated and, in some cases, competitive planning environment in which localities vie for development. Uncoordinated delegation also can result in exclusionary zoning practices in which localities can use their planning laws to "zone out" undesirable development and people. Cullingworth (1993, 14-16) refers to local planning in the United States as "piecemeal and disjointed". Aggravating this divisive geographic situation is the general tendency of the courts, in their efforts to ensure rights of individuals, to base their decisions primarily on constitutional principles such as "due process" and "just compensation" for the taking of private property rather than upon issues such as balanced development, regional equity and the preservation of natural resources.

However, the situation just described does not hold throughout the United States. Several states have adopted a policy framework to ensure coordinated local development. In the next part of this paper, we will compare the experience of Oregon which has such a framework and Ohio which does not.

The Oregon Approach

Oregon has gone further than any state in enacting legislation requiring the coordination of local land development. The Oregon approach is based upon the adoption of mandatory goals for local governments which must be incorporated into their comprehensive plans. A state agency, the Land Conservation and Development Commission, is empowered to develop statewide standards for local planning and to enforce "their incorporation in local plans and regulations by cities and countries" (Abbott, Howe, and Adler 1994, 49-50).

The Oregon approach has several innovative features. One requires that all local land use decisions must be based upon a locally adopted comprehensive plan. A second requires that local comprehensive plans be based upon state policy contained in 19 statewide planning goals (see Appendix). These goals pertain to all aspects of land development and resource conservation. Among the most significant are goals to preserve and maintain agricultural land, to conserve forest lands, to conserve open space and protect natural and scenic resources.

From the prospective of urban development two goals stand out. One concerns the need "to plan and develop a timely, orderly and efficient arrangement of public facilities and services to serve as a framework for urban and rural developments". Another strives to achieve an "orderly and efficient transition from rural to urban land" through the imposition of urban growth boundaries which separate urbanizable land from rural land. In effect, the growth boundary serves to stop unplanned urban sprawl. Boundary changes can only be permitted if it can be proven that developable land no longer exists within the bounded area. Extensions must take into account any or all pertinent aspects of the other 18 goals, including retention of agricultural land, compatibility of land uses adjoining each side of the revised boundary, and the "orderly and economic provision of public facilities and services" (Abbott, Howe, Adler 1994, 299-303).

How successful has the Oregon approach been? Evidence seems to suggest that growth pressures are less containable in small cities, but that the policy seems to be effective in larger cities. A recent study concluded that only 5 percent of urban growth in the Portland area occurred outside its growth boundary (Porter 1991).

Other states have been influenced by Oregon. Florida enacted compulsory requirements for comprehensive planning in each of the state's municipalities and counties. These requirements stipulate the need for conformity with state comprehensive planning guidelines and for coordination of development among all political units within counties. Florida has not adopted the urban growth boundary concept, however.

New Jersey adopted a statewide comprehensive plan in 1992. Among its features is a requirement to establish a set of development districts which are modeled on the Oregon planning system. Development priorities in New Jersey have been focused on older urban areas which have been most affected by outmigration of economic activity and people. Areas with significant growth potential, such as prime farmland, are placed in a lower development priority district. Forest and environmentally-sensitive areas are also given lower development priority status (Abbott, Howe, and Adler 1994, 231-233).

The Ohio Approach

In contrast to Oregon, where local planning and regional coordination of development is mandatory, Ohio is a "laissez-faire" state as far as public planning is concerned. Ohio has no statewide planning goals to provide a framework for local planning. Ohio's planning legislation is permissive in that localities are permitted, but not required, to create municipal or county planning agencies. Ohio law empowers local planning commissions to prepare "studies, maps, plans, recommendations and reports" (Section 713.02 of the Ohio State Code). A locality can adopt a comprehensive plan for the geographic area within its political jurisdiction, but there is no mechanism other than political persuasion to ensure that comprehensive plans of adjacent political jurisdictions are compatible. Ohio law also permits the adoption of zoning ordinances without any underlying comprehensive plan. Municipalities can thus control land development without reference to a plan or guide which sets forth local goals or policies for development. Change in this area may come from court decisions which in other states increasingly require that zoning be validated by a comprehensive development plan. In fact, in Ohio the reverse holds as a consequence of a 1979 Ohio Supreme Court case (Central Motors Corporation vs. City of Pepper Pike, 1979) in which the Court held that "a municipality has the discretion as to whether it will adopt a comprehensive zoning plan; failure to have a zoning plan which is separate and distinct from a zoning ordinance does not render a zoning ordinance unconstitutional" (63 Ohio Appeals 22nd 3).

The existing system creates, in contrast to Oregon, a competitive development environment in which zoning and planning are viewed as tools in the battle for land development. A speculative land development process arises when municipalities permit urbanization to occur before a requisite infrastructure is in place to support development. Thus, instead of coordinated growth, Ohio's metropolitan areas exhibit uncoordinated sprawl and premature development.

The Montrose area of western Summit County, Ohio is a developing *edge city* with significant office and retail activity. Not only is this new *edge city* competing for development with the old central city of Akron, it also evidences internal competition, because its area is apportioned among three separate political subdivisions, each competing with the others for development. An overall plan for the area, which would be required in Oregon, does not exist. As a result, traffic congestion and excessive land development are major problems. Relatively older and closer-in facilities within the *edge city* are already showing signs of decline, because the "pull" of newer facilities at the edge is strong and as leases expire, retailers and offices move to new sites.

Conclusion

In this study we have examined two very different planning systems. One has been highly centralized, while the other is constitutionally prohibited from centralizing. In Finland the national level of planning is dominant. The regional level is an administratively-created and a non-self-governing level, which exercises delegated authority from the central or national level. All local plans must ultimately be approved at the national level. The regional level has no inherent or constitutional authority and appears to function as a convenient intermediary agency between central government ministries and local authorities. The situation in Finland is in the throes of change, however, as strong currents for decentralization are growing, and as accommodations to the EU must be made.

The United States, in contrast, has no formal land use planning authority at the national level. Land use planning is exercised exclusively by the states. States vary significantly in the importance they attach to planning. Reasons for this are beyond the scope of this paper and relate in part to differing political cultures within states (Elazar 1966). Nevertheless, the impact of the differences is striking. Several states such as Oregon, Florida, New Jersey and Vermont have enacted comprehensive sets of policies concerning landuse planning and urban development; as comprehensive as any in Finland. However, other states such as Ohio, New Hampshire and Indiana have no statewide planning policy or tradition with the outcome that uncoordinated and competitive local level development is the norm rather than the exception.

References

Abbott, Carl, Deborah Howe, and Sy Adler, 1994, *Planning the Oregon Way: A Twenty-Year Evaluation*, Corvallis, Oregon: Oregon State University.

Blomstrom, Magnus and Patricia Meller (eds), 1991, *Diverging Paths: Comparing a Century of Scandinavian and Latin American Economic Development*, Baltimore: Johns Hopkins University Press.

Cullingworth, J. Barry, 1993, *The Political Culture of Planning: American Land Use Planning in Comparative Perspective*, New York: Routledge.

Elazar, Daniel, 1966, *American Federalism: A View from the States*, New York: Thomas Y. Crowell.

Hautamäki, Lauri, 1996, Aluepolitiikan ja aluesuunnittelun tilanne ja tulevaisuus Suomessa, (The situation and future of regional policy and regional planning in Finland), Vaasa, Finland, *Proceedings of the University of Vaasa, Report 7.*

Lai, Richard, 1988, *Law in Urban Design and Planning*, New York: Van Nostrand Reinhold.

Mäkinen, Marko, 1995, "A Theoretical Assessment Framework for the Regional Policy in the European Union", pp. 53-66 in *Regions and Environment in Transition*, Tempere, Finland: University of Tempere, Department of Regional Studies.

Mead, W.R. 1968, *Finland*, London: Ernest Benn Ltd.

OECD, 1983, *Regional Policies in Sweden and Finland*, Paris: OECD.

Palomäki, Mauri, 1994, "Transborder Cooperation over Quarken Strait between Finland and Sweden", pp. 25-27 in Werner A. Gallusser (ed), *Political Boundaries and Coexistence*, Basel: Switzerland. Proceedings of the IGU-Symposium.

Porter, Douglas R. 1991, "Reassessing Urban Growth Boundaries", *Urban Land*, 50:8:25-26.

Rose, Jerome G. 1989, "A Comparative Analysis of the Systems of Planning and Land Regulation in the United States of America and the Republic of Korea", *Boston University International Law Journal*, 7:35:35-59.

Saarinen, Oiva W. 1993, *Universities and Regional Development: The Case of Finland*, Sudbury, Canada: Laurentian University, Institute of Northern Ontario Research and Development.

Sipponen, Kauko, 1985, Maakunta hallintoyksikkönä. (Province as an administrative unit) in Jaakko Numminen and Jouko Hulkko, (eds), *Suomalaisuuden Liitto: Maakuntapolitiikka (Finnish Union: The Provincial Policy).*

Strong, Ann Louise, 1971, *Planned Urban Environments*, Baltimore: Johns Hopkins University Press.

Appendix

Oregon's Statewide Planning Goals

Goal 1: Citizen Involvement
   To develop a citizen involvement program that insures the opportunity for citizens to be involved in all phases of the planning process.
Goal 2: Land Use Planning
   To establish a land use planning process and policy framework as a basis for all decisions and actions to assure an adequate factual base for such decisions and actions.
Goal 3: Agricultural Land
   To preserve and maintain agricultural lands.
Goal 4: Forest Lands
   To conserve forest lands by maintaining the forest land base and to protect the state's forest economy by making possible economically efficient forest practices that assure the continuous growing and harvesting of forest tree species.
Goal 5: Open Space, Scenic and Historic Areas, and Natural Resources
   To conserve open space and protect natural and scenic resources.
Goal 6: Air, Water, and Land Resources Quality
   To maintain and improve the quality of the air, water and land resources of the state.
Goal 7: Areas Subject to Natural Hazards and Disasters
   To protect life and property from natural disasters and hazards.
Goal 8: Recreational Needs
   To satisfy the recreational needs of the citizens of the state and visitors.
Goal 9: Economic Development
   To provide adequate opportunities throughout the state for a variety of economic activities vital to the health, welfare, and prosperity of Oregon's citizens.
Goal 10: Housing
   To provide for the housing needs of citizens of the state.
Goal 11: Public Facilities and Services
   To plan and develop a timely, orderly and efficient arrangement of public facilities and services to serve as a framework for urban and rural developments.
Goal 12: Transportation
   To provide and encourage a safe, convenient and economic transportation system.
Goal 13: Energy Conservation
   To conserve energy.
Goal 14: Urbanization
   To provide for an orderly and efficient transition from rural to urban land use.
Goal 15: Willamette River Greenway
   To protect, conserve, enhance and maintain the natural, scenic, historical, agricultural, economic and recreational qualities of lands along the Willamette River as the Willamette River Greenway.
Goal 16: Estuarine Resources
   To recognize the unique environmental, economic and social values of each estuary and associated wetlands.

Goal 17: Coastal Shorelands

To conserve, protect, where appropriate develop, and where appropriate restore the resources and benefits of coastal shorelands, recognizing their value for protection and maintenance of water quality, fish and wildlife habitat, water-dependent uses, economic resources and recreation and aesthetics.

Goal 18: Beaches and Dunes

To conserve, protect, where appropriate develop, and where appropriate restore the resources and benefits of coastal beach and dune areas.

Goal 19: Ocean Resources

To conserve the long-term values, benefits, and natural resources of the nearshore ocean and the continental shelf.

3 Regional Planning and Development in Israel as Affected by the Peace Process

ELISHA EFRAT

The peace process between Israel and the Palestinians, which is now in a stage of bilateral and multi-lateral negotiations in different committees, and the signing of a peace treaty between Israel and the Hashemite Kingdom of Jordan in 1995, has affected the trends and the orientation of regional planning and development in Israel. While in the past, planning was mostly directed to internal and domestic issues, nowadays many planning initiatives address external problems, mainly those along the borders between Israel and the territories in Judea, Samaria and around Jerusalem, and those with Jordan.

The Palestinian State gradually crystallizing in Judea and Samaria, and the common interest of Israel and Jordan to develop their arid regions around the Dead Sea, the Arava and the surroundings of the Gulf of Elat and Aqaba, have given impetus to many new spatial ideas and regional plans. Following is a discussion of six of the most important of these ideas and plans.

The Return to the "Green Line"

The "Green Line" as a border between Israel, Jordan and the Palestinians was established after Israel's War of Independence in 1949, and separated Judea, Samaria and the Gaza Strip from the rest of Israel's territory. This line was closed by the establishment of the so-called "Nahal" settlements along it, populated by army units who combined security and military training with agricultural work. The line, nearly a thousand kilometers in length, has been reinforced by new development regions that were established parallel to it, such as Lakhish, Besor, Ta'anach and the Arava, mainly for the purpose of reinforcing border strips of land. The "Green Line" was very important as a border between Israel and the surrounding countries from 1948 to the outbreak of the Six Day War in 1967, which changed the region both politically and territorially.

In 1967, Israel unilaterally recognized the Jordan River as its eastern border, converted the armistice line in the Golan Heights into a military security border, and created a narrow security zone, between Israel and Lebanon, north of the international borderline of 1923.

The "Green Line" which had been delineated in 1949 as an armistice line, was actually a cease-fire line agreed upon by the surrounding Arab countries, and approved by military representatives of all the partners who participated in the 1948 War, with the mediation of UN observers. It expressed, more or less, the military situation of the positions that existed at the moment of cease-fire, but it did not contain any security elements which could promise full defense to Israel from Arab infiltrations and terrorist attacks. It was delineated at that time with insufficient data on the geographical characteristics of the regions through which it would run. South of the Hebron Mountains, for instance, it was marked on the high mountainous ridges of Dahariye and Eshtemoa, 500-700 meters above sea level; in the Jerusalem Mountains it dissected in a diagonal way, mountainous ridges 700-800 meters high; in northern Israel it bisected the Amir Ridge 400-500 meters high, and crossed the Gilboa Mountains in a longitudinal section. It was obvious that in such a twisted topographical condition the line would not be suitable for defense of Israel's sovereign territory. Furthermore, the "Green Line" enclosed in Judea and Samaria an area of 5,878 square kilometers. Every kilometer of border had to protect an average hinterland of about 23 square kilometers. This ratio demonstrated its great length relative to the territory it contained. The "Green Line" was also dissected by more than 50 wadis of different widths, five of them in the valleys of Bet She'an and Harod, four in the Valley of Eron, about 20 in the Judean Mountains, and all the rest in the southern Hebron Mountains. Most are dry streams used by the Palestinians as relatively easy infiltration routes, and because they are dry during most of the year, their effectiveness for that purpose is obvious (Figure 3-1).

In recent years, as terrorist attacks initiated by the extreme Arab political groups of "Hamas" and the "Islamic Jihad" became more frequent, the "Green Line", which after 1967 was abandoned according to the Israeli policy of "Greater Land of Israel", and which held that no inner borders would exist within the whole territory, had to renew its security functions for internal protection of Israel's population. The "Green Line" proved to be an imperfect border with many geographical shortcomings, which originate from the years 1948-1949, when it was first delineated, as well as the geographical changes which occurred along it since the Six Day War in 1967. Its weakness as a defense line for Israel may be emphasized in four main areas: 1) its excessive length; 2) its twisted delineation over difficult mountainous topography; 3) the

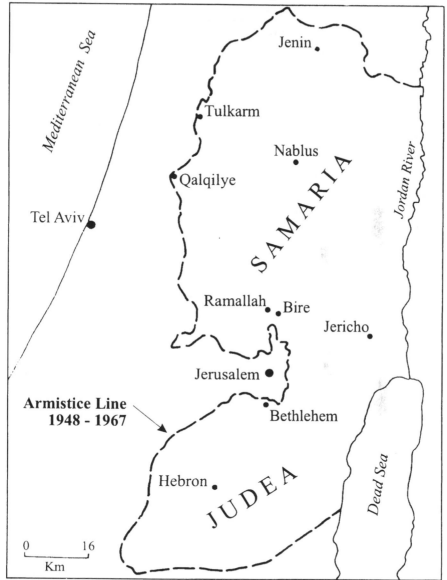

Figure 3-1   The Armistice Line ("Green Line") 1948-1967

large Arab population which resides on both sides of it, with the Palestinians to the east and the Israeli Arabs to the west; and 4) the relatively sparse Jewish population which settled adjacent to it over the years.

The Arab population residing in Judea and Samaria along both sides of the "Green Line" at the time when the line was marked, numbered about 100,000, while in 1996 the Arab population which lives in a distance of 2-3 kilometers east of the line, amounts to more than 200,000. On the west side, and in the same width, but in Israel's territory, live about 100,000 Arabs who have frequent contacts with the Arab people in Judea and Samaria. This is especially so in north and northwest Samaria, opposite the Hadera-Afula communications artery of the Eron Valley, and opposite the central coastal plain. A more substantial ethnic separation between Jews and Arabs along the two sides of that line exists in the Jerusalem Corridor, in the western Judean Mountains and in the southern Hebron mountains.

The Jewish population which resides adjacent to the "Green Line" in Israel's territory, in the same distances of 2-3 kilometers west of it, is only about 35,000, while those on the eastern side in Judea and Samaria number about 30,000. The population along the line is 300,000 Arabs compared with 65,000 Jews, excluding the Jewish-Arab concentrations in West and East Jerusalem. This is an advantage of 4.6 times in favor of the Arabs. The meaning of these basic deficiencies of the "Green line", which ought to secure Israel from terror in its sovereign territory, is very significant.

After the Six Day War there was a tendency in Israel to disregard the existence of the "Green Line". The development of new settlements along it was intended to erase its existence, but nowadays the "Green Line" has come to life again because of the new political circumstances. It is seen as vitally important to Israel, to secure its border with the future autonomous area in Judea and Samaria. The signing of an agreement between Israel and the Palestinians in 1994 regarding autonomy in the Gaza Strip and Jericho, naturally revived the problem of Judea and Samaria's borders in the future Palestinian State. This issue renewed the tension between the Israelis and the Palestinians in the region and sharpened the security and economic situation between Jews and Arabs. Of all the possible alternatives of spatial arrangements in the region, the Israelis now attach greatest importance to the "Green Line" as a security and ethnic border, because of the danger of terrorist attacks on Israel which may be initiated from the future autonomous territories.

The "Green Line" with its geographical and demographic complexities, is difficult to defend, unless it is replanned and fortified as a sophisticated and perfected security line. Such a step taken by the Israeli government would be not only very expensive, but also economically wasteful. Thus, it is critical to accelerate the peace process with the Palestinians, so that the "Green Line" will no longer be needed as a security line.

The late Prime Minister Yitzhak Rabin spoke in 1995 about the need to separate the Arab population in Judea and Samaria and the Jewish population in Israel. He appointed a committee of experts representing different Israeli security bodies to prepare a plan for a separation line east of the former "Green Line", but parallel to it, with all the needed installations to provide security for Israel. With this act Israel joined countries which have planned and constructed artificial borderlines to define and protect their territory. It may be assumed that the separation line which Israel delineated in Judea and Samaria cannot close the border hermetically because of the mountainous terrain and the strong inter-relationships which exist across the border. Israel needs working manpower and the Arab population seeks such places of employment. It is doubtful whether separation lines and artificial borders will be able to prevent totally immigration and commuting of population from neighboring areas which are economically unequal. It seems, that only a fundamental economic and political solution in the territories, which will create more places of employment for the local population, will ensure the effectiveness of the separation line between Israel and the Palestinians. No barriers, closure periods, guarding posts, patrols, watchdogs and electronic fences will solve the existing political problem. The new separation line between Israel and the Palestinians which has been established and partially fortified should be viewed as a temporary line to enable options for a final peace agreement on the borderline revisions of the future Palestinian State, with the hope, that "good fences make good neighbors".

The "Axis of the Hills" Plan

In response to Israel's policy to defend its eastern borderline with the Palestinian State, a new plan initiated by Israel's former Minister of Housing, A. Sharon, was presented as a measure to ease the absorption of the 1990 wave of Jewish immigration into Israel. In reality its aim was to create an irreversible border on the map representing new demographic, economic and political realities along the "Green Line". The government has been trying also to settle people more densely along the "Green Line" and "Judaize" the area. Both objectives were included in the original settlement plan named the "Axis of the Hills" Settlement Plan (Adiv and Schwartz 1992).

The plan focused on a narrow strip of land, 80 kilometers long, at the edge of the "Green Line". It is wedged between the West Bank and the eastern coastal plain, from the new southern town of Modi'in to the outskirts of Um el-Fahm in the north. The area, including the Arab town of Um el-Fahm and surrounding villages, is inhabited by 150,000 Arabs and 40,000 Jews, and a decision was made by the Israeli government to settle 350,000 Jews in this area over the next 15 years. The basic premise of the "Axis of the Hills" Plan is to create a string of settlements parallel to the heavily populated coastal axis, in order to ease population pressure on the coastal center, especially between Ashdod in the south and Hadera in the north. The plan also calls for creating an economic infrastructure alongside these settlements because the designated area lacks industrial or economic support systems. In addition, it has been decided to build a national road known as the Crossing Highway which will cross the country from its northern border to Beersheba in the south. No doubt, this plan has a political basis, and regardless of the type of peace agreement Israel may reach with the Palestinians in the future, these settlements, side-by-side with the Jewish settlements on the West bank, will become interdependent and strengthen the border (Figure 3-2).

According to the plan the "Judaization" of the area will occur in three stages, with the final stage reached in 2005 when it will have reversed the demographic composition from the present 82 percent Arabs and 18 percent Jews to 36 percent Arabs and 64 percent Jews. It will strive to settle the Jewish population between Kafr Qasem in the south and Um el-Fahm in the north. The backbone of this plan is the proposed Crossing Highway which will supply a necessary link between the settlements. The planners acknowledge that the decision to begin working on the Highway was motivated by the desire to implement the plan. Without some east-west arteries to link the settlements to one another, it would be impossible to implement this plan which includes the building of new cities and thousands of square meters of industrial structures.

The plan also envisages the building of four large centers which will offer the settlers commercial and industrial job opportunities, health services, cultural services, and a system of communications to the rural settlements to be built around them. The four centers in which the bulk of the work will occur are: Modi'in, Rosh Ha'Ayin, Kochav Ya'ir settlement cluster, and Harish in the Eron area. The settlement of Modi'in, expected to have a population of 160,000 by the year 2005, will be a new city along the Jerusalem - Lod highway. The population of Rosh Ha'Ayin was about 29,600 inhabitants in 1996 and is expected to increase to 50,000 by the year 2005. With a current population of 5,300, the Kochav Ya'ir settlement cluster, located between Taibe and Qalqilye, is projected to have a population of 20,000 by the completion of the plan. For the fourth center, the plan calls for the establishment of a new city, Harish, north of Baqa el-Gharbiye, with a projected population of 35,000.

Rural settlements built around each of the four urban centers will have between 1,000 to 4,000 residential units and will be connected by a new

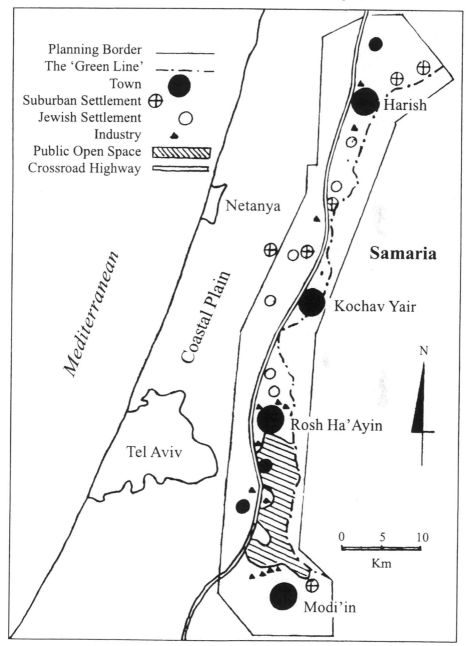

Figure 3-2   The "Axis of the Hills" Settlement Plan

network of roads. The planned network of roads will cut through land under the administrative jurisdiction of Arab villages. The plan also calls for constructing 14 industrial complexes spread over an area of 1340 hectares. One of these industrial complexes is planned for Rosh Ha'Ayin, covering 115 hectares and adjoining the houses of Kafr Qasem. Other industrial zones will be set up in Tzur Yigal, Tzur Natan, Taibe, Sha'ar Efrayim, Kokhav Hasharon, Baqa el-Gharbiye and Harish. The four urban centers also are expected to provide 25,000 new jobs to the settlers.

The plan, aimed at settling 350,000 Jews in the strip on the border of the West Bank, is being implemented at full speed. Three steps are currently being undertaken by the Ministry of Housing and Construction. First, the most recent, and apparently final, routing of the north-south highway has been determined and building the main intersection near Ben Shemen has been completed. Second is the formation of a regional council, Tal-Iron, representing the Jewish towns of Katzir and Harish. Last is a decision to reestablish a regional council consisting of seven Arab villages which lack any form of municipal representation such as a local council. The villages to be included are dispersed geographically. The inclusion of the villages in a single municipal framework will erode the few local services still provided through their *Mukhtar* or village leader. The planned council is unanimously opposed by the residents of the seven Arab villages who believe that it will restrict their independence.

Israel's Planned Crossing Highway

As a part of the national transport road network the Israeli government decided in 1992 to plan and build a Crossing Highway which will cross the country from northwestern Galilee and the Golan Heights through the eastern coastal plain and, from there, southward to Beersheba and the northern Negev (Figure 3-3). This highway has three main objectives: 1) to serve as a bypass road east of the Tel Aviv agglomeration and to facilitate quicker transport in a north-south and west-east direction in the central part of the coastal plain; 2) to enable dense Jewish settlement and development along the "Green Line"; and 3) to bring the Galilee and the Negev closer to the transport terminals in Haifa, Ashdod and Elat harbors and to the industrial and commercial centers and transport terminals in the Tel Aviv area. The Crossing Highway will be the main transport artery in the country, and aligned east of the population concentrations in the coastal plain, close to the foothills of Samaria and Judea. It is to be Israel's greatest infrastructure project for the next decade. Considered one of the biggest transport projects ever planned in the country, it should be linked

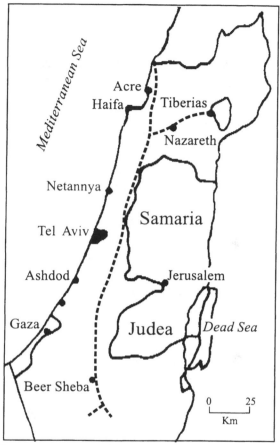

Figure 3-3   Route of the Crossing Highway

in the future to a larger transport system of the Middle East. The Crossing Highway will be similar in its function to the famous historic *via maris* which ran through the coastal plain and connected Egypt and Mesopotamia in ancient times. After peace in the Middle East has been achieved, it may also lead to Damascus in Syria or function as an international transport route between Cairo, Istanbul and Europe (Efrat 1994).

The length of the Crossing Highway will be approximately 230 km and will consist mainly of eight lanes, with a width of 64 m to 150 m. The central section of the highway, parallel to the Ashdod-Netanya axis, will be a toll road, the first highway in Israel built by private enterprise. It should be completed by the year 2000. The government recommended that private investors, possibly from abroad, finance and supervise the construction of this highway. In return, investors will receive revenues from the tolls. The financial investments in the first stages of construction are estimated to be approximately 450 million US dollars for the central section, 750 million for the southern one, and 650 million for the first north extension. The costs of construction and maintenance of the highway, as against the expected savings in vehicle operating costs and passengers' time, indicate an average total saving of about 250 million dollars (Ministry of Construction and Housing, 1991).

In order to execute the plan, a government company was founded to speed up approval of plans and the construction of the highway. Because the Crossing Highway will pass through lands owned by 75 agricultural settlements, the company will have to expropriate 4,000 acres of land for the road and its attached installations.

The geographical aspects of this Crossing Highway involve four main problems: traffic, land use, suburbanization and politics. The Crossing Highway will have a twofold aim: to intensify transport links between the northern and southern regions of the country, but also to function as an outer ring road for the metropolitan area of the Tel Aviv conurbation. Between 1967 and 1991 the number of cars in Israel increased by 475 percent from 193,000 to 1,075,000, or an average of 19.5 percent annually. Trucks and commercial vehicles increased by 251 percent from 47,000 to 165,000, an average of 10.4 percent per year (Central Bureau of Statistics, 1993). Assuming that in the next decade the number of vehicles in the country will double, the shifting of transport routes to the west of the Judean hills and to the east of the coastal plain is necessary to give priority to north-south traffic, and to encourage development in peripheral regions.

Regarding the land use problems it should be emphasized that the construction of this highway will create a loss of considerable agricultural land which is in very limited supply in Israel, and will do harm to many agricultural units which lie along the planned route.

It is feared that the Highway will encourage the trend of suburbanization and low-density housing within the outer ring of the Tel Aviv conurbation, a phenomenon which is against Israel's general population dispersal policy to weaken the large population concentrations in the central coastal plain and to give preference to housing and economic investment in peripheral regions of the country. This highway may attract more economic and transport activity in the outer areas of Tel Aviv instead of facilitating investments in the Negev and in Galilee.

The plans show that the Highway will be located very close to the former "Green Line" border between Israel and the occupied West Bank territories of Judea and Samaria. On the one hand it may perpetuate this boundary by strongly encouraging development on the Israeli side and increase the gap in living standards compared with the West Bank. On the other hand it could weaken economic interrelationships between the border area and Israel because not many lateral roads are planned in an eastward direction to link with Judea and Samaria.

Recent Partition Plans of Jerusalem

The city of Jerusalem is a site of demographic and physical competition between two populations, Israelis and Palestinians, each with the clear political purpose of holding and controlling the city and its environs. Planning and development under the current conditions initiated Israeli plans for the partition of Jerusalem. On October 1995 three maps, which were prepared by the Institute for Jerusalem Studies, describing alternative plans for the final political solution in Jerusalem, were passed by governmental officials to the Palestinian

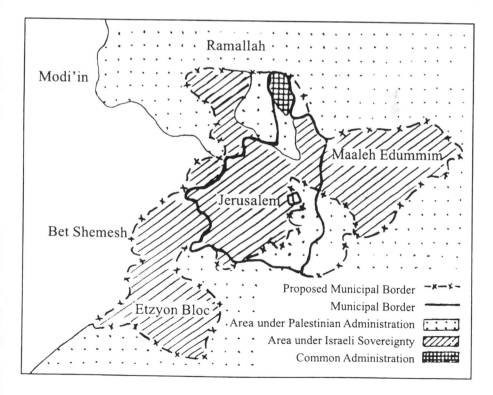

Figure 3-4   Extended Plan of Jewish Jerusalem, 1996

authority. In one of these plans the recognition of Palestinian sovereignty in East Jerusalem was clearly expressed, but with it also the annexation to Israel of some wide areas, including as the Etzyon Bloc, Ma'aleh Edummim, Givat Zev and Betar (Figure 3-4).

According to this plan five alternatives for Palestinian sovereignty in East Jerusalem were proposed. These varied from very limited to extensive territorial concessions:

- A limited area along the fringe line of Jerusalem municipal boundary will be connected by a strip of land to the Palestinian territories, and could be used in parts of it as a Palestinian government compound.
- Sovereignty in East Jerusalem, excluding its Jewish neighborhoods and the Jewish Quarter in the Old City. The other parts of East Jerusalem, the Old City, Mount of Olives, and the Town of David, which are the most important sites for the three religions, to have a special status of joint sovereignty or condominion.
- Sovereignty in East Jerusalem, excluding the Jewish neighborhoods and the Old City which will remain under Israeli control.
- Sovereignty in East Jerusalem, excluding the Jewish neighborhoods, the Jewish cemetery on the Mount of Olives, the Town of David, the Jewish and the Armenian Quarters in the Old City, West Jerusalem, and a strip of land connecting the Etzyon Bloc to Israel.
- Sovereignty in East Jerusalem, excluding the Jewish neighborhoods.

The advantages of these alternatives, from the point of view of Israel, are:

- Reconciliation with the Palestinian and the Muslim world, and a greater chance to achieve a long-term stable peace agreement.
- Preserving of Israel's sovereignty in the Jewish neighborhoods of East Jerusalem, including the Jewish Quarter, and control of a substantial part of East Jerusalem.
- Palestinian recognition of Israel's sovereignty over most parts of the city, including the new neighborhoods which were built after 1967 and which consist today of about 170,000 inhabitants.
- Improved chance to achieve a territorial exchange for the connection of Ma'aleh Edummim and Givat Zev to Jerusalem, and a connection of the Etzyon Bloc to Israel after Israel's surrender of all the Arab areas in East Jerusalem.
- No need for a physical division of Jerusalem and assuring free access to all parts of the city without disrupting its economic fabric, and no need to rule 160,000 Palestinians who live in East Jerusalem.
- Security in the city and its foreign affairs will be kept under Israel's authority.
- These proposals would enable a gradual move towards Palestinian sovereignty over a long-range period, starting from a limited joint government as an interim

arrangement to a larger share in East Jerusalem, excluding the Jewish neighborhoods as a final agreement.

The disadvantages of these alternatives from the point of view of Israel are:

- Two political capitals will function in Jerusalem, the Jewish nature of the city will be diminished, and its status as a Jewish property may be undermined.
- Palestinian sovereignty will be established on the Temple Mount which is a holy place for the Jews.
- Joint administration and policing will be complicated and create conflicts between the two peoples.

A second plan proposed by the same Institute took for granted the situation in the city which had existed since the Six Day War. The plan is based on the assumption that the municipal area of Jerusalem will remain under Israeli sovereignty. Changing areas by a mutual agreement between Israel and the Palestinians, for pragmatic reasons, will be possible as, for instance, improvement of security posts, or the improvement of territorial links for Jewish and Arab population in neighborhoods.

The advantages of this plan are:

- Protection of Jerusalem as a Jewish entity.
- Reinforcing the idea among the Israelis and the Jewish people that other alternatives may weaken Israeli sovereignty in Jerusalem, aggravate tension and violence between Jews and Arabs, and create breaches that cannot be healed.
- This plan may be viewed as a provisional step enabling a shift to other alternatives when the Palestinian entity proves its stability as a political and democratic body.

The deficiencies of this plan are:

- It is totally unacceptable to the Palestinians.
- It may be an obstacle to the Israeli-Palestinian negotiations.
- It may encounter American opposition
- It may create agitation and ignite a new uprising in the form of the 'intifada' (Noble and Efrat 1990).

A third plan is based also on the assumption that Israel will get exclusive sovereignty in Jerusalem within its present municipal boundaries. In the framework of a mutual agreement, exchanging of limited areas with those in Judea might be possible. A symbolic center of sovereignty for the Palestinians in the city might be approved. The Temple Mount will be under Jordanian administration; similar treatment will be accorded to the Church of Sepulchre, and to the Christian Quarter in the Old city; the Armenian Quarter will get a special status, as well as the space between the walls and to the near surroundings of the Old City.

The aim of this plan is to administer functional autonomy under Israeli sovereignty in all the quarters in East Jerusalem. Such an autonomy will be supervised by the Jerusalem municipality and will include among others the following: collecting of domestic taxes, and administration of borough councils with permanent staff-members for culture, education, sport, social services, gardening, health and religious services.

The idea of functional autonomy has been accepted in principle by different institutions and organizations which are involved in Jerusalem's political

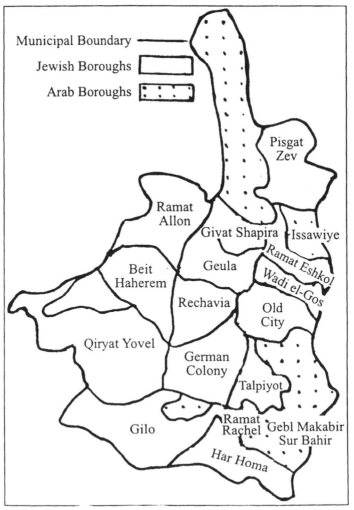

Figure 3-5   Planned Jewish and Arab Boroughs in Jerusalem

future. The premise is to convey some important authorities to borough administration, to give them policing authority and to encourage their independence from municipal administration. Borough administration would be established in all the parts of the city. In such a framework it will be possible to create a sub-municipality for the Old City with an inter-religious and international council (Figure 3-5).

This plan also recommended that the existing system of borough administration be dispersed to all the other neighborhoods in the city. The administration in the Old City would be established along ethnic and religious lines. The boundaries of each borough would be delineated by consultation with representatives of the residents in the area.

Functional autonomy to the boroughs may have a good chance of acceptance. The present situation that exists in the city is acceptable to the Palestinians, while a recognition of Palestinian sovereignty in East Jerusalem is not acceptable to many of the Israelis. But this plan is flexible and enables different kinds of arrangements between the peoples and it can be executed gradually according to future circumstances. The main deficiency of this plan is that if the rules are not strictly kept, it might cause a weakening of Israel's sovereignty in the city.

Without a comprehensive political plan agreed upon by both sides, and rapid systematic implementation of important aspects of it, Israel will not be able to safeguard the city as a capital. The reunification of the city in 1967 did not bring to an end the division between Jews and Arabs (Efrat and Noble 1988). The animosities are deep and have survived the recent geographical shifts. None of the different partition plans of Jerusalem will ensure normal functions of the city. It remains always a constant danger that Jerusalem might become a political arena where two peoples find themselves embraced in hatred with no logical solution.

Water as a Geopolitical Problem

More than five million inhabitants of Israel and about two million Palestinians suffer already from lack of water. When Israel's population reaches eight million in the year 2020, and the Palestinians have three million, the shortage of water will be even worse.

Water has been a touchy geopolitical issue in the Middle East. Israel faces a steadily worsening water shortfall and its neighbors face even worse. In the absence of any significant likelihood that peace will solve the distribution of water, an ever increasing possibility exists that force may be used to determine who has water and who is deprived of it. Israel is currently estimated to be using its water resources 15 percent faster than they can be replenished. Without creating new resources of water, together with equal and proportional distribution, the dilemma will remain unsolved. New resources could be created by circulation of drainage water, and by introducing advanced methods in agriculture which will decrease the consumption (Wolf 1995).

Israel's special water problem lies in the urgent need to recycle sewerage water which has become more salty and polluted over the years. Both Israel and the Palestinians should therefore invest all their knowledge and energy in purification of sewerage water, so that the water problem for agriculture will be less severe. In the Israeli urban and industrial sectors the average demand is 100 cubic meters per person per year, as against 35 in the West Bank and the Gaza Strip. In the next few decades the Palestinian demand for water will increase and become equal to that of the Israelis. The present water resources in Israel, including underground water, the Jordan basin catchment area, and recycled water, provide about two million cubic meters annually, which is not enough for future consumption. The resources will not increase amounts of water significantly, so that other artificial and sophisticated methods should be introduced.

Water experts realize that without new water resources in the area, wars may continue permanently. The water resources of Jordan can supply the needs of only 1.4 million inhabitants, while about four million people currently reside there. Water for irrigation has been diverted to urban consumption. To replace it, sewerage water is used on crops with a high health risk. Jordan, Judea, Samaria and the Gaza Strip have exhausted their efforts to ensure more water resources. The Jordanians feel that desalination of the River Jordan's water and the systematic development of the Jordan Valley on both sides of the border are the right solutions.

The Red-Dead Canal

One of the main objectives of the Jordanians is therefore the Red-Dead Canal, connecting the Red Sea with the Dead Sea, and the creation of hydro-electric energy. The Jordanians, as the Israelis, are concerned about the continuous process of drying up of the Dead Sea. Until 20 years ago its water table was 395 meters below the Mediterranean sea level, and today it is 401 meters. In another decade it may descend to 416 meters and the sewerage water flow in the area will increase. A canal connecting the Dead Sea with the Red Sea at the Gulf of Elat may control the water table of this lowest sea on earth. The

Jordanians feel, that the potential which lies in the elevation differences between the two bodies of water, 160 km apart, will be sufficient to produce from
the saline water, energy which could be used to desalinate the river's water for
both sides of the Jordan Valley. Before the peace agreement between Israel
and Jordan was signed, the borderline created topographical difficulties which
made such a project impossible. An open border between the countries and
mutual interests make it much easier now to supply the needs of a larger
population.

In addition to plans and policies for resources of energy, water and the
conservation of the Dead Sea's potential, another plan has been proposed for
a deep-water harbor at the outlet of the canal at the Gulf of Elat. Besides the
enlargement of commercial capacity of the two harbors of Elat and Aqaba, the
beaches between them would be designated for tourism and recreation. Tourists will be able to sail along the canal toward the Dead Sea and fishing will be
possible on both banks of that canal.

The characteristics of the Red-Dead Canal are: The cost of 50-60 cents to
desalinate one cubic meter of water; the period of building the project is 15
years; joint control by Israel and Jordan in supervising the project; the water
will be elevated from the Gulf of Elat to 220 meters altitude in open canals and
in tunnels; the project will diminish the costs of potash production in the Dead
Sea and may encourage tourism; and many new jobs will be created for thousands of people (Figure 3-6).

The Med-Dead Canal

Another project regarding the water problem in the area is the Med-Dead
Canal which would connect the Mediterranean Sea with the Dead Sea. It is a
project which has been discussed many times in the past, but because of political circumstances was not practical. With the acceleration of the peace process between Israel and the Palestinians this project returned to the agenda
again in different alternatives. A canal which will begin at the southern part of
the coast of Israel, between Ashqelon and the Gaza Strip, and be dug along the
northern Negev in a tunnel beneath the mountains, will lead water to the Dead
Sea and will create a waterfall supplying energy of 800 megawatts per year.

One alternative proposes to begin with the canal at the Qatif Bloc which is
the shortest length, but this will need the agreement of the Palestinians and
Jordan. Another alternative proposes to take the water from the Mediterranean at an outlet in the Gulf of Haifa and lead it through the valleys of Yisreel,
Harod and Bet She'an, and from there to let it fall by elevation differences

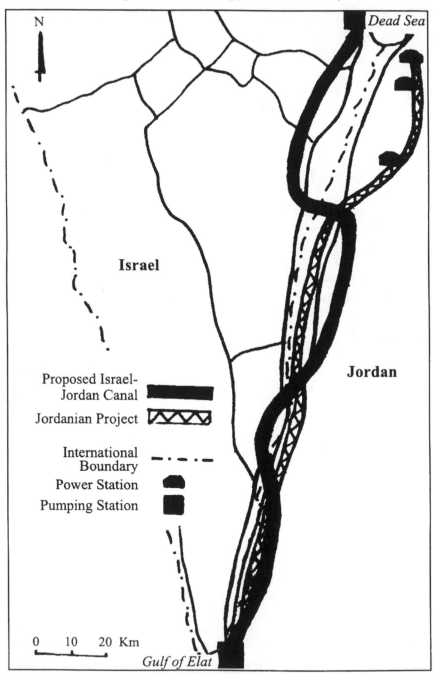

Figure 3-6   The Proposed Dead Sea - Gulf of Elat Canal

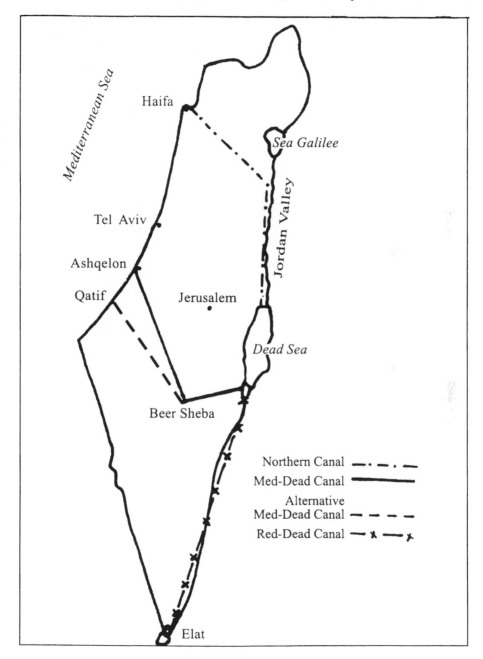

Figure 3-7   Alternatives for the Mediterranean - Dead Sea Canal

along the Jordan Valley into the Dead Sea while creating electric energy. Near the Sea of Galilee a desalination installation will be built to create 200 million cu. m. per year of fresh water. The desalinated water will flow to the Sea of Galilee and from there it will be distributed for irrigation and urban consumption (Figure 3-7).

This plan will require cooperation between Israel and Jordan, because at the desalination plant there will be a potential of 800 million cubic meters, 3/4 of which will be used by Jordan for irrigation along the eastern part of the Jordan Valley.

Development of the Arava Region between Israel and Jordan

A master plan for the development of the Arava region in the southeastern Negev has been prepared by the government of Israel with the aim of developing the potential of that region according to the options which exist there after the signing of the peace agreement with Jordan. The plan assumes an increase of tourists, and new markets for agricultural products and goods on both sides. Peace between Israel and Jordan provides opportunities for the region which will be developed as a crossroads between Israel, Jordan and Egypt. It is assumed that by the year 2000 there will be an investment equivalent to a hundred million dollars in this region.

The potential of the region has not been utilized in full because of climatic constraints, and remoteness from the central parts of the country and from the main centers of occupation. It is planned that by the end of the century the number of inhabitants in the Arava region will be doubled, together with an enlargement of the existing settlements and with an addition of 5,000 inhabitants. Two other settlements totalling 400 settlers will be established, in addition to the 20 which already exist.

The master plan envisages an economic structure in that arid region based on advanced technology, trained and qualified manpower, exploitation of natural priorities of the region, and creation of new jobs. The main targets of this plan are: increase of regional research and development in experimental stations; establishment of international centers for expertise in modern agriculture; development of desert tourism in the Arava mainly for archeological and geological expeditions; development of the Dead Sea region as a international center for healing tourism; common projects with the neighboring countries for desalination of water, development and exploitation of energy resources; the building of a common airport at the Gulf of Elat; the establishment of free commercial zones, and creating centers of employment and research institutions together with the neighboring countries.

The plan envisages that investments will come from abroad as a result of peace in the region, and the dynamism which the peace process creates in normalization between the countries. The demand for land in the central parts of Israel will increase land prices, and as a result a transfer of a part of the demand for land will go to the Negev and to the Arava. Also the freeing of agricultural land for housing in the central parts of the country will increase the demand for agricultural land in the south.

Solar radiation and geothermic water which are found in the Arava, and the fact that the region has an infrastructure, creates a potential of development of energy in modern ways. The potential of saline and geothermic water in the Arava fits desert and aquatic agriculture, and desalination of water may increase other uses.

The master plan proposes that the neighboring countries in the region organize and administer the tourist sites around the Dead Sea and the Gulf of Elat. In its details the plan proposes a link in tourism between the Negev and Petra, development of important tourist sites, a proposed railway line between the Dead Sea and Elat, and sites for border passages between Israel and Jordan.

Conclusion

Instead of solving substantive and physical problems within the territory of Israel, accelerated development and planning activity has been directed to sites along the "Green Line" and to the regions which border Jordan. The government gave greater impetus to the west-east direction, turning its attention to the occupied territories for reasons of security and economic development. The authorities were forced to find new responses to the new geographical conditions. The changed direction they took disregarded the principles that had underlined the building up of the country in the past. There is no doubt that the new regional plans were primarily influenced by political and security factors, but with the aim of mutual development in a background of peace. After the signing of the Oslo agreement between Israel and the Palestinians in 1993 and the peace treaty with Jordan in 1995, the gradual conveying of autonomy to the Palestinians and the stepwise withdrawal of Israel from the territories, the geopolitical map of Israel may undergo further changes in the near future, so that more new regional and development plans will be needed to solve problems and to establish peace in the region.

References

Adiv, A. and M. Schwartz, 1992, *Sharon's Star Wars: Israel's Seven Star Settlement Plan*, Jerusalem: Hanizotz A-Sharara Publ. House.

Central Bureau of Statistics, 1993, *Statistical Abstract of Israel*, Central Bureau of Statistics, Jerusalem, 43:513, 522, 531-532.

Efrat, Elisha, 1994, "Israel's Planned New 'Crossing Highway' ", *Journal of Transport Geography*, 2:4:274-277.

Efrat, Elisha and Allen G. Noble, 1988, "Problems of Reunified Jerusalem", *Cities, the International Quarterly of Urban Policy*, 5:4:326-343.

Ministry of Construction and Housing, 1991, "Highway No. 6", 2 vols, (in Hebrew).

Noble, Allen G. and Elisha Efrat, 1990, "Geography of the Intifada", *Geographical Review*, 80:3:287-307.

Wolf, A.T. 1995, *Hydropolitics along the Jordan River*, New York: UN University Press.

4 Regional Planning and Development as Affected by Tamil Insurgency in Sri Lanka

SHANTHA K. HENNAYAKE, BERNARD L. PANDITHARATNE

Post-independence development in Sri Lanka faced many external and internal challenges. Of the internal challenges Tamil insurgency has been the most damaging, both in spatial extent and in magnitude. Tamil insurgency increased in intensity from the early 1980s. Thus, it is also the longest continuing challenge to development in Sri Lanka. Tamil insurgency is also the most extensive in its damage, as its impact transcends economic development to other areas of life as well. Unlike other internal challenges to development, Tamil insurgency has led to the destruction of development achieved earlier.

A recent government publication summarized the impact of Tamil insurgency as follows: "It is the ethnic crisis which has evolved into the war which currently engulfs the North East. It has adversely affected every facet of Sri Lankan public life, seriously impairing every progressive move that we as a nation, have striven to make". As this document states, the areas most affected by the Tamil insurgency are in the Northern and Eastern Provinces (Figure 4-1). Jaffna peninsula in the north was most affected and was under the occupation of Tamil insurgents for over a decade. This chapter will pay special attention to the impact which Tamil insurgency has had on the regional development of Jaffna. The people of Jaffna, who enjoyed one of the highest levels of development before the inception of the insurgency, today as a consequence of it, experience the lowest quality of life of any group in the country. The ultimate irony of Tamil insurgency is that in promising a better future for the Tamils, it has delivered a worse present.

The Sri Lankan government has, after about 15 years, freed the Jaffna peninsula from the threat of Tamil insurgents and reestablished its authority. The government is engaged in a massive rehabilitation and reconstruction effort to improve the quality of life of people living in Jaffna. However, the Tamil insurgents have threatened to disrupt and sabotage government efforts. Henry Kissinger's famous statement is very valid here: "The guerrilla wins if he does

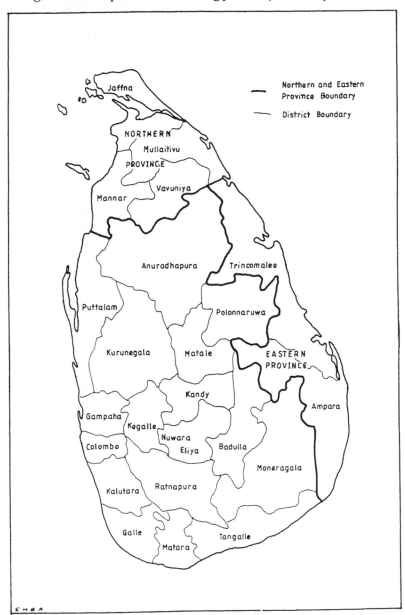

Figure 4-1   Sri Lanka: Provincial and District Map

not lose. The army loses if it does not win" (Athas 1996,7). The development efforts of the Sri Lankan government at the national level, and more particularly in Jaffna peninsula, will not be successful unless the Tamil insurgency is defeated fully. From the point of view of development, bringing an end to the Tamil insurgency should receive the highest priority in Sri Lanka today.

Regional Planning in Sri Lanka

Regional development is not a new experience in Sri Lanka. In ancient times, when accessibility was limited between different regions of the island, and a fair degree of regional autonomy was a fact of life, regional development occurred as a necessity. However, political and economic centralization during the colonial era, especially during the British period, effectively undermined any vestiges of regional autonomy that may have prevailed in the island. The linking of different parts of the country through a modern transport network and an efficient administrative machinery made the centralization very effective and strong. Except on a few occasions where the colonial rulers paid attention to regional problems in the periphery, such as rehabilitation of dilapidated irrigation reservoirs, no deliberate attempt was made for any kind of regional planning or development. Regional development did take place, but it was spontaneous and unintentional.

Post-independence Sri Lanka inherited from the colonial rulers a spatially uneven economic structure favouring Colombo (Samarasinghe 1982). Although the governments in independent Sri Lanka may have realized the need for a more spatially balanced economic development, they had to work through a machinery which was highly centralized, both institutionally as well as in thinking. Development and economic problems in post-independence Sri Lanka have always been conceptualized in national terms. Poverty, inflation, agrarian and agricultural problems, lack of industrialization, and rapid growth of population, are in fact national problems, but the regional variation in these issues did not receive adequate attention. In any case, unlike in many other Third World countries, the difference in the overall quality of life of people in different parts of the country was not extreme. Thus, explicit regional development planning and development was slow to come into the policy-making circles in the country. If there was any development activity in the periphery, it was not a product of deliberate regional planning, but it was a remote manifestation of national planning and development strategy.

The centralized political parties always gave prominence to national level economic and developmental problems. The masses readily identified with these issues as people shared these problems irrespective of their regional origin. The political manifestos of the various political parties had very little to say about regional problems and needs, although the individual politicians did emphasize very local needs and problems. In fact the centralized political parties can be seen as a machinery which overlooked and underestimated the regional development issues and needs. Chambers (1985, 3) argued in relation to peasantry in Sri Lanka that the "lack of an effective representation of the class or occupational interests of smallholders in the political process seemed to be associated with a tutelary, custodial or paternalistic attitude towards the 'peasantry' on the part of the political and administrative elite". To a large extent this argument is valid in explaining the lack of interest in regional issues as well.

The leftist political parties which were quite prominent in the early post-independence era highlighted the inequalities of the society. Based on a dogmatic acceptance of Marxist interpretation of class divisions of the society, these parties were oblivious to the regional disparities in the Sri Lankan economy. If at all, as in the case of the flight of the plantation labor, it was because, they were seen as the most exploited part of the proletariat. Categories such as rural, urban and plantation, which according to the Central Bank reports were used extensively, were conceptualized and treated by policy makers and the bureaucrats simply as sectors and not as regions. The spatial dimension of these categories were purely incidental and for some accidental.

An equally significant feature contributing to the lack of attention paid to regional planning and development is the general lack of demand from the periphery for such attention. Regional identities that prevailed in Sri Lanka during the pre-colonial era steadily lost their significance during the colonial period. Those that remained were very largely confined to the social and cultural spheres. During the post-independence era even the cultural and social significance of these regional identities have been eroding at an ever faster rate. As a result, regional identities never found their expression in the political arena. Sri Lanka has never had any significant regional political parties. A few such did contest at past elections, but their very poor performance was a clear testimony to the lack of regional identities. The Tamil parties, such as the Federal Party, Tamil Congress and Tamil United Liberation Front had their power base in Jaffna, but the "Tamil" element received much higher priority almost to the total neglect of the "regional" element.

At the same time, there has been a steady decline in the number of independent candidates contesting parliamentary elections. An even more significant development is the total elimination of independent members from the Parliament. These developments are a result of the rise of the national parties, and are indicative of the declining importance of region and locality. The decrease in number of independent candidates reduces the potential for the development of regional and local affinities.

The smallness of the country also contributed in many ways to undermine regional emphasis in development. In the first place, the interaction between the people of different parts of the country was quite prominent especially in the economic arena. This can be seen directly in the participation of people from one part of the country in economic activities in other parts. Indirectly, the regional economies are linked with each other, of course through Colombo. Thus, the fortunes of a region are not intrinsic to the region itself, but to the wider national economy.

Perhaps, more than anything else, the rise of non-regional affinities may have contributed in large measure to undermine the demand for regional development. The role of national level political parties is very significant. As two major political parties have grown in the post-independence era, people irrespective of their regional origin, began to think and behave as members of one or the other of the two parties. Consequently, regional issues and interest took a back seat in the political sphere. Agnew (1987) however, argues that increasing significance of national level politics or the "nationalization thesis" as he called it, can in turn spark regional and local reactions. To some extent the rise of Tamil ethno-nationalist politics can be explained from this point of view.

The rise of significant ethnic identities in Sri Lankan society may have been the most important force behind undermining regional interests. Heightened ethnic identity has led to political polarization of the two ethnic groups. The consequence of this is to ethnicize even non-ethnic processes such as development, migration, and education (Hennayake 1995). In the case of the Tamils, as a large number of them are concentrated in the northern and eastern provinces, their issues have been expressed as ethno-regional issues of those two provinces. Although a direct link between ethnicity and region may not necessarily be accurate, making such a link has been politically very effective for the Tamils. It is the political efficacy that led ethno-nationalists to invent a Tamil homeland from the lands of the northern and eastern provinces (Silva 1987). The concept of invention of a homeland is partially borrowed from Hobsbawm and Ranger (1983). Many of the cultural/political phenomena that appear to be ancient are in fact very recent creations. Although a *homeland* by definition should have an ancient origin, that of the Tamils is very recent going back only to mid-1970s.

For all the reasons outlined, regional planning and development did not receive sufficient attention in Sri Lanka until the 1970s. That decade is often considered as the "decade of the crises". The development thrust, which started immediately following World War II, was slowing down in the western world and in some areas of the developing world, especially in the former colonial countries in Asia, Africa and Latin America. A crisis was manifested in political, economic, socio-cultural and political problems experienced by almost all countries of the world, and Sri Lanka was no exception.

The impact of the economic problems was socially and spatially unequal and the segment of the population most affected by these problems was the rural poor. With population expansion, economic opportunities in the agricultural sector were quickly exhausted, and in areas where alternative employment opportunities were not created, the problem became serious. The people in these areas, especially if they felt subject to economic deprivation, increasingly came to realize that they could not depend on the government for their economic development. On the other hand, the government itself realized that national planning and development could not answer the problems in certain underdeveloped regions as these regions were linked to the core through a dependency relationship. Instead the government realized that the development issues and needs of different regions had to be addressed separately and deliberately. The strategy chosen by the government to tackle this problem was regional planning and development.

Deliberate regional planning and development thus became a prominent development strategy from about the middle of the 1970s. The rise of Tamil ethno-nationalist politics which blamed the government for "deliberate discrimination of Tamil areas", also put pressure on the government to implement regional development. Thus, the Tamil political institutions can be credited to some extent for highlighting the need for regional development as an important and necessary element in the overall development strategy in Sri Lanka. The early experiments in the 1970s in this direction gave only limited institutional and financial resources to the regions. Over the years, these resources were strengthened. The culmination of this process was the establishment of Provincial Councils within a framework of devolution of power introduced by the 13th Amendment to the Constitution.

From Development to Tamil Insurgency

Tamil insurgency in its present form is a guerrilla war waged against the Sri Lankan state to establish a separate state called "Eelam" in the areas within the northern and eastern provinces. Although Tamil guerrilla warfare has a recent origin going back only to 1983, the Tamil demand for political autonomy has a longer history.

Tamil ethno-nationalism first emerged in 1949 with the establishment of the Federal Party, which demanded that the northern and eastern provinces be made a State within a Federated Ceylon. The Federal Party manifesto argued

The Tamil speaking people in Ceylon constitute a nation distinct from that of the Singalese by every fundamental test of nationhood...which makes Tamil fully adequate for all present-day needs and finally by reason of their territorial habitation of definite areas.

However, the Federal Party did not enjoy the full support of the Tamil people. While keeping its original goal intact, the Party tried to enter into some agreements with the major political parties to achieve some degree of political autonomy, but all such attempts ended in failure. These failed attempts created a considerable degree of frustration among the Tamil political elite. Ethnic riots which arose subsequent to the passing of legislation which made Sinhala, the only official language in 1956, was a turning point. The mass of the Tamil people were aroused to make common cause with what had hitherto been confined only to the political elite.

After a period of relative silence in the 1960s when Tamil political parties supported the governing political party, Tamil concerns were heightened during the constitutional debates in 1972. Tamil political parties demanded a federal structure once again, but by this time the state was constitutionally defined as unitary, effectively making any demand for federalism extra-legal. Reacting to the lack of attention paid to their demands, the frustrated Tamil political parties were united for the first time and in 1977 they collectively formed the Tamil United Liberation Front (TULF), whose aim was to establish a separate state for Tamils.

TULF used extreme rhetoric in their election campaign and created unrealistic hopes among the young and unemployed segments of the Tamil population, who as a result of the campaign had become an anti-Sinhalese and anti-Sri Lankan group. The euphoria of the convincing victory by TULF in the areas they contested was soon lost when the youths realized that TULF could not deliver on its promises. Youths began to organize themselves in secret groups to nullify the writ of the Sri Lankan government in the northern province, especially the Jaffna peninsula. All democratic and guerrilla Tamil organizations have justified their demand for political autonomy on, among others, economic factors. The main issues were as follows:

• Jaffna has been deprived of government investments.
• University standardization discriminated against the Tamils.
• Colonization schemes were an "invasion" of the Tamil homeland.
• Tamils were deprived of government sector employment.
• The Tamil cultural landscape is becoming Sinhalized.

Before we analyze these claims, it is important to understand the development picture of the Tamil areas. No significant difference exists between

Tamil areas and the other areas of the country or between Tamils and Sinhalese in terms of development (Islam 1990, 15-30). Both Tamil and other areas have a wide variation in development. For example, the Jaffna peninsula is one of the most developed regions in the country with a level comparable with that of Colombo and Kandy measured in terms of a "quality of life" index. On the other hand, other Tamil areas such as Mannar, Kilinochchi and Mullaitivu are some of the least developed areas of the country ranking with the non-Tamil areas of Hambatoata, Moneragala and Ampara and Polonnaruwa districts. An overall assessment that Tamil areas are less developed than other areas is incorrect. Development and underdevelopment are found in all areas of the country.

Many economic problems are not limited to Jaffna Tamils, but are shared by most people living in the periphery irrespective of their ethnicity. As an ethnic group concentrated heavily in the periphery, (over 75 percent of the Tamils live in the northern and eastern provinces) the Tamils are affected by the general disadvantages of such location. Interpreting the problems of the periphery as affecting only the Tamils is incorrect (Hennayake 1991, 171-237).

No evidence suggests that the Sri Lankan governments have deliberately deprived the Tamil area in general, or Jaffna in particular, of government investments. The development of the service sector is a function of population density and thus Jaffna has always received a high share of state expenditure for services, followed by the coastal belt south of Batticaloa in the Eastern Province. The rest of the Tamil areas are very thinly populated and therefore the state investment in services in these areas have been significantly lower. The distribution of investment on infrastructure development also closely followed this pattern.

In contrast the state investment in agricultural development is very low in Jaffna as the potential for agricultural development projects is very limited there. However, the Eastern Province did receive significant government investment for irrigation rehabilitation projects from as early as the mid 1930s.

State investment in industrialization traditionally has been low in Sri Lanka. The greater emphasis paid to the agricultural sector, on the one hand, and the non-availability of industrial raw material, on the other, were the two main reasons for the lack of a sound industrial development. Of the government-sponsored large scale industries in Sri Lanka, those based on imported raw material were located in the vicinity of Colombo, and those based on local raw materials were located closer to their source. Thus two of the largest industrial establishments, cement and chemicals were located in the Northern Province. The Eastern Province housed sugar and paper industries. If any area in the country did not receive state investment on industries, it is the agricultural Dry Zone interior, where the population is largely Sinhalese.

Educational facilities in Jaffna peninsula, especially secondary school education, are among the best in the country (Silva 1977; Gunawardena 1984). The number of Tamil students entering university, especially in science-based programs such as medicine and engineering, has been exceptionally high. The share of Tamils in universities was much higher than their national population ratio or even the population ratio in the Tamil areas.

The ethnic and regional standardization of entrance marks introduced in the 1970s led to a decline of Tamil students entering the university system. One of the purposes of standardization was to give an opportunity for students from backward areas. Thus while the number of students entering the universities from major centers like Colombo, Kandy and Jaffna declined, those entering from backward regions including the Northern and Eastern Provinces increased rapidly. The impact of standardization was not confined to Tamil areas alone. After ethnic standardization was abolished in the early 1980s, the number of Tamil students entering the university remained well above their national population ratio (Committee for Rational Development 1984).

One of the issues that has been most vociferously presented by the Tamil ethno-nationalists is that the "Tamil homeland has been invaded by the Sinhalese colonization schemes". A careful analysis done at a smaller scale than that of the provinces, indicates that almost all colonization schemes were located in areas that have always had a Sinhalese majority and that they have not displaced Tamils.

The Impact of Tamil Insurgency on National Development

The problem of regional development cannot be discussed meaningfully by isolating it from national development. In Sri Lanka, planning and development are still very largely centralized despite the establishment since the late 1980s of the Provincial Councils. Planning for regional development, as well as the allocation of resources, is carried out by the agencies of the central government. Thus the fortune of the central government has a direct bearing on the regional development efforts.

Of various factors that have had negative impacts on national development the Tamil insurgency undoubtedly heads the list. Its effect has been devastating in a number of ways.

Destabilization of the economy is one of the most brutalizing effects of the civil war. A stable society and economy are essential requirements for sustained development. Since the early 1980s, Sri Lanka has been deprived of this essential requirement. The country has lost prominent statesmen because of assassinations and has been thrown into chaos and confusion on many occasions during the last ten years. Although the society has been resilient enough to bear such shocks, the actual loss and the extent of the damage will never be known. A number of development programs could not be sustained because the initiators and implementors were lost in these assassinations. The economy responds to politics very closely. Stable politics will ensure stability in the economy promoting extended periods of growth. However, unstable politics will deter investment as there is no reliable guarantee for the security of the investment, as well as a return on the investment. As a result, the country has been unable to maintain a steady rate of growth during the last decade. The roller-coaster of the Sri Lankan economy has manifested itself in other areas which will be discussed below.

The civil war has led to a general disruption of ordinary life and work in the country and the disruption has peaked during such occasions as massive bomb attacks and assassinations. The civil war has compelled the government to take precautionary security measures which have greatly inconvenienced and restrained the population. Road blocks, closing of roads, security booths at the entrances to government institutions, frequent checking of passengers travelling in public and private transport are some examples of how the war has impacted ordinary life. The landscape of the country has become militarized because of the omnipresence of the military throughout the country and especially in the major cities. Away from the center, in the border villages of the northern, eastern and north central provinces, the disruption to ordinary life has reached its peak. It is such a common occurrence that disruption to life has become ordinary. In many villages, daily life is patterned by regular migration to the jungle or to a common religious compound for the night and to the houses and the fields for the day. In some places the disruption is complete; people have lost not only their houses but entire villages and have been forced to live as refugees. Although it is difficult to quantify the disruption to ordinary life and the consequent uncertainty that has settled in the society, they nevertheless are significant in both monetary and human terms. With this type of uncertainty, and disruption, sustained development has become a difficult goal for Sri Lanka. It is almost unethical to preach development to the victims of the war.

Direct Cost of the War

The war is not just non-productive; it is destructive. Before the outbreak of the civil war, the Sri Lankan defense forces were small, maintained primarily as a ceremonial force to parade at Independence Day celebrations. Defense expenditure was one of the lowest in the world and its share in the government budget was negligible during this era. However, with increasing guerrilla activities since the late 1970s, defense expenditure has risen rapidly (Table 4-1).

Table 4-1 Sri Lanka: Increase in Defense Expenditure

Year	Rupees (Billions)	Percent of GDP
1977	0.4	1
1984	0.8	3.7
1993	8.0	-
1994	20.0	-
1995	28.0	4.7
1996	46.0	6.5

Source: Government of Sri Lanka

The expanding share of defense expenditure simultaneously reduces that of other sectors. For example, the defense expenditure was 12 percent and 20 percent of government expenditure in 1995 and 1996, respectively. A developing country such as Sri Lanka, which desperately needs to increase the quality of life of its people, cannot afford to spend 20 percent of its government expenditure on a war which brings no returns. One of the best investments that can be made to achieve sustained development is for education because it leads to the development of human resources of future generations. Yet in 1996, Sri Lanka spent four times more on defense than on education. The daily cost of the civil war in 1996 was over Rs. 125 million (US $2.27 million). This is a large sum of money using any standard. The basic infrastructure required to achieve sustained development is not sufficiently available in Sri Lanka; in fact it is a very poor nation. The transportation system in the country is very unsatisfactory. Many people still do not have access to purified water and electricity to their houses. In the cities, sewage and solid waste disposal systems are malfunctioning. Educational and health facilities are not adequate. Mere basic literacy, unsupported by professional training, is not useful to an economy. Vocational and professional training is in a poor state. Sri Lanka can no longer boast of a high quality of life - a legacy of past welfare investments by the state.

If one follows the simple logic of "cost incurred is a benefit foregone", the civil war expenditure has deprived the country of much needed capital for development to the tune of 125 million rupees a day in 1996. This is only the cost that the government has incurred. The overall cost borne by the entire society is much higher. A conservative estimate indicates that the country loses about Rs. 250 million a day as a result of the civil war.

The cost of the war is financed by taxing both people and industries. The government deducts 2 percent of the income from those earning between Rs. 15,000-30,000 and 3 percent from the higher income groups. In addition, the government has introduced a National Security (Defense) Levy of 4.5 percent in general and 2 percent for capital goods imported into the country.

The rising cost of the war means reduced investment in economically productive areas. The impact of such non-investment will only be felt by the country in the years ahead. It will postpone attempts to reach the Newly Industrialized Country (NIC) status in the near future. It is clear that Sri Lanka cannot maintain the present level of its economy, let alone reach NIC status with 20 percent of the government income wasted on the war.

Victims of the Civil War

The civil war has produced a very large number of refugees. The numbers increased to almost a million at the height of the Jaffna war in 1996. But as many of the Jaffna Tamils returned to their homes in 1996 defying the Liberation Tigers of Tamil Eelam (LTTE) ban, the number has declined to about half a million, including several special categories of refugees. Over 200,000 refugees currently live in refugee camps set up by government, and another 100,000 people have found refuge with their relatives. About 150,000 people have fled to South India and only a small number have returned. Over 150,000 Tamils have used the civil war to seek greener pastures in western countries. Over 15,000 fishing families have become refugees because the civil war totally halted fishing activities in the North. At its height, the civil war produced over 500,000 refugees

Unfortunately, no serious effort has been made to estimate the loss incurred by the refugees. A rough estimate indicates that it may be well over Rs. 50 billion. This is certainly a catastrophic loss for a developing country such as Sri Lanka.

All the Sinhalese and Muslims who lived in Jaffna prior to 1983 have become refugees, pushed out by the LTTE. A large number of Tamils also left Jaffna for Tamil Nadu and Colombo. In the border areas, many Sinhalese farmers have become refugees. Because refugees are unable to satisfy even basic needs, they become dependent. The country loses not only their economic contribution, but also it has to look after them. Thus, a refugee represents a double loss to the country. The long term impact of refugees is much greater especially because the younger generation is deprived of access to normal education, good health and sufficient nutrition. In refugee camps, the younger generation is brought up in a totally alien social environment, not conducive to a healthy development of a growing child. An entire generation of refugees has developed a hatred towards the larger society for allowing them to become refugees. The impact of the war will not end with the end of the war.

The civil war has resulted in the deaths of a large number of people and has made an even larger number physically handicapped or invalid. During the late 1980s, Sri Lanka ranked high in number of deaths caused by civil unrest. According to one estimate, more than 50,000 people have died since the separatist campaign erupted in 1972 (*The Australian*, 21/6/1996), and this may be a conservative estimate. Although data are not available on the number of people who have become physically handicapped as a result of the war, rough estimates put the number around 75,000.

The largest number of the casualties are from the productive age group (18-55 years). What is lost in the war is the future generation. With every life lost, a family loses part of its means of livelihood as well. Often when a person becomes physically handicapped, a whole family becomes dependent. Sustained development cannot be achieved by a society in which numbers of dependants are rising and in which its younger generation especially is lost to a senseless war.

Declining Foreign Investments

As a third world country Sri Lanka is forced to depend on foreign investment for much of its economic development. Foreign investment has come in many forms. Two of the most prominent sources were through tourism and direct investment by multi-national corporations. Since the early 1980s, international tourism had replaced exports of tea as the largest foreign exchange earner. Foreign exchange earnings depend on the presence of a stable economy and society. A country torn apart by civil war is seen as a dangerous place, both by people and corporations in most Western countries. In fact the United States has issued frequent travel advisories to its citizens who intend to visit Sri Lanka, not to travel to the northern and eastern parts of the island. Images of burning buildings, government security installations, dead bodies of innocent civilians, and confrontations between armed forces and guerrillas are distributed all over the world by international news media. Sri Lanka is portrayed as a very dangerous and unsafe place. Tamil propaganda has further intensified this negative image by alleging that the Sri Lankan government is fighting against an innocent ethnic group, who want to apply political pressure on the government.

Beginning in 1983, tourist arrivals steadily declined, although they have recovered slightly in the 1990s. Tourism lost its place as the largest foreign exchange earner by the early 1980s. Sri Lanka has yet to regain the peak tourist arrivals reached in 1982. As long as the civil war continues, even the best efforts to sell Sri Lanka as an attractive destination will not achieve much.

Sri Lanka regularly appeared in international news accounts because of ethnic riots and frequent bloody confrontations between the armed forces and guerrillas. Although the operational areas are located far away from Colombo, the economic hub of the country, it too is severely affected. Guerrillas have launched two major attacks on Colombo, both occasions leading to tremendous destruction. Guerrilla attacks on Colombo made most foreign investors think twice about investing in the country. Most important of all, assassinations of national and political leaders have frightened foreign investors, because such events have created doubt in their minds of the ability of the Sri Lankan government to provide security for their investments. It is impossible to estimate the potential loss of foreign investment as a result of these activities. In spite of the best efforts by the government to convince foreign investors of the merit of investing in Sri Lanka, the country is not attracting as much investment as possible. Among the factors responsible for this, the civil war is the foremost.

The Impact of Tamil Insurgency on Regional Development

If national development in Sri Lanka has been affected by Tamil insurgency, regional development in the Northern Province and especially Jaffna has been devastated. The LTTE was successful in bringing under their control the entire Jaffna peninsula, except for the military base in Palali, and much of the remaining parts of the Northern Province. In the Eastern Province their control was limited to the unpopulated jungles.

Before the Tamil insurgency emerged, Jaffna peninsula was the most populous part of the two provinces. Thus, except for the sandy beaches and the areas that came under flooding frequently, much of Jaffna peninsula was developed either as settlements or as a thriving agricultural region, specializing in vegetables, cash crops and paddy. Large tracts of paddy land existed in the adjacent mainland as well. Jaffna always had an agricultural surplus which was sent to markets such as Colombo and Kandy. Both provinces also had a thriving fishing industry and a significant part of the catch was sent to Colombo and Kandy markets.

Jaffna was also the location for several major industries, such as cement and chemicals. The Jaffna peninsula also had a thriving network of smaller, consumer-oriented industries catering mainly to the needs of the people in the peninsula.

The city of Jaffna was the fourth largest urban and service center in the island in 1981. Jaffna provided services to the peninsula and also acted as the collection center for agricultural produce for redistribution within the peninsula and to Colombo and other centers in the south. Jaffna was the most thriving commercial and financial center of northern Sri Lanka.

As the administrative capital of the Northern Province, Jaffna housed most of the government institutions. As the major service center, it also contained some of the best and most well equipped educational and health institutions in the country. Jaffna was the political and cultural center of the Sri Lankan Tamils and housed some of the most venerated cultural and religious institutions of the Tamils in Sri Lanka. Thus Jaffna, was seen as a secondary capital of the country by the Sri Lanka Tamils living in the north.

If any area in Sri Lanka can claim to have a regional economy, it is Jaffna. The relative spatial remoteness from the main economic hub of the country may be the main reason for this phenomenon. The economic history of Jaffna is also different from other parts of the island. Prior to the colonial era, no significant economic interaction existed between different parts of the island, and the Jaffna economy was very much isolated. During the colonial period, as the different parts of the island were connected through rail and road transport networks, Jaffna was slowly but steadily linked to the national economy in a number of ways. However, the remoteness of Jaffna remained, and thus restricted the regional economy of Jaffna.

One of the most significant aspects in Jaffna's becoming a source of human resources, started in the colonial period and continued into the post-colonial period. During the earlier period, missionaries developed very good educational institutions in Jaffna which became some of the best schools in the country. By the end of the colonial period, Jaffna was blessed with a large number of such educational institutions, whose graduates found employment in the colonial administration at various levels both in Sri Lanka and also in other colonies, such as Malaya. A large number of graduates were admitted to the university and subsequently found employment in prestigious positions in government sector institutions throughout the country. A part of their earnings was remitted back to Jaffna. This cash flow was so significant that the Jaffna economy was called a "postal-order economy" at one point. The process continues even today, although at a reduced level.

During the post-colonial period, the welfare policies of the government benefited the entire population of the country, but regional variation in this was clearly apparent. Colombo because of its privileged position in all spheres always enjoyed a higher level of development. Outside Colombo, there were a few regions that also enjoyed a fairly high level of development. Jaffna was among them. In fact as a region, Jaffna District was second only to Colombo in terms of overall development as evidenced from the quality of life index.

All that Tamil areas in general and Jaffna peninsula in particular, have achieved during centuries, was very largely lost in a period of less than fifteen years of civil war. The destruction of the Jaffna region was so immense that it may take several decades to bring it back to its position in the mid-1980s.

Destruction of Infrastructural Facilities

The destruction of infrastructural facilities has been an integral part of the strategy of the Tamil guerrillas. In one of the most recent bomb attacks, the LTTE destroyed the center of Colombo including the Central Bank of Sri Lanka and a few months earlier, they attacked and destroyed the petroleum storing facilities in the vicinity of Colombo.

The LTTE destroyed the industrial base in Jaffna by vandalizing the major cement, chemical and glass factories. The electricity grid to Jaffna from Vavunia and within Jaffna peninsula, and in many parts of the Eastern Province has been totally destroyed. Jaffna has been without electricity for the last thirteen years after the LTTE blew up the Chunnakam power plant. The railway and the main highway from Vavunia to Jaffna have been destroyed. Such destruction makes it very difficult to sustain current levels of production and services, let alone to provide an increase in production.

In the war zone, many houses, offices, schools, hospitals, and other private and government property have been destroyed. Jaffna hospital, for example, declined from 30 wards with 1015 beds to 17 wards with 380 bed by 1996. Similarly, of the 470 schools in Jaffna peninsula only 266 were functioning in

1996. The attendance of teachers was down to less than 40 percent and students 60 percent in Jaffna schools in 1996.

Towns, such as Kilinochchi and Paranthan, are damaged perhaps beyond repair. The destruction is so vast that the total cost of it may never be known. According to Dr. Michael Schubert of Medical Emergency Relief International, in Jaffna town, 60 percent of buildings were in need of repair. An estimated five percent of all buildings have been severely damaged as a result of the civil war. An entire regional economy has to be created anew in Jaffna.

Emergence of a War Culture

Sustained development requires an attitudinal change towards a condition of being altruistic, future oriented and socially responsible, but the civil war has created a culture that cultivates hatred, an emphasis on short term gains, and irresponsibility. "As wounded men may limp though life, so our war minds may not regain the balance of their thoughts for decades" (Colby 1926). The entire Sri Lankan society is now divided into two groups; those who instigate fear and uncertainty and those who live in fear and uncertainty. The civil war has made death and brutal violence against human beings such a common occurrence that Sri Lankans have become indifferent towards death. Incidents of children playing with toy guns and imitating the war, and looking up to the armed military men or guerrillas as role models, and elders settling their petty disagreements by killing each other have risen phenomenally in this decade. A war culture and its practice are visible not only on the war fronts, but also on the social front as well.

While the impact of the war culture is generally felt by the society at large, its impact in Jaffna is profound. Jaffna society has been transformed in a way that will take decades to correct. Parents of children in Jaffna schools found themselves helpless when the LTTE rebels forcibly conscripted the young pupils into their ranks. Many boys joining the LTTE are about twelve years old (UTHR 1990, 39). The attempts by parents to prevent the young from joining the LTTE could bring disastrous consequences to the child and the parents. Jaffna society is subdued, muted and totally suppressed with no room for free thinking. The society has turned into a totalitarian society by LTTE rebels. As G.L. Peiris (1996) pointed out "in their [LTTE] courts, boys of 21 and 23 years of age functioned as judges. People were sentenced to death. Those sentences were carried out" ("The Rising Price of War", *Ceylon Daily News* 01/07/1996).

Most families in Jaffna are in disarray with sons and daughters being forcefully inducted into the LTTE guerrilla force. In addition, over 2000 persons are detained in LTTE camps (UTHR 1990, 38). Some families have found that their relations and friends have fled to refugee camps in India, where the number of Sri Lankan Tamil refugees is over 150,000. According to UTHR, (1991, 5) "... the workings of the civil society have ceased to exist in a large category of instances". A generation is growing up without knowing that there used to be such things as post mortems, magistrate's inquiries and accountability before law and the humanity in this country has been devalued..."

Non-availability of a Large Part of Sri Lanka for Development

One of the biggest impediments of the civil war to sustained development is to make almost all the Northern Province and a fairly large part of the Eastern Province unavailable for the national development efforts of the country. Together these two provinces constitute about 20 percent of the land area and about 65 percent of the coastal belt. Some of the major minerals deposits (i.e. limestone, clay, illmonite) extracted for industrial purposes are located in these two provinces.

The Northern Province contributed about 4 percent of the national GDP, while the Eastern Province contributed about 7 percent in the late 1980s. However, since the late 1980s, most of the resources and production of the Northern Province have not been available to the national economy because the LTTE rebels have isolated it from the rest of the country. Although the Eastern Province did not entirely come under rebel control, many parts of the province were simply too dangerous to permit any development activity.

The Northern and Eastern provinces possess some of the finest beaches in the country with great potential for tourism. However, these beautiful beaches remain unutilized with the ruins of a few partially constructed hotels remaining as grim reminders of the impact of the civil war. All this is living testimony to the inability of the government to develop land and other resources in the LTTE-held territory.

Severance of the Link Between the North and Other Parts of the Island

Prior to the civil war, the economies of the northern and eastern parts of the island were linked to the economy of the rest of the island in numerous ways. Various industrial products (e.g. cement, chemicals, salt, paper) agricultural produce (paddy, chilies, tobacco, onions) and fish (fresh fish, dry fish, canned fish) were supplied by these two provinces to other areas. The severance of the link between the Northern Province and the rest of the country has stopped interregional trade and the Jaffna producers and traders have lost a lucrative source of income. The Northern Province, for example, at one point supplied over 50 percent of all the locally produced cement and chemicals, and about 40 percent of all fish. About 15 percent of all vegetables and 20 percent of other crops such as chilies and onions also came from Jaffna. After the rebels cut off Jaffna from the rest of the island, this economic interaction stopped. Amal Jayasinghe (1996, 2) has pointed out that the region has not contributed to the country's economy in the five years under rebel control. One of the reasons why Sri Lanka has failed to be self-sufficient in rice and has had to import a sizable part of its fish, vegetables and even fruit is because of this severance of economic interaction.

Increasing Spatial Unevenness in the Economy

The isolation of Jaffna from the rest of the island for over a decade and the inability of the government to function there prevented it from maintaining its services i.e., education, health, and infrastructure facilities such as communication, transportation and electricity. To make matters worse, the LTTE rebels have destroyed most of the infrastructure in the areas held by them. Because the LTTE has prevented the government from coming into the areas held by them, it could not initiate any new development activities there. The lack of infrastructure control of Jaffna by LTTE and especially the extortion of money from businessmen in Jaffna, prevented any new investment being made in Jaffna by private entrepreneurs. The destruction of existing facilities and the inability to start new development have increased the disparity between the LTTE-held parts and the rest of the country. In fact, what has been happening in Jaffna peninsula is not simply underdevelopment but literally "de-development." The net result is that "people are already starving, infants are dying due to want of milk powder, medicine is in short supply and medical aid is very minimum" (UTHR 1991, 53). The Jaffna economy and the standard of living today are at the lowest levels since independence.

The share of the industries in Jaffna peninsula is nil in 1996, while the share of the Northern Province and Eastern Province taken together is also almost negligible (Table 4-2). The rebel strategy of merely holding territory has completely destroyed the once vibrant economy of Jaffna and pushed it to the lowest level in the country.

Table 4-2   Sri Lanka: Regional Distribution of Industries

District	Industries under IDB	Total
Colombo	1211	1607
Gampaha	171	490
Kalutara	45	87
Galle	26	68
Kandy	30	65
Puttlam	17	55
Kurunegala	9	39
Matara	22	29
Hambantotta	18	27
Rathnapura	1	24
Nuwara Eliya	3	23
Badulla	9	22
Anuradapura	5	21
Kegalle	2	13
Matale	4	11
Ampara	4	7
Moneragala	1	7
Polonnaruwa	-	5
Vavunia	-	2
Trincomalee	-	2
Batticaloa	-	1
Total	1578	2605

All of this has led to severe economic problems in Jaffna. Because the trade between Jaffna and the rest of the country has virtually stopped, many consumer services and goods are simply not available to Jaffna residents. To curb the guerrilla activities the government banned a number of goods such as batteries, petroleum products, and agricultural chemicals. A black market emerged and together these two factors contributed to inflation in Jaffna (Table 4-3). Jaffna's standard of living which was one of the highest in the country in 1983 had deteriorated by 1994 to one of the lowest.

Table 4-3   Inflation in Jaffna, 1996

Item	Jaffna	Rest of the Country
Bicycle tire	1300	130
Red rice	100	20
Cabbage (1kg)	120	18
Potatoes	250	50
Kerosene Oil (bottle)	200	25

Source: Government of Sri Lanka

Cost of Rehabilitation

The first step in the rehabilitation of the region is to reconstruct what has been destroyed. Such rehabilitation and reconstruction will be a very substantial endeavour because it has to cover almost every aspect of life in Jaffna. The main components of the program include resettlement, relief assistance, bringing arable land under cultivation, resurrection of small and medium size industries, rehabilitation of roads and of Kankasanturai port, refurbishing schools and the Jaffna general hospital and peripheral hospitals, provision of telephones and of electricity, reviving fishing activities, and the re-establishment of the administrative infrastructure. The cost of the rehabilitation program proposed by the government is US $ 274 million or Rs. 13,000 million. This investment is intended to bring Jaffna back to what it was in the early 1980s. While the Jaffna economy as a result of the war is one and a half decades behind the rest of the country, all the new investment will not change the picture very much because it only replaces what has been lost in the war. It will take a very concerted effort and long years for Jaffna to catch up with the level of development in other parts of the country, especially Colombo.

The proposed program consists of two stages: stage 1, Resettlement and rehabilitation costing US$ 112 million, and stage 2, Economic development, infrastructure, and services costing US $ 162 million. Rehabilitation may not be as easy as the government is contemplating, as long as the LTTE remains a formidable force as it is in the Mullaitivu forests. The LTTE has already warned the government that they may not tolerate the rehabilitation program initiated by the government. It argues that "development and reconstruction (by government) are aimed at blunting the desire among the Tamils for (national) liberation. What is the need for development and rehabilitation after (your) rights, freedom, and self respect have been lost? The people have to be careful about development plans" (LTTE Pamphlet, Untitled, 1996). Such a pronouncement indicates that the LTTE may sabotage the rehabilitation program in Jaffna. As long as the LTTE is engaged is such destructive activities and sabotage, the Jaffna economy will continue to suffer and the region will fall behind the rest of the country.

Solutions

It is clear that Sri Lanka, or for that matter any country in the world, cannot conceivably think of development, let alone regional development while engaging in a protracted a civil war which consumes a large part of its natural, human and financial resources. It is quite evident, that if one is thinking of regional development of Tamil areas, the war must come to an end. Ending the war cannot be done by one party alone; it has to be negotiated between the two major forces involved, the government and the Tamil insurgency led by LTTE. Such negotiation may not be fruitful without a serious proposal towards devolution of power to Tamil regions, especially the Jaffna peninsula and its immediate vicinity, and a government committed to rehabilitate the war-torn areas and to implement the devolution in its true spirit. LTTE at the same time should assure that they will not further sabotage the development efforts by the government.

If Sri Lanka is truly concerned and interested in balanced development, then it needs to build the appropriate politico-social context within which it can be implemented. Creating and maintaining political stability is an urgent need. In the long run this can be achieved only if all Sri Lankans commit themselves to build a Sri Lankan state which can rise above the ethnically defined nations which have led to insurgency in the first place.

Acknowledgements

We wish to thank Dr. Nalani Hennayake for her comments on the first draft of this paper and Mr. S.M.B. Amunugama for preparing the map.

References

Agnew, John, 1987, *Place and Politics: The Geographical Mediation of State and Society*, Boston: Allen and Unwin.

Athas, Iqbal, 1996, "Return of the Enemy", *The Sunday Times* (July 21).

Budget Estimate of the Republic of Sri Lanka for the Financial Year from 01/01/1995 - 31/12/1995.

Colby, Frank M, 1926, "War Minds", *The Colby Essays*, vol 2.

Committee for Rational Development, 1984, *Sri Lanka: The Ethnic Conflict*, Delhi: Navrang.

Department of National Planning, 1995, *Public Investment*, Colombo: Department of National Planning.

Gunawardena, C. 1984, "National Integration of Sri Lanka: A Formidable Task before a Segregated System of Education", *University of Ceylon Review*, 1:4:40-46.

Hennayake, S.K. 1991, *Interactive Ethno-nationalism: Explaining Tamil Ethno-nationalism in Contemporary Sri Lanka*, Ph.D. Dissertation, Syracuse University, pp. 171-237.

Hennayake, S.K. 1995, "Ethnicism in the Studies of Ethnonationalist Politics in Sri Lanka", A paper presented to Ceylon Study Seminar, University of Peradeniya.

Hobsbawm, E. and T. Ranger, (eds), 1983, *The Invention of Tradition*, Cambridge: Cambridge University Press.

Islam, N. 1990, "Ethnic Differentiation, Relative Deprivation and Public Policies in Sri Lanka", *Canadian Review of Studies of Nationalism*, 17:15-30.

Jayasinghe, A. 1996, "Jaffna's Long Road to Rebirth", *Lanka Guardian*, 19:7:2.

Liberation Tigers of Tamil Eelam, 1996, Untitled Pamphlet, No place or publisher given.

Ministry of Justice and Constitutional Affairs, 1996, *Draft Provision of the Constitution Containing the Proposals of the Government of Sri Lanka Relating to Devolution of Power*, Colombo: SLMJCA.

Peiris, G. L. 1996, *Budget Speech*, Colombo: Government of Sri Lanka.

Recommendation on University Admission, 1983, Colombo: Center for Society and Religion.

"The Rising Price of War", 1996, *Ceylon Daily News*, 01/07/1996.

Samarasinghe, L.K.V. 1982, "Spatial Polarization of Colombo: A Case Study of Regional Inequality", *Sri Lanka Journal of Social Sciences*, 5:1:75-95.

Silva, C. R.de, 1977, "Education" in K.M. de Silva (ed) *Sri Lanka: A Survey*, Honolulu: The University Press of Hawaii.

Silva, K. M.de, 1987, *The Traditional Homelands of the Tamils of Sri Lanka: A Historical Appraisal*, Kandy, International Center for Ethnic Studies, Occasional Paper # 1.

Sumanasekara, H.D. 1983, "Measuring the Regional Variation of The Quality of Life in Sri Lanka", *Sri Lanka Journal of Social Sciences*, 2:1:27-40.

University Teachers for Human Rights (UTHR), 1990, *Report No. 5: August: A Bloody Stalemate*, Thirunelvely, Jaffna: UTHR.

University Teachers for Human Rights (UTHR), 1991, *Report No. 6: The Politics of Destruction and the Human Tragedy*, Thirunelvely, Jaffna: UTHR.

5 Population Dynamics and Planning: China and India

GEORGE M. POMEROY, ASHOK K. DUTT

The processes of development and urbanization in a less developed country are inextricably tied with the demographic patterns, dynamics, and prospects of that country. Success of a particular country along its chosen path to development depends in part on success in dealing with any population problems it may have. Though not precise and certainly indirect, relationships exist between population dynamics, such as fertility rates, and socio-economic development indicators such as GNP per capita (Findlay and Findlay 1987; Simpson 1987; Yaukey 1985). Strong correlations exist between measures of development and population. The World Bank Development Report (1980) commented that:

> Fertility is an area of human behavior where individual tastes, religion, culture, and social norms all play a major role. Yet evidence from large groups of people suggests that differences in fertility can be largely explained by differences in their social and economic environment.

Both India and China are particularly significant. First, there is a significance simply because of their massive populations, which together comprise nearly 38 percent of the world's total. Second, each has chosen distinctive population policies and distinctive paths to development. Each of these paths needs to be analyzed and evaluated in terms of successes and failures. In each country, the main concern is the coping strategies used in battling overpopulation defined as the mismatch between available resources and numbers of people. Population growth and size can inhibit development progress (Simpson 1987, 242). What may be learned from the cases of China and India can be applied elsewhere.

One important question is how have the chosen population planning policies and socio-economic development policies influenced population dynamics in each case? In this chapter, within the framework of a descriptive model, evaluation is made of each country's family planning policies, and certain development impacts on population dynamics. Selected demographic variables are compared through time with development indicators and the respective family planning policies adopted.

A General Framework

A number of theories have been put forth regarding demographic transitions. Included among these are classical demographic transition theory, wealth flow theory, rational decision-making theory, and the parental investment model theory (Kaplan 1994, 699). These theories attempt to explain the historical changes in demographic variables, particularly those relating to fertility change, both within and between societies. Typically, demographic transitions, notwithstanding the competing theories upon which they may be based, consist of four well-known stages (Figure 5-1). Stage one is characterized by both high birth and death rates, with population growing slowly. Stage two, a high growth stage, is characterized by rapidly declining death rates, and stable or slowly declining birth rates. Stage three, also a high growth stage, is characterized by low and more slowly declining or low, stable death rates. Stage four is characterized by both low, stable birth and death rates and population levels are stable. The authors of this chapter add a fifth stage, where the death rate stabilizes, and after reaching zero-level of population growth further lowering of fertility leads to population decline (Figure 5-1). Russia, Estonia, and Hungary may tentatively be placed in this category.

While the demographic transition model is based largely upon the historical context of population transitions in Europe, it has also been widely applied and fits reasonably well when considered in the context of demographic changes in other countries (Figure 5-1). However, some countries' demographic transitions cannot be easily fit into its framework (Johnston 1986). The demographic transitions of both China and India have typically been placed within the context of this standard and widely noted demographic transition model. These placements are commented upon later in this chapter.

A common approach in studying fertility is to work backwards "from the most proximate (immediate) cause to the most distant" (Yaukey 1985, 145). This procedure works from the most direct behavioral factors relating to childbirth to the more societal or cultural influences on those direct behavioral factors (Figure 5-2). The proximate causes of fertility are "those kinds of actions which have a direct impact on the reproductive process of women" (Yaukey 1985, 163). These variables may be grouped into three categories - intercourse variables, contraception variables, and gestation variables (Davis and Blake 1956; Bongaarts and Potter 1983).

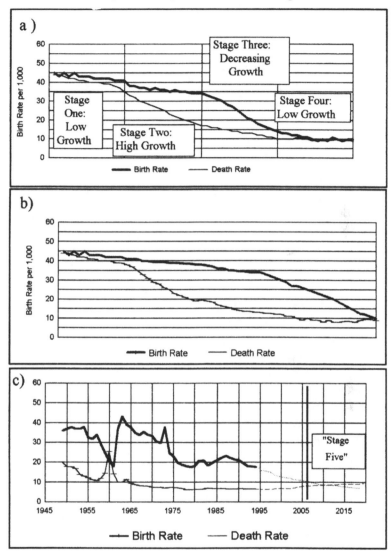

Figure 5-1   Demographic Transition Models

These proximate variables are, in part, derived from behavioral family decisions such as timing of marriage, the number of children to have, children's education, saving and consumption, and work time within and outside home (World Bank 1984, 66). These family behavioral decisions, in turn, are indirectly linked with, 1) socio-economic and cultural factors and, 2) governmental family planning efforts. It is at this point that distinctions may be made be-

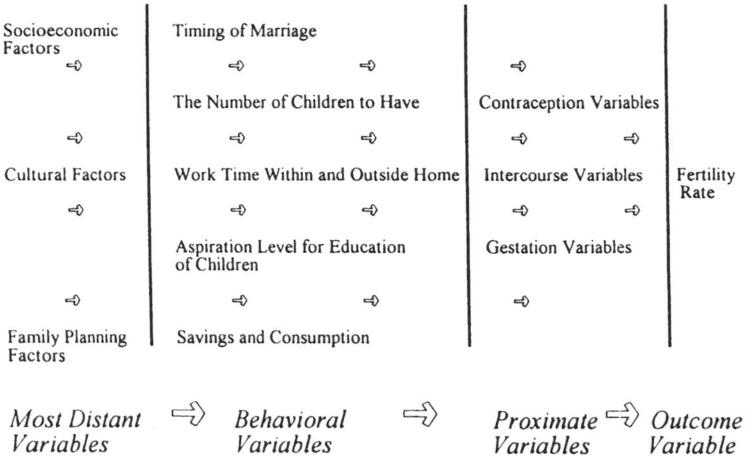

Socioeconomic Factors ⇨	Timing of Marriage ⇨ ⇨	⇨	
	The Number of Children to Have	Contraception Variables	
⇨	⇨ ⇨	⇨ ⇨	
Cultural Factors ⇨	Work Time Within and Outside Home ⇨ ⇨	Intercourse Variables ⇨ ⇨	Fertility Rate
	Aspiration Level for Education of Children	Gestation Variables	
⇨	⇨ ⇨	⇨	
Family Planning Factors	Savings and Consumption		

| *Most Distant Variables* | ⇨ | *Behavioral Variables* | ⇨ | *Proximate Variables* | ⇨ | *Outcome Variable* |

Figure 5-2   Variables Related to Fertility

tween population dynamics, socio-economic development, and family planning efforts. Distinctions between differing national contexts may begin to be made in terms of their impacts. A descriptive demographic model focusing on fertility (Figure 5-3) illustrates the proximate variables in relation to the more directly related intermediate variables and the less directly related socio-economic, cultural, political, and family planning factors.

Numerous social, economic, cultural and political factors have been cited as influencing fertility, as has the role of family planning in influencing fertility transitions (Rele and Alam 1993, 36; Simpson 1987; World Bank 1985). The socio-economic and cultural factors most often noted include urbanization, traditional orientation, religious beliefs and practices, per capita income, income distribution, levels of education, child and infant mortality, and the status of women. With all of these the relationships are broad and general; with connections between proximate variables being very indirect and irregular. Concerning family planning, governments work to "promote the use of contraceptives together with...policies enforced by disincentives such as taxation or housing penalties for large families or made attractive by privileges and facilities offered to the small [families]" (Simpson 1987, 54). The role of family planning in demographic transition may be considered in regard to a specific country, or even a specific political regime. As with the socio-economic factors, it is sometimes difficult to assess the impacts of family planning programs in more than general terms.

As mentioned earlier, the demographic structure of a country and certain dynamic aspects of that structure may be represented through descriptive models (Figure 5-3). As the element of interest typically is between demo-graphic variables and population growth, the focus of such descriptive models is on components of fertility. Fertility trends are the main items of interest and the critical point of policy intervention when discussing patterns and trends of population growth. Therefore, it is with the determinants of fertility that any descriptive model will be most concerned. It is here that the critical differ-ences between population dynamics in China and India are discernable.

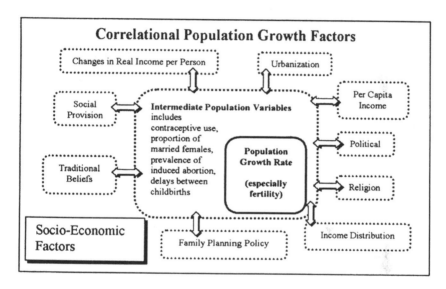

Figure 5-3   Correlational Population Growth Factor

The General Framework as Applied to China and India

Classical demographic transition theory suggests that declines in fertility arrive concomitantly with urbanization and industrialization. However, fertility de-clines have occurred in China, India, and other Asian countries without the large degrees of urbanization and industrialization that characterized fertility transitions in Europe. That is, the "rapid and sharp fertility declines in China [and other countries] occurred in the context of relatively modest socio-eco-nomic development and where the populations are predominantly rural" (Rele and Alam 1993, 36). India has undergone a substantial fertility transition also, though not as abrupt as that of China. At this point, selected aspects of the proximate, intermediate, and socio-economic factors may be discussed at length and specifically related to the cases of China and India. The overall populations of the countries and their respective growth rates, of course, must also be taken into consideration.

India's overall population currently stands at about 912 million with an annual rate of natural increase of 1.9 percent, while China's population numbers over 1.2 billion, with a rate of natural increase of 1.1 percent (Population Reference Bureau 1995). A comparison of indexed values (1967=100) shows that from 1967 to 1977 the populations grew at an equal rate (Figure 5-4). From about 1978 onwards, China's population growth rate has declined much more substantially than that of India. The demographic momentum as evidenced by birth rates does not reflect the changes in fertility that began to occur in the early 1970s. These changes in fertility are discussed further below.

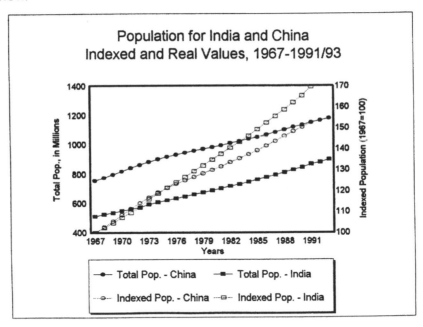

Figure 5-4   Population for India and China, Indexed and Real Values, 1967-1991/93

Selected Intermediate Variables, Proximate Variables and Total Fertility

As proximate variables impact fertility, it is appropriate to note total fertility itself. Fertility, defined as the average number of children born to a woman during her lifetime, has undergone substantial changes in both China and India in the period 1967-1991 (Figure 5-5). The fertility rate in India has declined steadily and slowly from 6.0 in 1967 to 3.3 in 1994. In China, the decline in the fertility rate has been substantial, especially in the period from 1968 (6.4) to 1980 (2.5). The rate has further declined to 1.9 in 1994. This decline represents perhaps the single greatest sustained change in fertility in any society over such a short period.

Declining infant mortality rates are considered a major determinant of fertility change. "As mortality declines, couples quickly realize that they do not need as many children as previously for a given number of them to survive to adulthood" (Caldwell 1993, 303). In addition, children with increased chances of survival receive more attention from their parents, with obvious implications for the child's health and education. "Lower mortality not only helps parents to

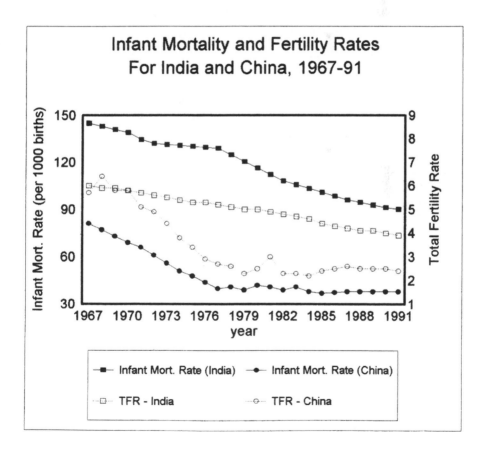

Figure 5-5   Infant Mortality and Fertility Rates for India and China, 1967-91

achieve their desired family size with fewer births, it leads them to want a smaller family as well" (World Bank 1984, 68).

Infant mortality rates are substantially different between India and China, as have been the trends in these rates (Figure 5-5). Infant mortality in India has remained high (92 infant deaths per 1000 live births in 1994), while China's has been relatively low (30), especially when compared to other developing nations (72). The indexed rate for China, like the fertility rates, declined substantially in the years 1967 to 1979 (81-39) and then leveled off after that. In India the infant mortality rates, too, show the same general pattern as the fertility rates, but its decline has been very gradual, dropping from 145 in 1967 to 129 in 1977. Since that time the rate of decline has become slightly more rapid.

Contraceptive use, its effectiveness, and its impact on fertility varies substantially across different nations, and sometimes within them. In the Less Developed Countries, two fifths of all child-bearing couples now practice birth control (Cleland and Hobcraft 1985, 115), due to demands related to development, better supply of modern contraceptives, and the influence of government programs in encouraging its use. A sharp contrast exists in the prevalence of contraceptive use between China, where 83 percent used them during the 1989-95 period, and India where the figure was 43 percent.

Contraceptive use in India varies widely in terms of acceptance, method, and promotion (Alam and Leete 1993, 160), though sterilization is also widely practiced as a birth control measure. Certain questions remain, such as "Why after almost two decades of efforts by family planning programmes, do the south Asian countries [specifically India] have such low levels of contraceptive use?" (Cleland and Hobcraft 1985, 125). In China IUDs (inter-uterine devices) are the most common contraceptive used, though sterilization is also important. Together these two techniques account for 88 percent of all birth control methods practiced. Contraceptive programs have been characterized as "coercive" (Alam and Leete 1993, 171). Table 5-1 shows contraceptive use and effectiveness in recent years. The rate of use in India is much lower than that of China, and it is also not increasing through time. Moreover, contraceptive use is more prevalent in urban areas of both countries compared to the rural areas. One-child families are most prevalent in urban centers of China.

Selected Socioeconomic Factors

Urbanization and declines in fertility rates can be broadly correlated (Simpson 1987, 46; Cleland and Hobcraft 1985, 212; World Bank 1984, 72). Urbaniza-

Table 5-1   Contraceptive Use Parameters

China			India	
Contraceptive Prevalence	Contraceptive Effectiveness		Contraceptive Prevalence	Contraceptive Effectiveness
-	-	1980	32.4	93.0
70.6	92.2	1982	-	96.0
74.0	93.7	1984	-	-
75.0	94.0	1988	43.0	-·
87.2	-	1990	-	-
89.4	-	1991	-	-
83.4	-	1992/3	40.3	-

Source: Sanderson and Tan, 1995:56.

tion, which in part is a measure of modernization and industrialization, influences income levels, economic opportunities, quality of life, and the cost of children. In all of these instances the urbanization impact is felt in terms of increases, such as increased incomes, for example. In addition, Cleland and Hobcraft (1985, 213) state that urban dwellers may have better access to contraception and may be more selective of marriage partners in their quest for upward mobility. The discrepancy between rural and urban fertility rates typically declines as overall fertility drops. Moreover, the urban middle- and upper-classes do not look upon children as a source of income. In the villages both the poor farmers and the landless laborers use child labor to augment family income. Although urbanization is occurring in both India and China, both countries still are largely rural (Figure 5-6). The percentage of people considered urban in China grew from about 17 in 1967 to 29 in 1994. In India in the same years, the percentage went from about 19 to 27, but the definitions of urban population differ in the two countries.

GNP Per Capita

Per capita incomes influence several proximate variables in a variety of ways. For example, both decreased breast feeding and shortening the time needed for dowry to be accumulated (each in turn is linked to higher fertility) can be correlated to increases in income (World Bank 1984, 69). In general, however, rising incomes lead to fertility decreases. People with higher incomes have less need for children as a form of "social security" and children may

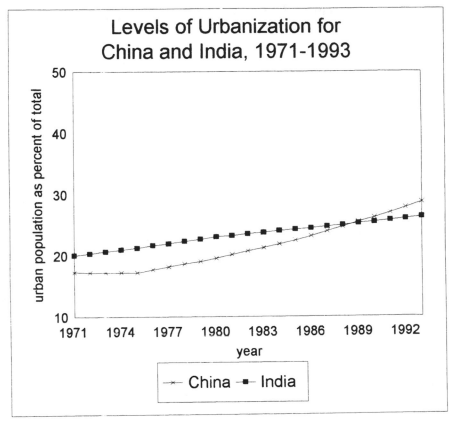

Figure 5-6   Level of Urbanization for China and India, 1971-1993

tend to compete with other demands, such as leisure activities, that one may be better able to afford.

The Chinese economy has been one of the world's fastest growing when measured in terms of Gross Domestic Product (GDP), growing 2.5 times, in real terms, in the period 1985 to 1995 (*Economist* 1996). Indexed values for per capita income (US$) show a moderate increase for the period 1967 ($100) to 1976 ($170), followed by more rapid increases in the period 1977 to 1981 ($320). A leveling of incomes then occurred, followed by renewed rapid increases from 1988 ($330) onwards to 1994 ($530) (Figure 5-7).

The Indian economy, too, has grown over the last 10 years, showing an increase in GDP of 64 percent. However, in per capita terms, economic growth has been less impressive. Increases in per capita GDP occurred at the same times as those in China, with the difference being in the magnitude of these

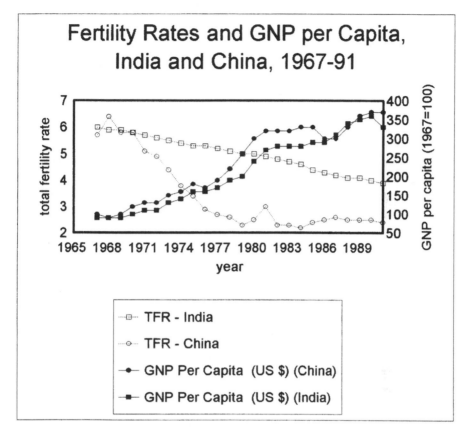

Figure 5-7   GNP Per Capita and Fertility Rate, India and China, 1967-91

changes. Per capita GDP grew from $100 (US) in 1967 to $320 (US) in 1994 (Figure 5-7).

The Role of Family Planning

Family planning has a substantial role in fertility changes (World Bank 1984). In many developing countries one can see the extent to which contraceptive use is reflected in birth rates and one may also see the sometimes significant interaction of family planning with other socio-economic or cultural factors (Simpson 1987, 52). Therefore, a greater need exists for governments in developing countries to be involved in population dynamics. It should be noted, too, that family planning goes beyond contraception to include incentives and disincentives to having children.

Family Planning in India

Indian central government officials have long recognized a problem of matching a large and growing population to limited resources. Most population planning efforts, like other development planning efforts in India, have been made largely in the context of the federal government's Five-Year Plans. Using the framework of the Five-Year Plans one may better understand policies and policy changes.

An official family planning policy was not immediately established in the aftermath of independence in 1947, as other more immediate concerns were being addressed. Also, Gandhi's teachings, which simply emphasized abstinence unless couples were attempting to have children were adopted as policy by default. In 1952, limited family planning efforts, which consisted of provision of information and supplies to couples attending clinics, were established (Alam and Leete 1993, 149). While India was one of the first nations to articulate support for a national policy, family planning efforts were not given priority at this time. Policy consisted mainly of providing information regarding birth control, though some limited emphasis was put on newly established programs which encouraged use of IUDs and male sterilization (Tirtha 1980).

The results of the 1961 census, however, aroused great concern because by then the population problem was very apparent, resulting from the slightly declining birth rate and the dramatically declining death rate. The net growth rate thus increased substantially. Since then, population limitation efforts have become a priority, because population growth was proceeding much faster than expected (Bouton 1987, 108). Central government authorities began allocating more resources and taking steps to implement an assortment of more substantial programs related to family planning and health care. These programs included an approach which relied on health care workers who promoted family planning.

Despite a greater application of effort and resources, especially in regard to male sterilization and adoption of IUDs, results of the population planning policies in the Third Five-Year Plan (1961-66) were disappointing. Efforts were redoubled and agencies re-organized in the Fourth Five-Year Plan (1969-74). Policy innovations were also made. Publicity of family planning was pursued through private advertising, subsidized production and sale of condoms was initiated, and training programs for doctors specializing in family planning were also begun (Tirtha 1980, 108). The Fifth Five-Year Plan was interrupted in June 1975 when an Emergency Period was declared by Prime Minister Indira Gandhi in response to political unrest. During the Emergency Period, civil liberties were curtailed drastically and the government took on the powers of a police state. At this time some persons were even subjected to forced sterilization.

Emergency status was rescinded when the Janata Party came to power in the elections in 1977. In the aftermath of these coercive policies all subsequent population planning efforts were looked upon with great suspicion by the Indian people. The backlash to events in the Emergency Period set back population planning to a great degree, especially for sterilization. With this backlash in mind, "family planning" programs were renamed "family welfare" programs. In the Sixth and Seventh Plan periods (to 1990) "emphasis of the programme shifted to promoting policies for controlling population growth through voluntary acceptance of a small family norm" (Alam and Leete 1993,149). The Eighth Five-Year Plan aimed at the unachievable goals of diminishing the crude birth rate to 2.6 and raising the "couple protection rate" to 56 percent by 1997. The following extract from a government publication describes the voluntary nature of the measures adopted (Government of India 1995,216):

> In keeping with the democratic traditions of the country, Family Welfare Programme seeks to promote on a voluntary basis, responsible and planned parenthood with two child norm, male and female, or both, through independent choice of family welfare methods best suited to the acceptors. Family welfare services are offered through total health care delivery system. People's participation is sought through voluntary agencies, opinion leaders, people's representatives, government functionaries and various other structures and influential groups. Imaginative use of mass media and inter-personal communication is resorted to for explaining various methods of contraception and removing socio-cultural barriers.

Overall, population planning in India has been characterized, with the one notable exception of the Emergency Period, by policies that encourage voluntary population controls, as opposed to coercive measures (Bouton 1987, 108). All population planning activities have been coupled with other development schemes. Throughout the post-Independence period, population planning has received an increasing budgetary commitment. As a large, secular democracy within a segmented and agrarian society, India encounters special difficulties when establishing population planning institutions and implementing policy. The special difficulties relate primarily to encouraging voluntary decisions in lowering birth rates. The programs have been particularly voluntary in nature from 1978 to the present (Figure 5-8).

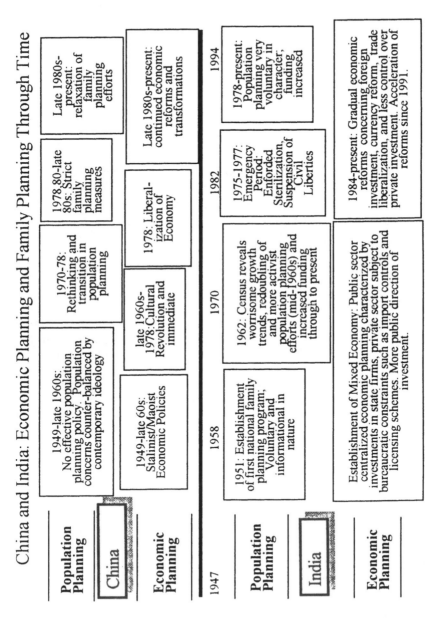

Figure 5-8   China and India: Economic Planning and Family Planning Through Time

Family Planning in China

With the victory of the Chinese Communist forces over the Nationalists in 1949 most of the efforts of the nation in the years following were directly related to establishing or rebuilding much of the war-torn nation, resulting in little attention, and even less focus, on problems related to population. The origins of China's family planning policies may be traced back to the mid-1950s when the results of the first census (1953) became known. The census revealed a population of 602 million, surprising political leaders including Mao Zedong (Tien 1991, 83). In fact, prior to the census, most leaders including Mao, held slightly pro-natalist views. Efforts, endorsed by the Communist Party leaders, were made to establish birth control, to initiate the manufacture of contraceptives, and to adopt other measures (Hong 1991, 43). The bureaucratic framework was being established at the same time as the short-lived Hundred Flowers Campaign in 1957 (Peng 1991, 20).

With the launching of the Great Leap Forward (1958-1960) and the turbulence that accompanied it, birth control efforts never became fully implemented. Nonetheless, restrictions on the use of certain contraceptives were permanently relaxed (Tien 1991, 175). The famines which occurred in 1959 and 1960-61 were exacerbated by Great Leap Forward policies. Birth rates dropped precipitously and death rates skyrocketed during this period of upheaval. Upon the rescinding of the Great Leap Forward policies and reestablishment of order, birth control efforts again were established. This time policies sanctioned further government interference in personal family decisions and the campaign emphasized the importance of family planning for the nation at large, as opposed to appealing to family interests alone. Focus upon urban areas, use of subsidies, and support programs (beyond the mere provision of birth control devices) became more widespread. However, just as the first campaign had been interrupted by political upheaval so was this one--this time by the Cultural Revolution. Much more sweeping, the social and political anarchy of the Cultural Revolution (1966 to 1970) brought a virtual halt to any family planning efforts.

As the chaos of the Cultural Revolution abated in the early 1970s, the gravity of the population dilemma was again recognized, especially by Premier Zhou Enlai (Peng 1991, 23). Population control efforts were undertaken with a sense of great urgency and seriousness, becoming yet more extensive and intrusive. Zhou initiated programs with material incentives and sanctions, in addition to the provision of maternal and infant welfare. Like efforts prior to the Cultural Revolution, attention was focused on urban areas. Ideological campaigns associated with the slogans such as "later, longer, fewer" concerning marriage, birth intervals, and number of children; and "one child is not too few, two children are just right, but three children are too many" were started at this time.

Upon consolidation of power by Deng Xiaoping in 1978, efforts towards population control were even further strengthened. The Dengist long-term economic development strategy was premised on the idea that population levels could not exceed a certain point—1.2 billion in the year 2000 (Peng 1991, 24). The one child per couple campaign, which started in 1978, relied on intensive publicity campaigns, persuasive measures, and some coercive measures (such as the requirement that married women practice contraception). The rigorous new program better recognized where further drops in fertility might be achieved and focused resources on these areas.

Since the late 1980s, the rigor of population control programs has been slightly relaxed. Exceptions or relaxations in policy are being allowed in certain rural areas and among the minority populations. The one child policy in some cases has been adjusted in some cases to allow for a second child in "only girl households" (Peng, 1991, 26). Economic liberalization has also, to a small degree, diluted governmental powers and degree of control. The enormous changes due to rapid development in some portions of the country have taken away (to a limited degree) the government's control in private affairs.

Family planning efforts in China, like economic policies, may be divided into four distinct policy periods (Figure 5-8). Between 1949 and 1969, family planning efforts were largely ineffective; with population concerns counterbalanced by contemporary ideology. In the period 1970-78 rethinking occurred with a fundamental transition in population planning (Tien 1991; Hong 1991). Strict family planning measures were implemented during this time period. From 1978 until the late 1980s even more strict family planning measures predominated, notably the one child policy. From the late 1980s to the present some relaxation of family planning efforts has been evident. One feature that has always characterized the Chinese program and contributed to its success is its decentralized policy implementation (Peng 1991, 16).

Demographic Models

Demographic transition models, as discussed earlier, may be broken into four stages depending on changes in birth rates, death rates, and growth of population. As mentioned in the discussion of the situations in China and India, classic demographic transition theory posits that these demographic changes arrive concomitantly with socio-economic development. Considering that it is a model based largely on Western historical experience, the theory has certain shortcomings when applied to the developing world. India, China, and many other nations throughout Asia, have achieved lowered fertility rates while still having low per capita incomes and lesser urbanization (Caldwell 1993, 301). While classical demographic transition theory could have predicted the fertility declines in a few of the East and South-East Asian countries , it could not have predicted the decline in others and "the sharp and rapid fertility declines in China...occurred in context of relatively modest socioeconomic development..." (Rele and Alam 1993, 36). This is also noted by Caldwell in comments on the fact that a radically different fertility situation exists than that predicted, particularly as both China and India are low income countries with low fertility (Caldwell 1993, 301).

Given the enormous size of its population, China's low fertility rate (1.9) and death rate (7 per 1000), should not be considered an exception to general rules of demography. After all, it is difficult to consider any characteristic of 21 percent of humanity as an anomaly! The pervasive efforts of China's family planning programs have been key factors in population control in that country. Peng (1991, 63) states:

> There is no doubt that the very successful family-planning is one of the major factors responsible for the rapid fertility decline in China. Since the early 1960s, control of population growth has been strongly advocated by the Chinese leadership, no matter who has been in power. With the shift of the Party's priority from political struggle to economic construction, more and more efforts have been made to bring down the birth rate and modernize reproduction. The regional differences in the timing and pace of the transition are to a great extent a result of the differentials in the implementation of the local family-planning programme.

In discussing the changes brought about by intensive government programs in China, Caldwell suggested that "the same policies, if sustained, could have reduced India's fertility to replacement level some time in the 1980s, at least in large parts of the country..." (Caldwell 1993, 311). Such policies are likely only possible in a state such as China, with an historical record as a unitary state and its Confucian attitudes. Given India's status as a democracy, as well as its historical political and cultural diversity, such sustained policies are an impossibility. Even in the short-lived Emergency Period, public reaction to enforced sterilization was notable and lasting backlash against sterilization occurred. The levels of coercion used in China, if applied to India, would likely only serve to topple governments, as they in part did in 1977.

Conclusion

In India, the history of development and population dynamics since independence, has been characterized by a very gradual transition towards lower birth rates and fertility rates, and slow to moderate, stable economic growth. These gradual transitions have been punctuated by several subtle changes in economic policies and family planning strategies. In China, since 1949, policy history can be broken into four distinctive periods, all characterized by dropping fertility levels (especially in comparison to other developing countries, including India) and, since 1978, rapid economic growth.

Though both countries have undergone significant socio-economic change, these changes have not reached the point where they are responsible for any appreciable drop in birth rates. Of the socio-economic factors, the discrepancy in infant mortality is one that may be best linked to the differences in fertility rates between the two nations. Both India's modest declines in the birth rate and the remarkable decline of the same in China are due in large part to the nature of family planning programs in each. India with its more subtle and voluntary programs has been less successful than the highly energized and sometimes coercive programs in China. India can achieve more successes in its family planning efforts, not only by its general economic uplift, but also by creating conditions (such as lowering of infant mortality, raising the proportion of literacy among young women, and making family planning techniques more readily available) on which contraceptive prevalence may augment.

The differences between the demographics of India and China discussed here bring to attention certain implications for demographic transition theory. Noting the current status of each country in terms of the standard demographic transition model, it is evident that China's position is unique, as is India's but to a more limited degree. A low fertility rate has been achieved in a largely rural, low income country. At the same time death rates are very low and characteristic of those found in the more developed countries. It is possible that this "revolution in Asian fertility" which is especially notable in China needs to be accommodated by a new demographic model. Or, perhaps China itself, due to the extreme changes induced by its population policies, may be placed in a "fifth stage" of demographic transition.

The United Nations projects that worldwide more people soon will be living in urban areas than in rural. As much as 61.2 percent of the population will live in urban areas in 2025. By then India's and China's urban proportions will also rise to 45.2 percent and 54.5 percent, respectively (United Nations 1992). Such a rise in urban population will also be accompanied by a drastic change in the use of family planning techniques, rises in per capita GNP, and lowered rates of infant mortality. These two areas of Asia are at the threshold of very important changes in the socio-economic spheres.

References

Alam, Iqbal and Richard Leete, 1993, "Variations in Fertility in India and Indonesia", pp 148-172 in Leete, Richard and Iqbal Alam, (eds), *The Revolution in Asian Fertility: Dimensions, Causes and Implications*, Oxford: Clarendon Press.

Davis, Kingsley, and Judith Blake, 1956, "Social Structure and Fertility: An Analytic Framework", *Economic Development and Cultural Change*, 4:4:211-235.

Bongaarts, John and Robert G. Potter, 1983, *Fertility, Biology, and Behavior*, New York: Academic Press.

Bouton, Marshall M., (ed), 1987, *India Briefing, 1987*, Boulder, CO: Westview Press.

Caldwell, John C. 1993, "The Asian Fertility Revolution: Its Implications for Transition Theories", pp. 299-316 in Leete, Richard and Iqbal Alam, (eds), *The Revolution in Asian Fertility: Dimensions, Causes and Implications*, Oxford: Clarendon Press.

Chinese Academy of Social Science (CASS), 1989, *Information China: The Comprehensive and Authoritative Reference Source of New China*, vol. 2, Oxford: Pergamon Press.

Cleland, John and John Hobcraft, (eds), 1985, *Reproductive Change in Developing Counries: Insights from the World Fertility Survey*, New York: Oxford University Press.

Coale, Ansley J. 1984, *Rapid Population Change in China 1952-1982*, Washington, D.C.: National Academy Press.

Dutt, Ashok and Frank Costa, 1977, "Necessary Inputs for Reducing Population Growth in Bangladesh", pp. 153-163 in S. Parvez Wakil, (ed), *South Asia: Perspectives and Dimensions*, Ottawa: Canadian Association for South Asian Studies.

Economist, The, 1996, "Emerging Market Indicators", (March 16) 108.

Findlay, Allan M. and Ann Findlay, 1987, *Population and Development in The Third World*, New York: Methuen Press.

Galenson, Walter, (ed), 1993, *China's Economic Reform*, South San Francisco, CA: The 1990 Institute.

Goldscheider, Calvin, (ed), 1992, *Fertility Transitions, Family Structure, and Population Policy*, Boulder, CO: Westview Press.

Greer, Harold C., (ed), 1991, *China: Facts and Figures Annual*, vol. 14, Gulf Breeze, FL: Academic International Press.

Hinton, Harold C., (ed), 1986, *The People's Republic of China, 1979-1984: A Documentary Survey*, Wilmington, DE: Scholarly Resources.

India, Government of, 1995, *India 1995: A Reference Manual*, New Delhi: Ministry of Information and Broadcasting.

Johnston, R.J., (ed), 1986, *Dictionary of Human Geography*, 2nd ed, Oxford: Basil Blackwell.

Kaplan, Hilliard, 1994, "Evolutionary and Wealth Flows Theories of Fertility: Empirical Tests and New Models", *Population and Development Review*, 20:4:753-793.

Lee, Chung H. and Helmut Reisen, (eds), 1994, *From Reform to Growth: China and Other Countries in Transition in Asia and Central and Eastern Europe*, Paris: OECD.

Leete, Richard and Iqbal Alam, (eds), 1993, *The Revolution in Asian Fertility: Dimensions, Causes and Implications*, Oxford: Clarendon Press.

Li, Cheng-jui, 1992, *A Study of China's Population*, Beijing: Foreign Languages Press.

Liu, Zheng, Jian Song, et al, 1981, *China's Population: Problems and Prospects*, Beijing: New World Press.

Overholt, William H. 1994, *The Rise of China: How Economic Reform is Creating a New Superpower*, New York: W.W. Norton.

Peng, Xizhe, 1991, *Demographic Transition in China: Fertility Trends Since the 1950s*, Oxford: Clarendon Press.

Population Reference Bureau, 1995, "1995 World Population Data Sheet", Washington: Population Reference Bureau.

Rele, J.R. and Iqbal Alam, 1993, "Fertility Transitions in Asia: The Statistical Evidence", pp. 15-38 in Leete, Richard and Iqbal Alam, (eds), *The Revolution in Asian Fertility: Dimensions, Causes and Implications*, Oxford: Clarendon Press.

Sanderson, Warren C. and Jee-Peng Tan, 1995, *Population in Asia*, World Bank Regional and Sectoral Series, Washington: The World Bank.

Scherer, J.L. 1981, *China Facts and Figures Annual*, Gulf Breeze, FL: Academic International Press.

Shirk, Susan L. 1994, *How China Opened Its Door: The Political Success of the PRC's Foreign Trade and Investment Reforms*, Washington: Brookings Institute.

Simpson, E.S. 1987, *The Developing World: An Introduction*, New York: John Wiley.

Singh, Ajay Kumar, 1989, *Indian Population and Fertility Behavior*, Delhi: Prabhat Prakashan.

Sudan, Falendra K. 1992, *Demographic Transition in South Asia*, New Delhi: Anmol Publications.

Tien, H. Yuan, 1991, *China's Strategic Demographic Initiative*, New York: Praeger.

Tirtha, Ranjit, 1980, *Society and Development in Contemporary India: Geographical Perspectives*, Detroit, MI: Harlo Press.

United Nations, 1992, *World Urbanization Prospects: 1992 Revision*, New York: United Nations.

Wang, Hong, 1991, "Population Policy of China", pp. 42-68 in Wang, Jiye and Terence H. Hull, (eds), *Population and Development Planning in China*, Sydney: Allen and Unwin.

Wang, Hong, 1993, *China's Exports Since 1979*, New York: St.Martin's Press.

World Bank, 1980, *World Development Report 1980*, New York: Oxford University Press.

World Bank, 1984, *Population Change and Economic Development*, New York: Oxford University Press.

Yaukey, David, 1985, *Demography: The Study of Human Population*, New York: St. Martin's Press.

6 Transportation, Regional Development, and Economic Potential in Mexico

DAVID J. KEELING

As countries and regions continue to move toward economic integration within the context of the evolving global economy, geographers and others have focused increased attention on the role of transport and communication in facilitating the regional development and integration process. Transport is a catalyst for interactions over space and through time among individuals and communities, and between them and the physical and cultural environment. The intimate relationship among transport, space, and time provides the basis for the widely accepted proposition that transport is a fundamental component of regional development and integration (Voight 1984; Dugonjic 1989). Regional transport services and infrastructure function as the lifeline for economic and social interchange. They enable those in local and regional economies to interact with other parts of the global economy.

Analyses of transport and communication's role in the regional development process typically have concentrated on changes at three different scales: global, supra-state regional, and sub-state regional. Each scale of connectivity has implications for the ability of regions and countries to interact and compete in the global economy. Within and between these different scales, transport spaces and actors are inextricably intertwined with transport processes and infrastructure. In turn, these processes have distinct impacts at myriad scales. For example, research on global transport networks has helped our understanding of the relationship between world cities and their regional hinterlands (Keeling 1995), while studies of global telecommunications systems have shed light on the articulation of local, regional, and national economies in the global system (Heldman 1992; Kellerman 1993). At the supra-state regional scale, Europe has dominated recent examinations of transport's role in the regional development process (Noam 1992; Gibb 1995; Ivy 1995). However, recent passage of the North American Free Trade Agreement promises to stimulate research on the transport-regional development nexus in Mexico, Canada, and the United States (ECLAC 1995). Over the long term, hemispheric integration will demand greater attention to the importance of transport links between contiguous and non-contiguous countries.

The most typical framework for analysis is the substate regional scale. Considerable literature exists on the role of transport and communication in shaping regions within political states, from the pioneering work of Taaffe et al. (1963) in Africa to more recent work on regional development in such diverse areas as Indonesia (Leinbach 1983), Ecuador (Rudel and Richards 1990), and Argentina (Roccatagliata 1993). Yet despite this body of research, few refinements have been made in recent years to basic theories about the role of transport in regional development (Simon 1996). Analyses of the impact of transport on regions continue to concentrate primarily on ways in which transport and communication infrastructure might influence regional economic change within a national context (Vickerman 1991). Part of the problem, of course, is the dominance of the relationship between economics and national regional development in the plans, policies, and strategies of most governments. As a consequence, little empirical evidence explicating the ways in which different types of infrastructure exert spillover effects, their impact on regional development from a non-economic perspective, or their supra-state regional implications has been forthcoming. A notable exception is Gibb's (1994) recent edited volume on the Channel Tunnel. We know even less about how and why transport networks evolve and change in specific regions. Nonetheless, empirical analyses have proved that an integrated interregional transport and communication network does not automatically encourage capital flows from rich to poor regions. Nor does it guarantee a reduction in regional inequalities, either at the national or supra-national scales (Blum 1982; Dugonjic 1985)

In addition, despite tremendous theoretical and empirical advances in our understanding of how local, regional, and global socio-economic spaces are evolving and coalescing, progress on examining the role of transport in shaping and structuring inter-regional spatial relationships has been slow. The lack of an adequate theoretical framework of transport's role in regional and global development has inhibited research on the changing circumstances of cities, regions, and states in the contemporary global economy. As Nigel Thrift (1986,62) argued, few researchers have seemed willing to "come out of their national shells and take the wider view which would enable them to understand what is going on within their own countries". Latin America, in particular, has suffered tremendously in recent decades from a lack of focus by governments, planners, and academics on the relationship between transport, urban growth, regional development, and global economic change. Arguably, intra- and interregional connectivity in Latin America has improved only marginally during the twentieth century. Constraints on the flow and interaction of people, goods, and information within Latin America remain at the top of a long list of regional development problems facing planners and bureaucrats at both the national and supra-national levels.

As geographers and others continue to explore the "local-global" continuum of socio-economic relationships and the impact of the globalization phenomenon on peoples and places, a much clearer understanding is needed of the links between transport, regional development, and "local-regional-global" restructuring. This chapter summarizes ongoing research on this important theme in the growing debate over regional development issues. Of particular concern are the nature and rates of socio-economic development in disparate regions, both within a country and across international political boundaries, and the role of relative regional accessibility in shaping transport investment decisions. First, I develop a framework of analysis for examining the relationships between the main components or "pillars" of regional development. Second, using Mexico as a case study, economic potential analysis is used to illustrate the importance of transportation in the regional development process. The chapter concludes with a critique of existing transport-based regional development strategies in Mexico.

A Framework of Analysis

The contemporary global economy has been described as a historically unprecedented phenomenon. According to Amin and Thrift (1992), during the 1970s and 1980s an important shift occurred in the capitalist system, with a transition from an international to a truly global economy. Four characteristics define the contemporary global economy. First, industries now function on a global scale through the medium of global corporate networks. Second, oligopolistic, progressively centralized power has increased, helping to create important nodes in the global system. Third, new forms of joint ventures, subcontracting, and other types of networked organization and strategic alliances are contributing to the ongoing process of corporate decentralization. Finally, there is a new, more volatile balance of power between corporations, local governments, and national states. These new power relationships have resulted in the increased prominence of cross-national issue coalitions, where "fragments of the state, fragments of particular industries, and even fragments of particular firms" are united in a global network (Amin and Thrift 1992, 574-5). Although debate continues over the precise nature of the new global economic order, most observers seem to agree that a reconstituted economic land-scape is emerging, along with a realignment of class forces.

In recent decades the state's relative dominance in global affairs has given way to new space-time patterns where both the multinational-global and the regional-local scales have become more prominent. Corporate, financial, and political activity at the global scale still play crucial roles in structuring daily life. However, geographers and others are finding that regional and local responses and restructuring processes have become equally important. Certainly, local and regional processes always have played a crucial role in structuring daily life, yet until recently they generally were given short shrift in the development of meta-theories and models about global change. In the 1990s, the forces of decentralization, devolution, and localization are counter-balancing the processes of globalization. As a result, the local-global continuum of social and economic relationships is realigning itself over space and through time, with major impli-cations for regions and states. Yet treating global and local activities as con-tradictory forces is much too simplistic an approach. In practice, these activi-ties are generating complex and dynamic tensions within and between social divisions and structures inside national states, tensions that are creating distinct spatial patterns of development, growth, and change.

Moreover, within the capitalist global economy, certain regions and subre-gions have emerged to function as significant focal points for transport, com-munication, and international economic activity. These regions, and more spe-cifically their dominant cities, also serve as concentration points for global fi-nancial activity and can be distinguished by certain characteristics such as a concentration of multi-national headquarter functions, a flexible and mobile labor force, major cultural infrastructure, and substantial flows of goods and information (Friedmann and Wolff 1982). In countries such as Mexico, the regional context is being restructured to enhance these global characteristics and to capture the benefits of participation in the emerging system of world cities and regional trading blocs. Mexico, for example, is playing a key role in the development of the North American Free Trade Area (NAFTA). If inter-national and national processes are impacting differentially on different kinds of regions in Mexico, then these processes need to be identified, measured, and analyzed in order to shed meaningful light on the spatial implications of restructuring.

The interplay between national uniqueness, regional processes, and global interdependence in Mexico can be synthesized in a basic schematic of regional growth and change (Figure 6-1). The diagram functions as a framework for analyzing regional restructuring within the context of global economic pro-cesses. A thematic-spatial approach treats the core components of regional growth and change as an interactive, integrative whole. This approach suggests that a highly generalized set of central relationships is at work in and among regions of many scales. Place, capital, and labor function as central pillars in local, regional, and global economic systems. Interaction between these three components is inextricably intertwined with, and shaped by, social, political, cultural, and environmental forces. Changes in the ethnic and demographic character of place, for example, directly affect the interaction between place and the other pillars of development—labor and capital. Circuits of capital change as places become more impoverished, and the dynamics of labor change as local and regional populations age or relocate.

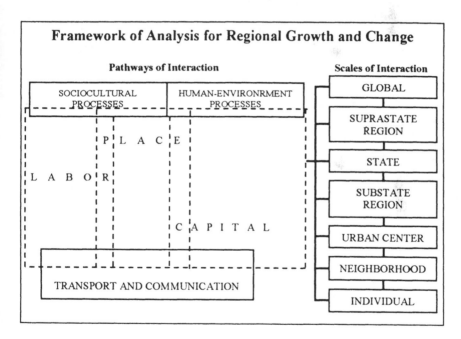

Figure 6-1   **A Framework of Analysis for Regional Growth and Change**
Source: After Keeling (1997).

The constantly expanding and contracting interface between the diagram's components, represented by the dashed lines, determines the strength of the relationship and its possible impact on a region. Weak interfaces between labor, capital, and place can serve to inhibit the development of relationships within the world economy and with the hierarchy of regions and urban centers. If global or national capital has a weak interface with place, theoretically little regional restructuring is likely to occur. For example, in Mexico the southern states of Chiapas and Oaxaca both are substantial areas, with about 2.6 and 2.7 million inhabitants respectively, yet they exhibit little physical evidence of a strong interface with the circuit of global or regional capital. These regions are impoverished, underdeveloped, and isolated from the main pathways of regional-global interaction, although international tourism has become increasingly important in Oaxaca and coffee cultivation has expanded significantly in the agricultural economy of Chiapas.

Most important of all, however, are transport and communication, which function as the foundation upon which rest the central pillars of the regional development process. Transport and communication facilitate the movement of people, goods, and information in and among regions and are necessary, though not sufficient, components of regional and global socioeconomic genesis, growth, and change. Weak linkages between transport, communication, and any of the three central pillars can have a deleterious spatial impact on regional development and can influence the process of growth and change in negative ways. Weak transport links to and between regions can inhibit the spread of economic potential from core or nodal areas and can isolate potential economic activity in underdeveloped areas. Economic potential analysis, therefore, can serve as a useful tool for highlighting the relative importance of transport services and infrastructure between and among regions and urban centers at both the state and suprastate levels.

Regional Development and Economic Potential in Mexico

For many Latin American countries, the gloomy prognosis that the 1980s would be a "lost decade" for socio-economic development proved to be true. Rampant inflation, continued political upheaval, increasing social inequalities, and deteriorating infrastructure throughout the region retarded the development process and widened the gap between rich and poor. In urban Mexico at the beginning of the 1990s, for example, 35 percent of all urban income went to the richest 10 percent of the population, while the poorest 40 percent received only 17 percent of the total urban income (Krooth 1995). Moreover, deteriorating socio-economic conditions in more isolated areas such as Chiapas and Guerrero, underscored by political violence, have focused increased attention on spatial dichotomies in the regional development process.

As the early 1990s unfolded, Latin American countries found themselves in a desperate struggle to redefine their role in a rapidly evolving and sophisticated global system. New regional economic alliances, the end of an ideologically bifurcated world order, and the spread of free-market capitalism and democracy have forced countries such as Mexico to reappraise their policies, economies, and territorial spaces. One of the Mexican government's key strategies was to negotiate participation with Canada and the United States in the North American Free Trade Agreement (NAFTA). The implementation of NAFTA in Mexico has begun to change the country in profound and fundamental ways. New social, political, and economic spaces are emerging to redefine Mexico's role in the region and in the world. At the same time, global, hemispheric, and regional forces are changing the map of Mexico at sub-regional, state, and local scales.

Tensions, strengths, and weaknesses in Mexico's bi-directional links along the local-global continuum of economic, social, and political relationships are having a direct impact on the country's patterns of growth and change. For example, tensions exist between Mexico's regional integration strategies and domestic social policies; witness recent political and economic discord in Chiapas. Strengths are evident in Mexico's development of regional economic ties, and weaknesses continue in the country's transportation links to both internal and external regions. To understand more clearly the role that Mexico is likely to play in emerging regional, hemispheric, and global networks, we need to examine in more detail the impact of these policies and linkages over space and through time. Geographers, in particular, are extremely interested in how changes in these linkages are reshaping patterns of interaction between people and the environment at the local and regional levels. Especially important are the problems created by uneven socio-economic development among a country's constituent regions and the role that transport and communication play in facilitating regional accessibility and mobility.

As the foundation for the three major pillars of regional development, transport and communication are critical to growth and change in Mexico. Transport and communication services and infrastructure facilitate and condition interaction between people, institutions, labor, capital, and the regional environments. Mexico's geography, however, has presented obstacles to economic integration and regional development throughout the country's history. Primarily a mountainous environment, Mexico lacks internal navigable waterways and suffers generally from high transportation costs associated with infrastructural development and service provision. Since the nineteenth century, transport routes have been fragmented in nature, primarily north-south in orientation, dominated by the urban-industrial core of Mexico City, and geared to carry export products from the interior to the global economy. In the more remote areas of the North, the Southeast, and the coastal regions, transport provision historically has been inadequate for regional development. Moreover, successive transport policies failed to address the lack of connectivity with neighboring countries and served to reinforce Mexico City's dominant position within Mexico.

Mexico's Mesa Central, or core region, long has dominated the socio-economic dynamics of the country. The twelve states that comprise the Mesa Central today account for over 60 million of Mexico's 95 million inhabitants, or approximately two-thirds, on less than 18 percent of the national territory. These states historically have provided over 60 percent of the country's gross domestic product and they are endowed with over one-third of Mexico's railroads and paved federal highways (Table 6-1). Compared to Mexico's three peripheral regions, the Mesa Central zone enjoys a much higher transport density ratio. Migration patterns also have favored the Mesa Central over other regions, although *maquiladora* (assembly plants) activity along the U.S.-Mexico border has begun to attract increasing numbers of migrants in recent years. Moreover, current privatization of the transport and communication sector and other socio-economic changes wrought by NAFTA are beginning to realign internal spatial arrangements in Mexico, with important consequences for the locational utility of regions and places.

Table 6-1　Gross Domestic Product, Population, Land Area, and the Density Percent of National Routes in Mexico

Political Division	GDP US$ (millions)	1991 Population	Area (SqKms)	National Routes (Kms)[a]	Density Percent[b]
Mesa Central					
Colima	913	419,439	5,191	542	10.44
Federal Cap.	36,477	19,150,275	1,479	464	31.37
Guanajuato	5,64	3,542,103	30,491	2,389	7.84
Hidalgo	2,908	1,822,296	20,813	1,824	8.76
Jalisco	11,578	5,198,374	80,836	3,322	4.11
Mexico	19,475	11,571,111	21,355	2,205	10.33
Michoacán	4,283	3,377,732	59,928	3,615	6.03
Morelos	2,181	1,258,468	4,950	732	14.79
Puebla	5,304	4,068,038	33,902	2,533	7.47
Querétaro	2,242	952,875	11,449	954	8.33
Tlaxcala	976	665,606	4,016	946	23.56
Veracruz	9,706	6,658,946	71,699	4,708	6.57
Total	101,685	58,685,263	346,109	24,234	7.00

Table 6-1   continued

The South

Campeche	3,802	592,933	50,812	1,642	3.23
Chiapas	3,322	2,518,679	74,211	2,801	3.77
Guerrero	3,217	2,560,262	64,281	2,329	3.62
Oaxaca	2,927	2,650,232	93,952	3,773	4.02
Quintana Roo	1,232	393,398	50,212	1,009	2.01
Tabasco	3,172	1,299,507	25,267	872	3.45
Yucatán	1,991	1,302,600	38,402	1,791	4.66
Total	19,663	11,317,611	397,137	14,217	3.58

The Northeast

Coahuila	5,103	1,906,119	149,982	3,753	2.50
Nuevo León	10,816	3,146,169	64,924	2,721	4.19
San Luis Potosi	3,158	2,020,715	63,068	2,795	4.43
Tamaulipas	4,683	2,266,677	79,384	3,078	3.88
Total	23,760	9,339,680	357,358	12,347	3.45

The Northwest

Aguascalientes	1,249	684,247	5,471	566	10.35
Baja	5,141	1,703,571	143,396	3,204	2.23
Chihuahua	5,550	2,238,542	244,938	5,064	2.07
Durango	2,250	1,384,518	123,181	3,368	2.73
Nayarit	1,244	846,278	26,979	1,165	4.32
Sinaloa	3,834	2,367,567	58,328	2,159	3.70
Sonora	4,702	1,799,646	182,052	3,600	1.98
Zacatecas	1,754	1,251,531	73,252	2,182	2.98
Total	25,724	12,275,900	857,597	21,308	2.48
Grand Total	170,832	91,618,454	1,958,201	72,106	3.68

Sources: INEGI (1986-1988); Pick and Butler (1994).

[a] National routes include total railroad kilometers in service and the total kilometers of federal highways.

[b] Density ratio is the total kilometers of national routes to the land area in square kilometers expressed as a percent. Theoretically, the higher the density ratio, the greater the accessibility to the national surface transportation network.

Changes in transportation infrastructure over the next decade will reshape Mexico's connectivity levels at a variety of different scales and will play a major role in the country's regional development. For example, the level of connectivity experienced by Mexico along the local-global continuum is crucial to the restructuring of relationships between Mexico, its neighbors, and important regional urban centers within the country. Transport and communication also help to shape the response of individuals and institutions to the forces of change. The long-term success of regionalization and globalization forces in Mexico and the country's incorporation more fully into the world economy depend, in large part, on the government's response to increased demands for transport and communication services and infrastructure. An analysis of regional economic potential can provide good data and a spatial rationale to support government, planning, and private transport investment strategies for future development.

Economic Potential Analysis

The concept of economic potential measures the nearness or accessibility of a specific volume of economic activity to a particular point or region. Its genesis lies in the concept of regional potential and was first used in an analysis of population distribution (Stewart 1947). Between the 1950s and the 1980s, researchers such as Harris (1954), Clark (1966), and Rich (1980) refined the regional potential concept in analyses of industrial location, focusing specifically on the market or economic potential of regions. Most recently, Gibb and Smith (1994) used economic potential analysis to challenge the generally accepted belief that the benefits generated by the Channel Tunnel linking Britain and France would be confined to the southeast region of England and to the Nord-Pas de Calais region of France.

The formula for measuring a region's accessibility to economic activity, as given by Rich (1980), is:

$$P_i = \sum_{j=1}^{n} M_j / D_{ij}^{\propto}$$

where P_i = the potential of region "i"

M_j = a measure of the volume of activity in region "j"

D_{ij} = the measure of the journey distance/time or transport cost between regions "i" and "j"

$\propto$ = the distance/time/cost exponent.

When summed for all "n" regions in the study area, the potential value for region "i" in units of economic activity per unit of distance, time, or cost is yielded. Regions with the highest potential values theoretically have access to more economic activity within a specified distance than regions with lower values. High relative accessibility confers comparative advantages within the region concerned by reducing the time and distance costs associated with moving products, people, inputs, and information. In contrast, more inaccessible regions are handicapped by comparative disadvantages in the form of higher time and distance costs, which include the perception of regional inaccessibility. Thus, economic potential analysis provides useful evidence of how and where transport improvements theoretically could be the most effective.

The Methodology of Economic Potential Applied to Mexico

The major methodological issues presented by economic potential analysis include scale of analysis, the definition and measurement of economic activity, the distance matrix, the problem of tariff barriers, and the estimation of a region's self-potential. The complete study area comprises the 90 states and provinces, including two federal districts, that make up the North American Free Trade Area, with the "island" provinces of Newfoundland and Prince Edward Island combined with Nova Scotia for transportation purposes. The Mexican portion of the study area, the focus of this essay, comprises 31 states, including the Federal District of Mexico City. The two states of Northern and Southern Baja are combined into one. The NAFTA region contains approximately 380 million people and is one of the most urbanized in the world. There are over 40,000 cities and towns in the NAFTA region, including 280 cities with populations exceeding 100,000 and about 50 metropolitan areas with populations surpassing one million.

In each state, a major town or city was chosen as the nodal point for that state, and the volume of economic activity in the state was allocated to that location. For Mexico, gross domestic product (GDP) values based on data provided by Mexico's National Statistical Institute (INEGI) were used to measure the mass M_j term, expressed in millions of U.S. dollars converted from Mexican pesos at 1988 average exchange rates. At the time of the study, data for 1988 were the latest available for reliable GDP values for each state in Mexico. GDP values are regarded widely as the best available accurate measure of the volume of economic activity in regions.

All nodes in the study area were connected to neighboring nodes, including, where necessary, links to nodes in non-contiguous regions. The nodal matrix for Mexico, including cross-border connections to U.S. nodes, comprises 57 primary links, with a mean distance between each node in the matrix of 441 kilometers. The nodal matrix for the entire NAFTA region has 196 primary links, with a mean distance between links of 661 kilometers and a standard deviation of 530 kilometers. Thirty-seven percent of the 196 primary links had distances between nodes above the mean distance for the entire study area. The shortest link between two Mexican nodes (Puebla and Tlaxcala) was 61 kilometers, with the longest link 966 kilometers between Tuxtla and Acapulco.

The major empirical work in calculating potential economic values centers on the calculation of the distance, time, or transport cost between nodes. This involves selecting appropriate cost or distance measures and calculating the necessary transport time, cost, or distance exponents. Meaningful transport cost data for Mexico are extremely difficult to obtain and the considerable time and problems involved in gathering such data could not be justified in terms of providing any appreciable difference in the final potential results. Therefore, the present study uses the most direct road or rail distances between the regional nodes. A Thomas Cook *Overseas Timetable* (1994) and a North American road atlas provided the base data required to build a distance matrix. The distance term D_{ij} represents potential interregional rail freight journey times. Manufacturers are concerned primarily with the time needed to ship their goods, especially in the contemporary economic environment of "just-in-time" inventory practices, not with the distance over which those goods have to travel. Moreover, in most urbanized, industrialized countries, transport costs have become a relatively minor component of the final cost of delivered goods and services. However, the importance to regional socio-economic development of the accessibility and mobility of people should not be overlooked. Efficient, cheap, rapid, and high-quality passenger services and infrastructure are fundamental to the regional development of Mexico and are a major component of the present study.

A further consideration in using the economic potential model is the issue of a region's or state's self-potential, the contribution to the potential of region "*i*" of its own mass value. In other words, how can the special case of D_{ii} in the array D_{ij} be quantified? Various approaches to this problem have been tried in past research. Rich (1980) has argued that the clustering of economic activity around the major urban center of most officially-defined regions supports the use of the following formula to calculate D_{ii}:

$$D_{ii} = 0.5 \ \sqrt{\frac{\text{area of the region}}{\Pi}}$$

General Hypothesis

Globalization of the region's economy, coupled with the integration processes engendered by the NAFTA, is having a differential effect on cities and regions. The purpose of this study is to analyze which regions and states within Mexico could benefit theoretically from enhanced accessibility to economic activity occurring throughout the NAFTA area. In other words, what portion of, for example, California's economic activity or Ontario's economic activity theoretically could accrue to Mexico's southern region and its constituent states as a consequence of improvements in transport services and infrastructure? Within the context of economic potential analysis, the following general hypotheses can be stated. First, basic improvements in transport services and infrastructure are likely to endow the Mexico-U.S. border regions and cities with higher gains in absolute and relative accessibility than Mexico's core region. Benefits from economic liberalization strategies, including *maquiladora* activity, duty-free policies, and proximity to the U.S. marketplace, already favor Mexico's northern regions.

Second, southern areas such as Chiapas and Campeche are likely to experience only moderate increases in absolute and relative accessibility. With the exception of several locations that have specific competitive advantages in terms of the regional and global economy (particularly tourist zones), most of the southern region of Mexico has been excluded from the economic integration process. For example, the Chiapas conflict that began in January 1994 is driven not only by traditional isolation from the national economic core, but also by a strong perception that the NAFTA, globalization, and structural adjustment programs are leaving the southern region behind. Third, the core region of Mexico centered on Mexico City is likely to experience the lowest gains in absolute and relative accessibility. The megacity region already dominates the Mesa Central or core area and is reasonably well connected to most areas of Mexico. Moreover, metropolitan Mexico City is losing jobs to the seven states and their major cities that surround the capital.

Finally, two time scenarios are incorporated into the economic potential calculations for the entire NAFTA study area. The base scenario assumes an average node-to-node journey time of 80 kilometers per hour. This scenario recognizes existing road and rail conditions in Mexico, accounts for the difficult physical environment, and draws on field experience of internal journey times. Scenario one also assumes an overall improvement in the operating conditions of Mexico's major highways and railroads. In the second scenario, average node-to-node journey times are calculated at 160 kilometers per hour. These times would result from the introduction of an integrated, multimodal

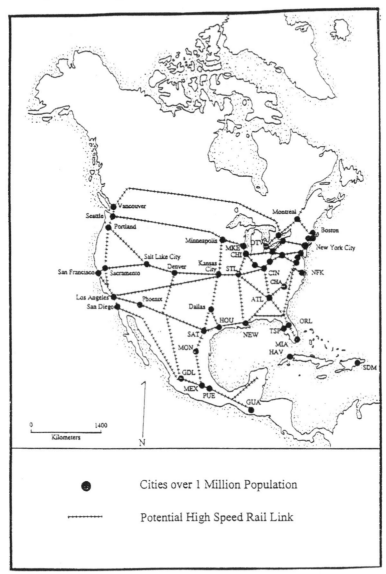

Figure 6-2   An Idealized High-Speed Rail Network for the NAFTA Region

national road and rail network between Mexico's major cities, allowing for rail running speeds up to 200 kilometers per hour along rehabilitated intercity pathways (Figure 6-2).

Economic Potential Study Results

The general hypotheses posited that with basic improvements in transport services and infrastructure the more peripheral regions of Mexico could experience higher gains than the core areas in absolute and relative accessibility. The initial results for scenario one (80 kph) show that the smallest potential increases in economic potential are experienced by four of the twelve states that comprise the urbanized, industrialized core: México, the Federal District, Jalisco, and Veracruz (Table 6-2). Overall, potential increases are highest in the northern states of Aguascalientes, Nayarit, Zacatecas, and Durango, These states traditionally have lagged behind in social and economic development and long have been considered more underdeveloped than Mexico's core states. Not surprisingly, the largest potential increases in Mexico are found in the two core states with the smallest share of national GDP: Colima and Tlaxcala. Their small areal size and their proximity to the dynamic activity concentrated in the Greater Mexico City region gives them a distinct advantage over more peripheral areas. The only other states to record potential gains in excess of 100 percent are Morelos and Querétaro in the Mesa Central region, and Quintana Roo in the southern part of Mexico (Table 6-2).

As expected, states with above-average shares of national GDP experienced the highest percentage of self-potential increase. In other words, the Federal District gained only 30.43 percent of its potential increase from outside; 69.57 percent of the District's total potential increase came from its own economic potential. In contrast, states with below-average shares of national GDP experienced the lowest percentage of self-potential increase. Durango, for example, one of Mexico's poorest states, gained 98.49 percent of its total potential increase from outside the state. With the exception of the Federal District and the state of México, no state recorded a self-potential increase 10 percent or greater.

At the national level, Mexico's total potential economic increase from transport improvements overall favors the Mesa Central region, with the Northwest, the South, and the Northeast gaining far less. However, when total potential economic increases are compared to national GDP as a percentage, the Northwest with an 81.6 percent increase in GDP ranks the highest, followed by the South, the Northeast, and the Mesa Central. Moreover, in terms of potential benefit from outside Mexico, the states peripheral to the Mesa Central region recorded the highest gains. South of the Mesa Central, for example, 90 percent of Quintana Roo's economic potential is generated from the United States and Canada, with other peripheral states such as Aguascalientes, Yucatán, Chiapas, and Tamaulipas recording high economic

Table 6-2   Absolute and Relative Economic Potential Values for Mexico: Scenario One—80 Kilometers per Hour

Political Division	GPD (millions of US$)	Self Potential	Percent Self	Total Potential (millions)	Percent Increase
Aguascalientes	1,249	89.978	3.78	2,381	190.64
Baja	5,141	72.197	1.46	4,954	96.36
Campeche	3,802	89.965	6.07	1,478	38.87
Chiapas	3,322	64.850	4.07	1,595	48.01
Chihuahua	5,550	59.636	2.44	2,446	44.08
Coahuila	5,103	70.072	2.79	2,509	49.17
Colima	913	67.390	3.32	2,031	222.43
Durango	2,250	34.092	1.51	2,251	100.07
Federal Cap.	36,477	5043.833	69.57	7,250	19.88
Guanajuato	5,642	171.824	7.24	2,374	42.08
Guerrero	3,217	67.476	3.76	1,794	55.77
Hidalgo	2,908	107.192	4.34	2,468	84.86
Jalisco	11,578	216.557	9.26	2,339	20.20
Mexico	19,475	708.697	22.96	3,087	15.85
Michoacan	4,283	93.040	4.24	2,196	51.28
Morelos	2,181	164.853	6.35	2,595	118.99
Nayarit	1,244	40.276	1.96	2,052	164.93
Nuevo León	10,816	225.738	7.76	2,908	26.88
Oaxaca	2,927	50.782	2.81	1,807	61.74
Puebla	5,304	153.189	7.12	2,152	40.58
Querétaro	2,242	111.426	4.81	2,317	103.34
Quintana Roo	1,232	29.238	2.10	1,395	113.22
San Luis Potosi	3,158	66.873	2.89	2,312	73.22
Sinaloa	3,834	84.421	4.04	2,091	54.54
Sonora	4,702	58.604	2.43	2,412	51.30
Tabasco	3,172	106.119	6.09	1,742	54.91
Tamaulipas	4,683	88.388	3.59	2,465	52.64
Tlaxcala	976	81.900	3.56	2,301	235.74
Veracruz	9,706	192.763	9.01	2,139	22.03
Yucatan	1,991	54.030	3.90	1,387	69.64
Zacatecas	1,754	34.464	1.43	2,408	137.29
Total	170,832	8499.593	11.24	75,636	44.28

potential from outside Mexico. In contrast, the Federal District gains only 23.1 percent of its economic potential from outside the country, while the state of México gains approximately 53 percent externally.

Scenario two is based on transport network improvements, including a high-speed rail system, and presumes average point-to-point journey speeds of 160 kilometers per hour (Table 6-3). At 160 kph, many of the southern states could experience increases in their economic potential of over 100 percent. Several of the poorer north-central states such as Aguascalientes, Nayarit, and Zacatecas could conceivably enjoy a rapid rate of socioeconomic development with improved transport links.

As Table 6-3 suggests improvements in transport connectivity to attain the running speeds used in the analysis could generate acceptable economic returns and would provide a strong boost to the depressed economies of the traditionally peripheral states.

Economic potential analysis does have several theoretical and practical weaknesses, and a number of other inputs to the model could have been considered: for example, population dynamics, levels of industrialization, existing external trade links, decentralization policies, and infrastructure plans. Nonetheless, at a very general level, the results suggest a strategy of spatial development that should focus more on the northern and southern regions than on the core. Certainly the northern states have a greater potential to capture economic activity across the international border than do the southern states.

A further benefit to Mexico's development that transcends purely economic growth lies in the ability of transport improvements to reshape individual and community perceptions about place and region. Mexico City, for example, is perceived by many as the sophisticated, industrialized, and civilized heart of the nation, whereas interior states often are viewed as isolated, backward, uncivilized, and barbaric. Rural-urban migration patterns of the past half-century have been driven, in part, by this perception of the core region's attractiveness, as well as by the location of major industries in and around the Federal District. Growth pole strategies in Mexico designed to encourage development outside the Mesa Central mostly have been a miserable failure, in part because little attention has been paid to the practical and perceptual issues of accessibility and mobility. Transport improvements that embrace a broader regional approach could not only revitalize stagnant and depressed economies, but they could also help to reshape traditional perceptions about the quality of life in Mexico's poorer states and to improve individual and community mobility.

Table 6-3   Absolute and Relative Economic Potential Values for Mexico: Scenario Two—160 Kilometers per Hour.

Political Division	Increase in State Potential		Percent Increase Relative to Mexico DF
	Millions of US$	Percent	
Aguascalientes	4,762.170	381.28	32.84
Baja	9,907.310	192.71	68.32
Campeche	2,955.467	77.73	20.38
Chiapas	3,189.460	96.01	22.00
Chihuahua	4,893.004	88.16	33.74
Coahuila	5,017.880	98.33	34.60
Colima	4,061.587	444.86	28.01
Durango	4,503.124	200.14	31.05
Federal Cap.	14,500.834	39.75	100.00
Guanajuato	4,748.550	84.1	32.75
Guerrero	3,588.193	111.54	24.74
Hidalgo	4,935.436	169.72	34.04
Jalisco	4,678.653	40.41	32.26
Mexico	6,173.275	31.70	42.57
Michoacan	4,392.434	102.56	30.29
Morelos	5,190.373	237.98	35.79
Nayari	4,103.362	329.85	28.30
Nuevo León	5,815.344	53.77	40.10
Oaxaca	3,614.138	123.48	24.92
Puebla	4,304.882	81.16	29.69
Querétaro	4,633.916	206.69	31.96
Quintana Roo	2,789.625	226.43	19.24
San Luis Potosi	4,624.725	146.44	31.89
Sinaloa	4,182.397	109.09	28.84
Sonora	4,823.855	102.59	33.27
Tabasco	3,483.666	109.83	24.02
Tamaulipas	4,929.800	105.27	34.00
Tlaxcala	4,601.657	471.48	31.73
Veracruz	4,277.056	44.07	29.50
Yucatan	2,773.219	139.29	19.12
Zacatecas	4,816.259	274.59	33.21
Total	151,271.651	88.55	

Transport Policies in Mexico

Transport policies over the past three decades have been driven by two major factors: the development of regional planning strategies designed to deconcentrate industrial and manufacturing activities away from Mexico City, and the management of transport first through direct government intervention and more recently through privatization policies. However, Mexico's transport policies have not responded to current theories about the role of transport in regional development. The idea that transport's role is merely to facilitate economic interaction at the global and interregional levels has become embedded in transport policy planning. Present regional development and integration strategies make the same mistake as past policies of focusing solely on global and inter-regional links. Little attention has been paid in these policies to intra-regional and local connectivity.

Furthermore, transport planners and policy makers have elected to address transport issues and problems on a modal basis or from a modal perspective. The development of an integrated, multi-modal approach has not been forthcoming. The potential cost and service benefits from focusing on the coordination of multi-modal transport have been largely ignored. Throughout Mexico, air, rail, bus, and road freight networks operate independently of each other, with little spatial or temporal coordination at any level of the system. Vertical or horizontal modal integration in transport and communication networks is relatively unknown.

In terms of modal split, trucking is the most important mode of merchandise transport in Mexico, accounting for approximately 75 percent of all cargo movements. The national railroad system handles about 15 percent of the total cargo moved (Fernández 1995). Moreover, approximately 90 percent of all interstate passenger traffic is handled by bus services. Deregulation of the trucking and bus industries at the beginning of the 1990s has changed the dynamics of cargo and passenger movement within Mexico. Thousands of new trucks and buses have taken to the country's highways, exacerbating the problem of very unsafe and poorly maintained roads. Major routes are congested and accidents are frequent, even though privatization and deregulation policies have encouraged the introduction of new toll roads.

Approximately 5,800 kilometers of highways were constructed between 1988 and 1994, at a cost of over US$15 billion, with most of these new roads run by private consortiums (Emmons 1996). Concessions were granted originally for only 10-15 years after the construction date. Consequently many of the companies set outrageously high tolls in order to recoup their investment as rapidly as possible. For example, before the late-1994 devaluation of the peso, a 400-kilometer trip from Mexico City to Acapulco cost the equivalent of US $63.00 for a passenger vehicle. Tolls in March 1996 on the new freeway between Mexico City and Veracruz for a passenger car cost about US $20 for the 450-kilometer journey, with tolls for large trucks costing upwards of US $60. Many smaller trucking companies cannot afford the cost of these tolls and continue to use the *carretera libre*, or non-toll roads, many of which are poorly maintained and require longer journey times.

Deregulation also has induced a major change in the geography of trade flows. International trade routes such as Mexico City-Monterrey-Laredo, Guadalajara-Manzanillo, and Mexico City-Veracruz have experienced dramatic increases in traffic volumes, whereas certain internal routes such as Guadalajara-Monterrey and Guadalajara-Mexico City have experienced decreases in traffic (Fernández 1995). Mexico's national airline system also has been privatized and the railroads currently are undergoing privatization.

Another regional development and planning problem in Mexico is that transport policies have responded to a conception of the interior that emanated from strategies and policies formulated in the Federal District. These policies have treated regions such as the Yucatán or the Northwest, for example, as functional regions with homogeneous characteristics, rather than as historically defined regions with temporally and spatially complex characteristics. Policy priorities have been driven by an emphasis on state and corporate action rather than by considerations of distinctive regional economic, social, and cultural forces. Thus, policies frequently are reactive rather than proactive, and the federal government has had to respond to regional inequalities instead of preventing them from occurring. Conceptualizations of non-core regions also have been driven by the implicit assumption that these regions should be like Mexico City or, more preferably, like the United States rather than by the realities of spatial conflict and struggle within the regions themselves.

Transport's role in Mexico's regional, hemispheric, and global development should include an explicit acknowledgement of the relationship between transport, people, and places. Transport networks cannot be planned without an understanding of the cultural inputs that help to define the network. Why do people choose to travel to a particular location? Where are needed services located? What are the time and cost relationships between communities, individuals, and the needed services? And what are the ideologies that drive the provision of the infrastructure needed to facilitate interaction? The role of transport in regional development, however, cannot be measured by transport mode and provision analyses alone. Applying a gravity model, for example, to the cities of Monterrey and San Antonio may indicate that a certain amount of interaction is likely to occur between them. Yet without a detailed analysis of the economic, social, political, and environmental circumstances of each city (in other words, the cultural factors), such models have little value.

Mexico's transport policies in the 1990s are aimed at privatizing and deregulating the transport arena. Airline, railroad, and telephone networks have been sold piecemeal to private concerns without any overarching national development plan. The government believes that supply and demand mechanisms will act as a development agent and thus, beyond the privatization strategies, no federally driven national, integrated, multimodal transport policies are needed. A long history of government financial and management ineptitude in the transport sector lends credence to the popular belief that private enterprise will provide a better quality of service, while relieving the financially strapped federal and provincial governments of fiscal responsibility for transport provision. Although short-term benefits are beginning to be felt from the transport privatization process, the long-term consequences may be disastrous for Mexico's goal of creating a dynamic, unified, national economy and society able to compete at the regional and global levels.

Conclusion

The results of the economic potential analysis for Mexico demonstrate that the advantages offered by transport improvements in the NAFTA region are not necessarily confined to the already dominant Mesa Central states. However, inadequate supporting transport infrastructure and services are likely to limit any potential benefits from transport improvements and to have some serious repercussions for Mexico's more isolated regions. In light of the country's recent move toward neoliberal economic restructuring policies, there appears to be some potential for meaningful transport improvements in the near future.

However, the nature and evolution of the privatization and deregulation of Mexico's transport systems will determine the character, shape, and efficiency of the country's transport networks at both the national and international scale for the remainder of the 1990s and beyond. Some critics of Mexico's neoliberal economic strategies have argued that the government is repeating the mistakes of the past by allowing the fate of the nation's transport systems to be determined by foreign interests. Moreover, transport planners have been criticized for focusing on the ideological aspects of privatization and ignoring the very real problem of developing a multi-modal, integrated transport and communication system capable of carrying Mexico into the next century.

Although it is still too early to determine if private sector ownership of Mexico's transport infrastructure will lead to increased investment and improved linkages to interior regions and cities, many analysts, including this one, are extremely doubtful. If Mexico's more isolated regions are to realize the potential advantages that transport improvements will offer (for example, reduced journey times, lower transport costs, changing perceptions of the interior, and enhanced intra- and inter-regional mobility), governments must take the initiative in providing the policies, funding, and motivation for transport infrastructure improvements. As this analysis has pointed out, the non-core regions of Mexico could benefit tremendously from increased accessibility within the NAFTA region. However, absent any serious government action in the transport arena, Mexico's citizens, industries, and businesses will miss some of the opportunities presented by the NAFTA and many areas will almost certainly be further peripheralized in the emerging regional and world economies.

References

Amin, Ash and Nigel Thrift, 1992, "Neo-Marshallian nodes in global networks", *International Journal of Urban and Regional Research*, 1694:571-587.

Blum, U. 1982, "Effects of transportation investments on regional growth: A theoretical and empirical investigation", *Papers and Proceedings of the Regional Science Association*, 49:169-184.

Clark, C. 1966, "Industrial location and economic potential", *Lloyds Bank Review*, 82:1-17.

Dugonjic, V. 1985, *Transport Policy and the Regional Aspects of Economic Development in Yugoslavia*, Cambridge, Mass: MIT, Department of Urban Studies and Planning, Working Papers.

Dugonjic, V. 1989, "Transportation: Benign influence or an antidote to regional inequality?" *Papers of the Regional Science Association*, 66:61-76.

ECLAC (ed), 1995, *Trade Liberalization in the Western Hemisphere*, Washington, DC: Economic Commission for Latin America and the Caribbean (ECLAC).

Emmons, William M. 1996, "The Mexican Toll Roads Program", pp. 281-316 in Ravi Ramamurti (ed), *Privatizing Monopolies: Lessons from the Telecommunications and Transport Sectors in Latin America*, Baltimore: Johns Hopkins University Press.

Fernández, Arturo M. 1995, "Deregulation as a source of growth in Mexico", pp. 311-332 in Rudiger Dornbusch and Sebastian Edwards, (eds), *Reform, Recovery, and Growth: Latin America and the Middle East*, Chicago: The University of Chicago Press.

Friedmann, John and Goetz Wolff, 1982, "World city formation: An agenda for research and action", *International Journal of Urban and Regional Research*, 6:3:309-344.

Gibb, Richard (ed), 1994, *The Channel Tunnel: A Geographical Perspective*, New York: Wiley.

Gibb, Richard, and David Smith, 1994, "The regional economic impact of the Channel Tunnel", pp. 155-176 in Richard Gibb (ed), *The Channel Tunnel: A Geographical Perspective*, New York: Wiley.

Harris, C. 1954, "The market as a factor in the localization of industry in the United States", *Annals of the American Association of Geographers*, 44:315-348.

Heldman, R.K. 1992, *Global Telecommunications: Layered Networks, Layered Services*, New York: McGraw Hill.

INEGI. 1986-1988, *Anuario Estadístico*, Mexico City: Instituto Nacional de Estadística, Geografía e Informática (INEGI).

Ivy, Russell L. 1995, "The restructuring of air transport linkages in the new Europe", *The Professional Geographer*, 47:3:280-287.

Keeling, David J. 1995, "Transport and the world city paradigm", pp. 115-131 in Paul L. Knox and Peter J. Taylor, (eds), *World Cities in a World-System*, Cambridge: Cambridge University Press.

Keeling, David J. 1997, *Contemporary Argentina: A Geographical Perspective*, Boulder, CO: Westview Press.

Kellerman, Aharon, 1993, *Telecommunications and Geography*, New York: Wiley.

Krooth, Richard, 1995, *Mexico, NAFTA and the Hardships of Progress*, London: McFarland.

Leinbach, Thomas, 1983, "Transport evaluation in rural development: An Indonesian case study", *Third World Planning Review*, 5:23-35.

Noam,E. 1992, *Telecommunications in Europe*, Oxford: Oxford University Press.

Pick, James B. and Edgar W. Butler, 1994, *The Mexico Handbook: Economic and Demographic Maps and Statistics*, Boulder: Westview.

Rich, D. 1980, *Potential Models in Human Geography*, Norwich: University of East Anglia Press.

Roccatagliata, Juan A., (ed), 1993, *Geografía Económica Argentina*, Buenos Aires: El Ateneo.

Rudel, T. and S. Richards, 1990, "Urbanization, roads, and rural population change in the Ecuadorian Andes", *Studies in Comparative International Development*, 25:3:73-89.

Simon, David, 1996, *Transport and Development in the Third World*, London: Routledge.

Stewart, J.Q. 1947, "Empirical mathematical rules concerning the distribution and equilibrium of population", *Geographical Review*, 37:4:461-485.

Taaffe, Edward J., Richard L. Morrill, and Peter R. Gould, 1963, "Transport expansion in underdeveloped countries: A comparative analysis", *Geographical Review*, 53:4:503-529.

Thomas Cook, 1994, *Overseas Timetable*, Peterborough, UK: Thomas Cook Publications.

Thrift, Nigel, 1986, "The geography of international economic disorder", pp. 12-67 in Ronald J. Johnston and Peter J. Taylor, (eds), *The World in Crisis?* New York: Blackwell.

Vickerman, R.W. 1991, "Other regions' infrastructure in a region's development", pp. 126-141 in R.W. Vickerman, (ed), *Infrastructure and Regional Development*, London: Pion.

Voight, F. 1984, "Transport and regional policy: Some general aspects", pp. 3-16 in W.A.G. Blonk, (ed), *Transport and Regional Development*, Aldershot, UK: Gower.

7 Shifts in Slum Upgrading Policy in India with Special Reference to Calcutta

ASHOK K. DUTT, ANIMESH HALDER, CHANDREYEE MITTRA

Though formation of slums takes place in cities all over the world, the Indian city slums are characterized by a total lack of upward mobility and persistence of poverty. Twenty-eight percent of the population of the million-plus cities of India live in slums. The four metropolitan cities, Bombay (Mumbai), Calcutta, Delhi and Madras (Chennai), have a third of their population in slums (Ribeiro, 1985). The proportion of people living in slums declines as the city-size declines, confirming that the larger cities make more intensive use of space. An Anova test shows that between city-group variance is statistically significant (Table 7-1). Moreover, the larger cities have more industrial and service employment opportunities, which attract rural migrants in large numbers who settle in the slums. It is the agriculturally productive Indo-Gangetic plain and the East coast region (Figure 7-1), where rural population densities are high and where a great deal of "rural push" is generated. Urban slum population proportions are very high in cities of these areas (Mukhopadhyay and Dutt, 1993; Dutt and Mukhopadhyay, 1997). This chapter analyzes policy shifts in slum upgrading, first for the nation and then for the city of Calcutta, with a case study of three slums of Calcutta to show how the vicious cycle of poverty persists.

Slum Policy and Five Year Plans

A significant policy change in the 1960s initiated efforts for slum improvement. The 1950s bulldozing policy aimed at clearance of the slums and of building new housing. Unfortunately, the new properties generally housed the middle class and the rich, while the original, poor slum dwellers were pushed to other slums or forced to form new ones. The Indian First Five-Year Plan (1951-56) clearly indicated that its housing scheme was "for slum clearance and rehous-

Table 7-1 India: City Size and Population Residing in Slums, 1991

City Size Category	Count	Average(%)
Above 4 million	6	33.4
1,000,000 to 3,999,999	17	26.9
500,000 to 999,999	27	24.0
200,000 to 499,999	76	23.6
100,000 to 199,999	155	23.7
Anova Results		
F-statistic	4.45	
critical value of F	2.4	
p-value	0.001	

The between group variance is statistically significant.

Source: Ribeiro (1985); 1991 Census of India

ing of slum dwellers" (First Five-Year Plan Progress Report 1954, 259). The Second Five-Year Plan (1956-61) proposed "slum clearance and houses for sweepers" (Second Five Year Plan 1956, 161) and allocated 17 percent of its housing budget for the purpose. The Plan also acknowledged that compared to 1951 the shortage of housing (mainly urban) could double by 1961. The Third Plan (1962-67) combined both slum clearance and improvement strategies. Improvement was proposed only on selected sites and the local authorities (municipalities or public corporations) were empowered to carry it out. These local bodies were, however, incapable of properly carrying out the slum improvement programs, resulting in inaction in most cases (Roy 1994, 29). The Fourth Plan (1969-74) started to change strategies. It subsidized urban housing schemes and called for both "slum clearance and improvement" (Fourth Plan n.d., 401) and it was proposed that 69,556 tenements be built under this scheme by 1968-69. Such a low number of proposed housing also points out that the urban slums were given very low priority. By 1972-73, schemes were introduced to improve slum environments by providing basic services, such as sewerage, drainage, safe water supply, paved streets and streetlights. The schemes covered eleven cities of the country that each had population of over 0.8 million (Fifth Plan, II 1973, 257). Thus, the concept of slum improvement, not clearance, took definitive shape. Nonetheless, slum bulldozing policy per-

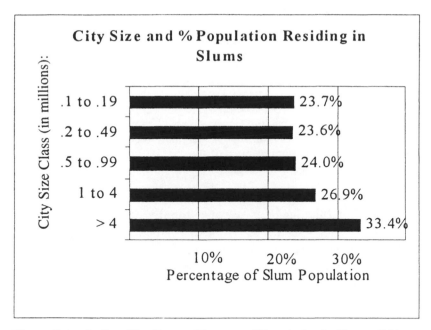

Figure 7-1   India: City Size and Percent of Population in Slums, 1991
Source: Riberio, 1985 and Census of India, 1991.

sisted in many cities including in Delhi, where thousands were uprooted from their inner city slum locations during the Emergency Period (1975-77).

The Fifth Plan (1974-79), under the sub-title of "Environmental Improvement Scheme in Slum Areas", made it clear that the practice of slum "clearance" would be replaced by slum improvement.

> The existing scheme of environmental improvement in slum areas...would be continued... The scheme is proposed to be extended to all towns with a population of 3 lakhs [300,000] and above and in addition to one town in each State where no such town exists. The provision being made would make it possible to cover a population of nearly 6 million slum dwellers. With the expenditure already incurred in the Fourth Plan, all slum areas in such towns as are not scheduled for clearance and/or otherwise amenable to improvement would stand covered. (Fifth Five-Year Plan, II 1973, 259)

The Plan also allocated one third of its urban development budget to the States and Union Territories for environmental improvement of slum areas.

The Sixth Five-Year Plan (1980-85) acknowledged that slums were caused by poor environment, such as poor drainage and sewerage and lack of sanitation. Therefore, environmental improvement was paramount for uplifting a slum. The "strategy of attempting massive relocation of slums in urban areas" was discarded once and for all, while the Plan advanced the following arguments for slum improvement:

> ...Relocation not only involves substantial hardship to those affected in terms of loss of easy access to employment centres and other amenities, but results in unnecessary destruction of existing housing capital, however sub-standard it may be. It is, therefore, important that substantially increased investments be made in the environmental improvement of slum areas. Low cost sanitation and drainage are key areas of much needed investment in the slums of our cities. (Sixth Five-Year Plan 1981, 392)

The Plan proposed that the facilities—"water supply, storm water drainage, paving of streets, street lighting and provision of community latrines"—be provided "to all urban areas irrespective of the size of the city/town" (Sixth Plan 1981, 396). The Sixth Plan's complete aversion to massive relocation was a reaction to the unpopular slum clearance efforts of Sanjay Gandhi in Delhi during the black days of the Emergency Period (1975-77).

The Seventh Plan (1985-90) reiterated government's role in housing stating that major effort for new housing had to come from the private sector, while the direct public sector involvement was for the weaker section of people and for slum improvement. In order to benefit more, the slum dwellers' allocated amount of the Environmental Improvement of Slums (EIS) Program was increased; 9 million slum dwellers were to be covered. The Plan recognized that legal safeguards be provided so that slum dwellers were not displaced after the implementation of the EIS Program.

> The EIS programme has to be continued with greater vigor and steps should be taken to provide security of tenure to the slum dwellers so that they may develop a stake in maintaining and improving their habitat (Seventh Five-Year Plan 1985, 299).

The Eighth Plan (1992-97), like the previous plans, expanded its schemes for housing different income groups. It introduced a new housing and shelter upgrading scheme for the urban poor living in cities of 100,000-2,000,000 size. The Eighth Plan emphasized "total sanitation" rather than adopting piecemeal environmental sanitation projects.

> The concept of "total sanitation"—covering primary health care, water availability, women's welfare, immunization and provision of sanitation facilities, all linked to cleanliness as a basic human need — will be emphasized. Every effort will be made to adopt a low-cost approach, employing technical and scientific know-how

and the experience already gained by several non-governmental organizations in this regard (Eighth Plan, II 1992, 381).

Also the global idea of decentralization has been related to upgrading schemes water supply component.

> In order to reduce costs, it is desirable that the programmes of drinking water supply and sanitation, both in rural and urban areas, are implemented in a more decentralized manner with the involvement of the people and local institutions at all states — planning, project formulation, execution, operation and maintenance and monitoring and evaluation. (Ibid.)

In pursuance of further decentralization, states have been charged with the main responsibility for housing the weaker section of the people.

> The State would address specifically to the needs of the houseless, poorer households, SCS/ST [Scheduled Castes and Scheduled Tribe], women and other vulnerable groups. The long term objective is provision of "Shelter for All". (Eighth Plan, I 1992, 15)

In spite of the efforts made by the central and state governments of India, the slum problem has not ameliorated. Access to sanitation remained non-existent for 54 percent of urban people in 1993 (World Bank 1997, 115). In 1990, 48.8 million people were living in urban slums, 40 percent of them in the million-plus cities.

> In 1991, the National Buildings Organization estimated that there was 10.4 million backlog of housing in urban areas. (Eighth Five Year Plan 1992)

Why Slums Continue?

There are several reasons why slums and the associated problems continued to aggravate during the last five decades. First, though the urban population has increased only from 17 percent in 1950 to 26 percent in 1990, during the same 40-year period 155 million people were added to the urban areas, making a total of 216 million in 1990. The supply of decent urban housing was either too limited or beyond the reach of most new migrants to city, thus giving rise to sub-standard housing with very few sanitary facilities. Second, most of those who came to the city had few skills and little education and ended up in low-paying jobs, generally in the informal sector. They could not afford rents for decent or semi-decent housing and therefore, they settled in to one-room slum tenements. Third, the numbers increased so rapidly that the civic authorities and others in charge of administration could not keep pace with the increase, leading to deterioration of civic and sanitary facilities, especially of the slums. Fourth, though India is a democracy and slum dwellers have voting rights in the civic, state and national elections, their strength in decision making remains largely ineffective because they are the poorest in the city and compared to rural poor their numbers are much less. Often, they are manipulated by political parties through intimidation, bribery and false promises. Fifth, rampant corruption results in pocketing all or part of the money allocated for slum improvement by civic and government officials, contractors and a few "leaders" of the slum. Sixth, a developing country like India has a government that has a limited amount of funds but an enormous amount of problems. Therefore, only a small share of the public budget is allocated for slum improvement. Seventh, as most migrants to the city lack upward mobility in their income they remain mired in the slums for good.

Slum Policy in Action in Calcutta

When the poor migrated to colonial Calcutta as servants of the British families the first group of slums was formed. This trend continued when the wealthy Indian landlords and trades people settled in Calcutta as they also needed servants. The servants lived in sub-standard houses not too far from their masters' homes. With the growth of industries from the early 19th century, low paid workers swarmed into the city and they lived near the factories creating additional slums. The British policies towards slums were initially directed against the spread of infectious diseases and fire, as both originated there. In order to eradicate slums from the British quarters, a Lottery Committee was formed in 1817 which bought slum lands in the British areas for clearance. Starting in 1850 steps were taken to make the city fire-safe. The slum huts, made of thatch and straw, were most susceptible to fire, which consumed rich peoples' housing too. By the end of 1860s a law was passed, which resulted in thatch roofs being replaced by tiles and straw walls by mud walls. No wonder Calcutta slums to the present generally have tiled roofs. At first the southern side of the city, where the British lived, was made fire-safe. Gradually, such safety expanded into the native quarters in the north (Roy 1994).

Except for the levee adjacent to Hooghly and other distributaries of the Ganges, most land in Calcutta and the Calcutta Metropolitan District is low-lying and susceptible to water-logging during the rainy seasons. Many areas turn into swamps. As much as 59 percent of the slum dwellers, according to a 1996-97 Calcutta Metropolitan District Authority Survey, live in waterlogged areas after heavy rains. Therefore, one way the city folk, particularly the poor migrants, could construct houses was by cut, fill and build; they excavated a tank, piled the dirt on the banks, and created artificial highlands for settlement. This process left the slums with a large number of tanks in the middle of their habitation. These eventually became multi-purpose tanks, where families took baths, utensils and clothes were washed and from there water was brought to the homes for cooking and drinking. These tanks, never cleaned or disinfected, became a breading ground for cholera. Calcutta earned the nickname, "Cholera Capital of the World", as the disease became endemic to the city. In order to rid the city of deadly cholera, the Calcutta Corporation in the 1870s launched a drive to clean and/or fill the tanks in the slums. Filling of tanks had disastrous environmental consequences as it deprived a) the people of a rare open space, b) the presence of a safety valve, which acted as a water storage during monsoon, and c) the slum dwellers of a nearby bathing and washing place.

During the 1880s, a significant sanitary improvement occurred in the slums as a result of the efforts initiated by Henry Harrison, Chairman of the Calcutta Corporation. Harrison

> promised to reform the existing drains, pipes, sewer and filthy privies with necessary arrangement of water supply. To keep a track on the condition of the bustees, the number of members of the staff were increased and individual overseers were made responsible for each bustee. Under his initiative, almost all the bustees were connected with the drainage system; most of the open drains were converted into sewers and covered with footpaths; many tanks were filled and cleaning services were set up. (Roy 1994, 11)

The Municipal Act of 1899 was intended to empower the Corporation to prepare a "Standard Plan" for the slums, in which improvements were standardized and it was made obligatory for the slum landlords to pay for them. The powerful, rich landlords lobbied against the Act, which was modified to the extent that it became impotent and inoperative.

The Calcutta Improvement Trust (CIT), formed in 1911, selected large areas, not built with permanent *pucca* (solid, all weather) structures within the city, for the purpose of residential development to house upper middle class and the rich. In doing so, a large number of slum huts were razed to the ground and their dwellers had no other alternative but to move to the fringes of the city. Thus, a structured and continuous slum clearance activity was in effect in Calcutta, until the Calcutta Metropolitan Planning Organization (CMPO) was formed in 1961.

Calcutta Plans for Slum Improvement

Even after India's independence in 1947 the colonial policy of slum clearance not only continued to exist, but thrived. Calcutta Slum Clearance and Rehabilitation of Slum Dwellers (CSC/RSD) Act was adopted by the West Bengal legislature in 1958, in which clearance and rehabilitation were subsidized by the government. The subsidy amount was so inadequate that the Act remained ineffective. Moreover, in accordance with the provision of the Act,

> the slum may be acquired and the huts may be demolished, provided alternate accommodations are offered to the occupier of such huts within a radius of one mile from the slum area. The intention of the "one-mile clause" was to minimize the hardships of the displaced slum dwellers, many of whom work within the immediate vicinity of their huts. Area restrictions of this type have seriously hampered slum clearance, since in most cases no suitable vacant land is available at moderate cost within a mile from a bustee [slum]. Area restrictions do not guarantee, moreover, that the displaced bustee dwellers will be able to maintain their existing occupations in the new location. (First Report 1962, 47)

Traditionally, the slums in Calcutta have an atypical land tenure system in which a landlord leases his land to a hut owner; he in turn builds one-room tenements for renting to slum dwellers. Thus, this three-tier Thika tenant system complicated any improvement action in the slum. In view of land tenure complications the CMPO also proposed to explore

> the possibilities of evolving a method which will permit the executing agency, appointed by government, legal right of entry to the slums to effect the improvement programme; and also provide for legal powers to operate and maintain the services and to ensure recovery of the unsubsidised portion of the cost without unnecessary disturbance to the landlord or hut owners. (First Report 1962, 48)

For the first time, the CMPO called for slum improvement with the installation of basic sewer and water lines and the provision of water stand pipes and sanitary latrines and baths. This was considered feasible as costs were low and affordable. Such a policy of improvement was the result of a marriage between western and eastern ideas, which evolved in the CMPO, after a thorough discussion and analysis between the Ford Foundation consultants and the Indian planners. Bannerjee and Chakrovorty (1994, 77) call it a "successful marriage of minds between the hosts and visiting experts" and a two-way street.

In spite of such marriage of ideas, the remnants of "clearance and rehousing" continued to attract. Both Western and Indian planners had some inclination towards such a scheme. In the Western cities urban renewal projects emanated from clearance or bulldozing efforts; the Ford Foundation consultants were too familiar with those activities; many of them had planned such activities in the United States. The Indian staff of CMPO, consisting of sanitary engineers, architects and physical planners as well as sociologists, geographers, and economists, were also quite familiar with the clearance idea. The Chief Executive Officer, Mr. P.C. Bose was one of the sanitary engineers. It was, therefore, not surprising when CMPO, in 1962, also proposed a conventional slum clearance and rehousing scheme in their Manicktala Work-cum-Living Project. Thus, in the early 1960s, CMPO's slum policy was still in a transition stage; the emphasis was on improvement, but the clearance idea was not discarded.

The Basic Development Plan (BDP) elaborated the ideas of the First Report, stating that the slum improvement program ought to be divided into two parts. First, was clearance and new construction, and second, was improvement of the existing slums. It was the second that was emphasized. For clearance, the recommendation was that the displaced slum dwellers must be re-housed, whereas the improvement program that was planned to cover 57 percent of the slum population during a five-year period, 1966 to 1971, gathered momentum. The massive slum improvement program in Calcutta was designed for eight large areas, mostly outside the old city, bounded by Circular Roads (Acharya -P.C. Roy and J.C. Bose Roads) (Figure 7-2).

The program also catered for social needs such as community facilities, schools, hospitals, parks, and children's playgrounds. The main improvement program had eight components: i) an adequate water supply--clean and safe--by installing tube wells, pump houses, and chlorination; ii) community water taps and baths, which included one water tap for every 100 persons; iii) a sanitary sewer system to include necessary sewers; iv), sanitary latrines (toilets) on the basis of four per 100 persons, or, one sanitary latrine for every hut. All service privies (toilets built on raised platforms, under which night soil accumulates in a bucket, and then is transferred into larger vessels for disposal by scavengers (*Mathers*), who belong to Scheduled Caste or "untouchables") were to be gotten rid of; v) stormwater drainage to prevent flooding and unsafe or unsanitary conditions, especially during monsoon periods; vi) concrete pavements to supplement the storm drainage system, in order to provide decent pathways; vii) lighting at regular intervals and at crosswalks; viii) tanks, found in almost all slums, needed to be treated and made health hazard free. All tanks were earmarked for preservation.

BDP's slum improvement project became a model. It was modified later on, but its basics remained intact. This model has been followed in other cities

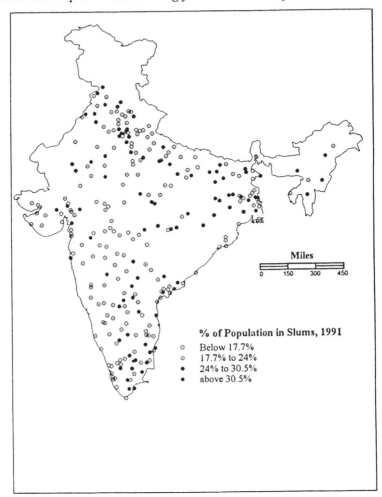

Figure 7-2   India: Percent Slum Dwellers to Total Population for Cities of Over
100,000 in Population
Source: see Figure 7-1.

of India and provided a new solution for slum improvement that is economically feasible. Based on this model, the Calcutta Metropolitan Development
Authority (CMDA), formed in 1970, launched an "Environmental Hygiene"
Program for Calcutta. The CMDA recognized that the contributing factors
for environmental degradation were: a) the existence of "service privies" in
the slum areas, b) absence of public conveniences and c) the habit of the rural
migrants to defecate wherever they could. The CMDA plan consisted of two

main components--service privies to be converted into sanitary latrines and provision of public convenience facilities. The specific CMDA project was to be completed by 1974.

> This project aims at replacing the service latrines in the Calcutta city area with sanitary ones of proper design. Since such service latrines are situated both in sewered and unsewered areas, it has been proposed that in sewered areas the service privies will be replaced by sanitary latrines connected to the nearest sewer. In the unsewered areas septic tanks will be built to treat the sewage for individual latrines. (Calcutta Development 1972, 41)

It was estimated that by 1986, 65 percent of the slums were covered by slum improvement schemes, designated as the "Slum Improvement Model" (Bagchi 1987, 598). This model did not mean simply the conversion of privies, but was a comprehensive sanitary model, with much broader consequences.

> ...it was ultimately decided to go for limited slum improvements. This was effectively a "Sanitation Model" aimed at providing basic infrastructure facilities to slum dwellers without attempting to provide conventional housing to the target group. The ingredients of the action programme included conversion of service latrines, providing potable water supply connections, surface drainage facilities, construction of paved roads and path-ways, arranging street lighting and providing garbage vats and dustbins in adequate numbers in slum areas. This model of improvement of living conditions of slum dwellers recognized slums as a part of the city housing stock occupied by people who had been priced out of all conventional forms of housing. This model did not involve interference with the rights and interests of the land owners and thika tenants and physical shifting of slum dwellers further from the place of their employment. The fact that this model of development was comparatively less expensive and improvements in living conditions could be done in stages were considered to be its strengths. (Chakrabarti and Halder 1991, 8-9)

Though Pugh (1989) considered the slum improvement schemes of CMDA to be one of its success stories, Charkrovorty and Gupta (1996) estimated that there had been a significant decline in slum improvement development expenditures from 24 percent to 12 percent between the 1978-82 and 1983-92 periods. Whatever the cause, such lowered expenditures ignore the facts that any improvement done in the slums needs maintenance, one-third of the slums are still unimproved, and a constant influx of migrants continues to inflate slum population every year. Moreover, the one-room tenement dwelling (even with sanitary improvements) is still inadequate and must not be taken as permanent. The next stage of improvement must plan for the addition of greater floor-space for the slum dwellers and in-house toilet and piped water facilities.

Anatomy of Calcutta Slums

A Slum Survey based on 20 slums was conducted by the CMDA in 1989. Another survey based on 4500 samples was conducted by the CMDA in 1996-97. This latter survey covers about half of the 138 census wards of Calcutta, which reveals the most recent spatial pattern; eight wards do not have any slum population while 5 have between 1 percent to 5 percent; only one ward has 100 percent of its population in slums while 4 have between 90 percent to 99 percent and 13 between 60 percent to 89 percent. In general, it is the area north and east of old Calcutta, surrounded by the Circular Road, that has the greatest concentration of slum population. The following figures taken from the two surveys provide a profile of Calcutta slums:

- 39 percent of Calcutta's population live in slums.
- 54 percent of the households consist of nuclear families (husband, wife and minor children), while extended families (with adult children and relatives) form 33 percent of the total households.
- family size averages 5.12 persons; 35 percent of the families consist of 6-10 members, and 4 percent have over 10 members.
- the Bengali-speaking families are 22 percent of the total slum population; 56 percent are Hindi speaking, and 21 percent Urdu speaking.
- of the total population 54 percent are male, and 70 percent of male earners and 67 percent of female earners are in the 15-44 working age group.
- 4 percent of the dwellers stayed in the slum for 0-5 years, 16 percent for 6-15 years, 29 percent for 16-30 years, 42 percent for 30 plus years and 9 percent do not have any record.
- 7.25 percent of the total slum population fall in the 0-4 age-group, 24.5 percent in 5-14 age-group, 51.64 percent in 15-44 age-group, 11.09 percent in 45-59 age-group and only 5.67 percent in the 60- plus age-group.
- of the adult population 53.8 percent never married, 41.77 percent married and 4.43 percent are widowed or divorced.
- 17.9 percent illiterates, 20.7 percent having less than primary schooling, 20.2 percent primary school, 20.7 percent middle school, 15.1 percent secondary school. Only 5.2 percent are graduates.
- 4.04 percent employers, 41.35 percent employees, 32.26 percent self-employed, 19.59 percent unemployed and 2.76 percent household labor.
- the monthly per capita income is less than Rs. 1200 (US $33) for 53.7 percent of the slum dwellers, while 11.6 percent earn more than Rs. 2600.
- 25 percent of the migrants to the slum remit money to their native villages at least once a year and 24 percent visit their respective villages on a regular basis.
- Over half of the slum dwellers (53 percent) have a daily caloric intake of less than 2100.

Poverty Level Per Capita Income

The question of what should be regarded as the nationally desirable minimum level of consumer expenditure was considered in 1962 by a Government of India Study Group. Dandekar and Rath (1971) defined poverty as based on the income required to procure food (primarily food grains) which would ensure a minimum level of caloric intake per capita. They estimated a caloric intake of 2150 per day to meet the minimum nutritional requirement, and they also, calculated the poverty line in India at the level of Rs. 22.50 per month per capita in urban areas and Rs. 15 per month per capita in rural areas at 1960-61 prices. Modifying Dandekar and Rath's study, the Indian Planning Commission estimated the poverty line for per capita calorie intake in India to be 2400 calories per person per day in rural areas and 2100 calories per person per day in urban areas, which at 1984-85 prices amounted to Rs. 107 per capita in rural areas and Rs. 122 per capita in urban areas; this was called the "Poverty Line Approach". The CMDA survey analysis, by incorporating the Consumer Price Index of 1989, inflated the 1984-85 urban poverty line per capita income to Rs. 180 (CMDA 1991).

The Nature of Migrants

People who live in slums are mostly rural migrants. Four types of rural/urban migrations were identified in a study in West Java and Jakarta. They are commuting, circular, seasonal and permanent (United Nations 1989, 5). In analyzing Calcutta we have not identified commuting as a form of migration because a commuter does not stay in the city during the night. Four types of rural migrants have been classified: permanent, long-term circular, seasonal and occasional migrants.

The migrants who move to the city and become permanent dwellers are permanent migrants. They form the stable population of the slum.

The long-term circular migrants move to the city for all or a large part of their working life. They also become a stable population for a long time; they save money while working in order to live the remaining part of their life from that saving when they return to their native villages. They are often replaced by their sons and other relatives who continue to live in the slum.

The seasonal migrants come to the city in particular seasons. They are short-term circular migrants. These are landless workers or poor peasants who come to the city during lean agricultural months in search of work. Such migration occurs during the time when there is no labor demand in the villages, which occurs between harvesting and sowing. The seasonal migrants return to their villages with the increase in labor demand during harvesting and sowing months. This process goes on almost every year. The construction industry, especially, makes use of such seasonal labor in the city.

The occasional migrants gravitate to the city following any disaster or unusual catastrophe in the village, such as drought, famine and flood. Most of them go back to the village after the mitigation of the disaster.

The migrant population identified in this study is that of either permanent or long-term circular migrants. Most are long-term inhabitants of the slum.

Methodology

The following analysis is based on a survey conducted by the Calcutta Metropolitan Development Authority (CMDA) in 1989 with 100 percent household coverage. Animesh Halder, a co-author of this chapter, directed the survey. Chandreyee Mittra, another of the chapter's authors, conducted interviews with the slum dwellers in 1995. The survey questionnaire related average monthly household income, per capita income, per capita caloric intake, yearly remittance to native place, number of visits to native place, per capita monthly expenditure on food, and distance traveled by the head of the household to work.

Rural/urban linkages play a dominant role in attracting new migrants to the city. The rural migrants residing in urban areas maintain ties with their native villages by remitting money and by visiting them once or twice a year. The long-term circular migrants remitted money frequently, indicative of their strong ties with their native places. The permanent migrants reflected very weak remittance patterns that implied severing of native place ties. The data for three slums of Calcutta: Paikpara, Narkelbagan and Betbagan (Figure 7-3) were analyzed using frequency analysis and cross tabulations.

Analysis and Findings: Three Calcutta Slums

Of the three slums studied, the Narkelbagan and the Paikpara slums, are located north of the city center, while the Betbagan slum is to the east of the city center. All three slums are about 100 years old and are sited at the periphery of the-then built-up area of the city. They originated just outside the-then municipal boundary. With the evolution of the city and the extension of the built-up area, the slums continued to exist while other land uses grew around them.

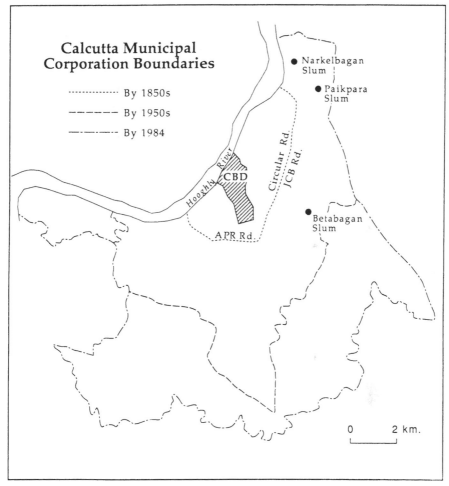

Figure 7-3 **Calcutta: City Boundaries in 1850s, 1950s and 1984 and Location of Three Study Slums and the CBD**

Paikpara

The Paikpara *bustee* (slum) was a Bengali-speaking slum of 300 households, 66 hutments and 1225 people in 1989; located north of the original municipal limits of old Calcutta, but now within the Calcutta Metropolitan Corporation (CMC) boundary. The slum is mostly formed of migrants from the rural areas of West Bengal and Bangladesh, both Bengali-speaking areas (Figure 7-4).

Initially, these rural migrants were employment propelled; moving out of the village by the "push factor"-lack of rural employment. Continued lack of employment prospects prompted a large number of them to close down their establishments at their native places and set up permanent homes in the Calcutta slum. The majority are permanent migrants to Calcutta. About 68 percent of the slum dwellers are below the poverty line; with monthly per capita incomes between Rs. 180 and Rs. 400 (US $10 - US $22). Most of the Paikpara inhabitants work in the informal sector as vendors and hawkers, and some stitch clothes at home for very low wages.

The Paikpara slum shows a relatively low incidence of remittances and visits to native places, indicative of a low-level of connection with their rural homes. Only 8 percent of the migrants remitted money to their native places and only 1.3 percent visited their native place in the year of the survey (Figure 7-5). For all practical purposes, they are permanent migrants to Calcutta.

Because of their low income, intra-urban migration is not possible. Forty-eight percent of the Paikpara slum dwellers have a per capita income of less than Rs. 180 per month, while 80 percent earn less than Rs. 300 (Figure 7-5). Eighty two percent have middle school or less than middle school education (Figure 7-6), and can only obtain low paying jobs in the informal sector. Their diet is low calorie and they have very little money to spend on housing and schooling of children (Figure 7-7). The children on reaching working age join the unskilled workforce in order to augment family income. When these children grow up and become adults the whole cycle is repeated again (Dutt, Mittra and Halder, 1997).

Narkelbagan and Betbagan

Narkelbagan was a Hindi-speaking slum of 400 households, 25 hutments with 977 people in 1989. Betbagan, also Hindi-speaking, was a slum of 513 house-holds, 32 hutments and 1359 people in 1989. The inhabitants have migrated from the districts of neighboring Bihar, and more distant Uttar Pradesh (UP), about 400 miles to the northwest (Figure 7-4). Their migration pattern is different from Paikpara. Over 90 percent in Narkelbagan and 85 percent in Betbagan are Hindi-speaking (Figure 7-4), and they show a process of long-term, circular migration for a segment of their inhabitants, while another segment is in the permanent category.

A large number live below the poverty line, in Narkelbagan, 48 percent, and in Betbagan 33 percent. The remittance pattern and visits to native place reveal that these slum dwellers have very strong ties with the native villages. At Betbagan 51.2 percent and at Narkelbagan 24 percent remit money to their villages; 55.2 percent at Betbagan and 24 percent at Narkelbagan visit their native place once or twice a year (Figure 7-5). Their migration to Calcutta follows a linkage chain through a village acquaintance or relatives. (Dutt, Mittra and Halder, 1997).

Mukhopadhyay (1990) in a study of a predominantly Hindi-speaking slum of Calcutta found that the Hindi-speaking people maintained strong kinship ties within their individual clan groups. New migrants preferred to come to that slum because of the concentration of Hindi-speaking people. Yadav (1989) in a survey of Delhi's rural migrants, also observed similar patterns. The migrants to Narkelbagan and Betbagan, like those in Paikpara, have very low earnings which do not allow them the choice of moving to better housing.

There is, also, a process of long-term circular migration among a part of these Hindi-speaking slum dwellers. The process of such migration and the cycle of poverty are also intertwined. Most of the long-term circular migrants live and work in Calcutta as long as they are physically fit. They also transfer parts of their incomes to their native places. Though many of them return to their native places when they are old and live on savings made earlier, still others live on in the slums. Many of these dwellers have been living in the slum for decades. When members of the present workforce become old and return home, their sons and nephews replace them. The households thus continue to exist, but the individuals change.

Cycle of Poverty in Operation

The Calcutta slum dwellers are mostly spatially immobile as the cycle of poverty engulfs them. These slum dwellers make meager money because they have low education and they lack technical training; making them eligible for only low paying jobs mainly in the informal sector. These slum dwellers live in rented houses for which they have to pay a monthly rent between Rs. 20-25 (under US $1.50). As the rent, food and clothing expenses consume most of their income, they are left with very little expendable money. They eat low calorie food and very few of them send their children to the secondary school. The question of moving to a place which is better than their one-room tenement accommodation does not arise. They cannot afford the cost of such housing. They move to other locations only when a) they have new employment at a distant place, b) there is a possibility of earning more money, or c) the slum is bulldozed. Proximity to work place and amount of rent are deciding factors for the choice of slum locations (Dutt, Tripathi and Mukhopadhay, 1994).

The migrants came to the locations outside the municipal limits and remain settled there due to lack of upward mobility. After settling in those areas initially, they usually become spatially immobile. With expansion of the city's

LANGUAGE

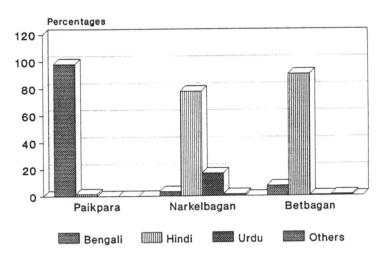

Bengali   Hindi   Urdu   Others

PLACE OF ORIGIN

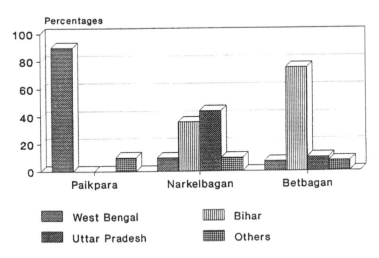

West Bengal   Bihar

Uttar Pradesh   Others

Figure 7-4   **Top: Language Composition of Three Slums, Percentage to Total Slum Population**
Bottom: Place of Origin of Three Slums in Percent to Total Slum Population
Source: CMDA 1991.

MONTHLY PER CAPITA INCOME

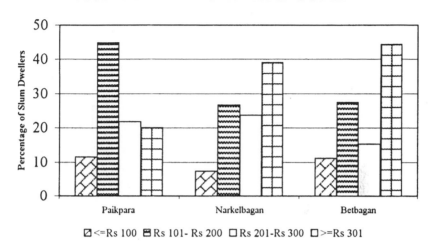

☑ <=Rs 100   ☒ Rs 101- Rs 200   ☐ Rs 201-Rs 300   ☐ >=Rs 301

REMITTANCE AND VISITS TO NATIVE PLACE

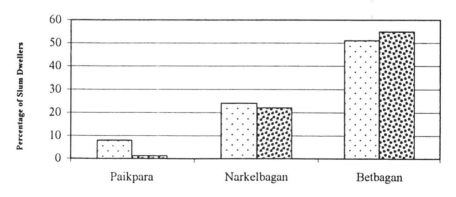

☐ Remittance ☒ Visits to Market Place

Figure 7-5   **Top: Monthly Per Capita Income of three Slums in Percent to Total Earning**
Bottom: Remittance and Visits to Native Place by the Slum Dwellers of the Three Slums in Percent to Total Slum Population
Source: See Figure 7-4.

RELIGION

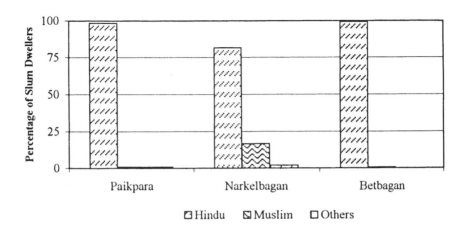

EDUCATION

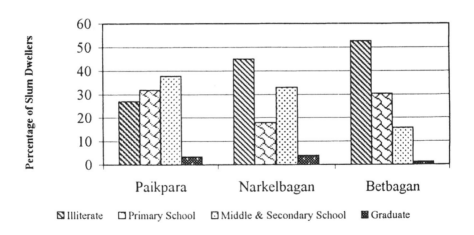

Figure 7-6   Top: Religious Composition of Three Slums in Percent to Total Slum Population
Bottom: Educational Levels of the Three Slums in Percent to Total Slum Population
Source: See Figure 7-4.

PER CAPITA CALORIE INTAKE

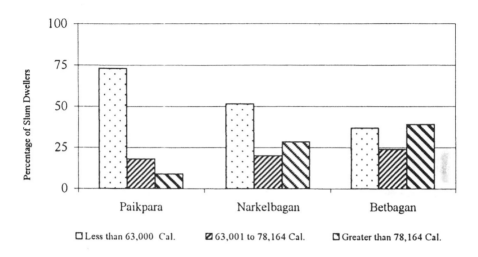

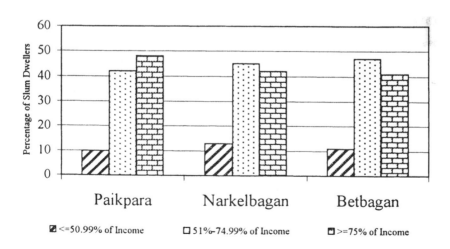

Figure 7-7   **Top: Per Capita Caloric Intake of the Three Slums in Percent to Total Slum Population**
   Bottom: Expenditure on Food of Three Slums in Percent to Total Expenditure

Source: See Figure 7-4.

built-up area, some slums are obliterated, but the three slums of Paikpara, Betbagan and Narkelbagan, maintained their existence as islands of poverty and depression.

Two descriptive models showing the comparative characteristics of the three slums sum up their situation (Figures 7-8 and 7-9). Many of the out-of-state Hindi-speaking migrants are long-term circular migrants. They consider themselves to be "aliens" in the Bengali-speaking city of Calcutta. Their slum home is a little "UP/Bihar village" in Calcutta. They continue to live in the city as long as they are physically fit enough to earn; eventually returning to their village in their old age. The first generation Hindi-speaking migrants generally go to their village for securing a bride or for attending specific rituals. The Bengali-speaking migrants, on the other hand, fit very well into the culture of Calcutta and essentially become permanent dwellers. As they became a part of the city, they and their offspring repeat the cycle of their stay in the slum for generations (Dutt, Mittra and Halder, 1997).

Conclusion

During the last fifty years, the slum upgrading policy in India has undergone significant changes. Policy moved from earlier "clearance" effort of the 1950s, to a "transition" stage in the 1960s, and then to the "improvement" and "sanitary environment" model of the 1970s and thereafter. Such changing policy implications were experienced in Calcutta, too. As a matter of fact, in 1962, CMPO was the pioneer in initiating the idea of slum improvement. Most slums of Calcutta have been improved and made sanitary as a result of the persistent work of the CMDA since the early 1970s. As a result of these endeavors, some basic infra-structure facilities have been instituted in the slums, but several, such as, accommodation (one-room tenements), very low levels of per capita income, low caloric intake for a sizable population and illiteracy or semi-literacy continue to plague the Calcutta slums. The studies of three Calcutta slums—Paikpara, Betbagan and Narkelbagan—confirm that a "cycle of poverty" exists because the poor migrants continue to remain poor all their life without any possibility of upward mobility.

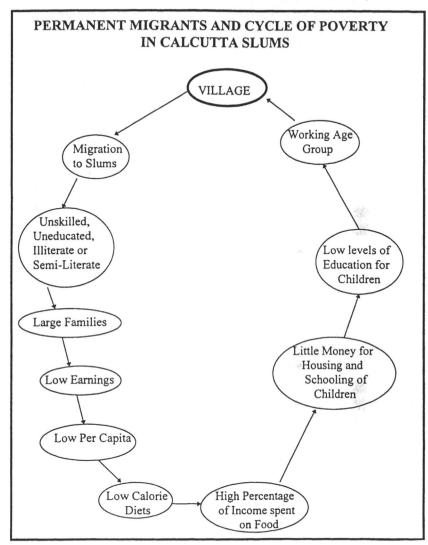

Figure 7-8   **Cycle of Poverty in Calcutta Slums and Permanent Migrants**
Source: Prepared by A.K. Dutt

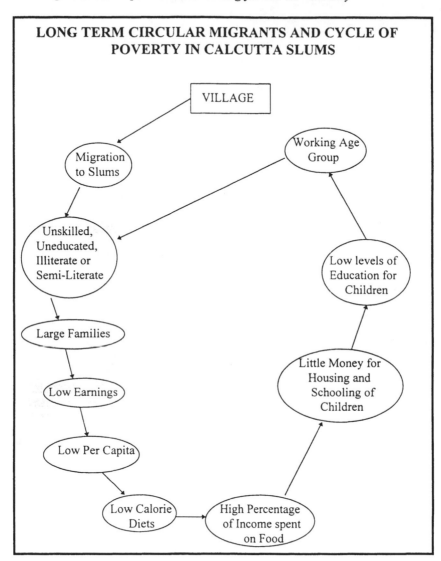

Figure 7-9   Cycle of Poverty in Calcutta Slums and Long Term Circular Migrants
Source: Prepared by A.K. Dutt.

References

Banerjee, Tridib and Sanjoy Chakravorty, 1994, "Transfer of Planning Technology and Local Political Economy: A Retrospective Analysis of Calcutta's Planning", *Journal of the American Planning Association*, 60:1:71-82.

Calcutta Metropolitan Planning Organisation (CMPO), 1963, *First Report: 1962*, Calcutta: CMPO.

Calcutta Metropolitan Development Authority (CMDA), 1972, *Calcutta Development: Programmes and Projects*, Calcutta: CMDA.

Chakraborti, Asok M. and Animesh Halder, 1991, *Slum Dwellers of Calcutta: Socio-Economic Profile, 1989-90*, Calcutta: Calcutta Metropolitan Development Authority (mimeographed).

Chakravorty, Sanjoy and Gautam Gupta, 1996, "Let a Hundred Projects Bloom: Structural Reform and Urban Development in Calcutta", *Third World Planning Review*, 18:4:415-431.

CMDA. 1991, Socio-economic Survey of Calcutta Slums, 1989 (based on 100% sample of 20 slums), Calcutta: A CMDA Report, (mimeographed).

CMDA. n.d. Socio-economic Survey of Calcutta Metropolitan Area, 1996-97, Calcutta: A CMDA Project.

Dandekar, M. and Nilakantha Rath, 1971, *Poverty in India*, Bombay: Indian School of Political Economy.

Dutt, Ashok K. and Anupa Mukhopadhyay, 1997, "Geographic Patterns Of Indian City Slums", pp. 325-362 in Rais Akhtar, (ed), *Contemporary Approaches to Indian Geography*, New Delhi: APH Publishing Corporation.

Dutt, Ashok R., Chandryee Mittra and Animesh Halder, 1997, "Slum Location and Cycle of Poverty: Calcutta Case", *Asian Profile*, 25:5:414-425.

Dutt, Ashok K., Suprabha Tripathi, and Anupa Mukhopadhyay, 1994, "Spatial Spread of Daily Activity Patterns of Slum Dwellers in Calcutta and Delhi", pp. 309-326 in Ashok K., Frank J. Costa, Surinder Aggarwal, and Allen G. Noble, (eds), *The Asian City: Processes of Development, Characteristics, and Planning*, Dutt, Dordrecht: Kluwer Academic Publishers.

Mukhopadhyay, Anupa, 1990, *Spatial Analysis of Daily Activity Patterns of a City Slum in Calcutta*, Unpublished Master's Thesis, University of Akron.

Mukhopadhyay, Anupa and Ashok K. Dutt, 1993, "Slum Dwellers Daily Movement Pattern in a Calcutta Slum", *GeoJournal*, 29:2:181-186.

Planning Commission, Government of India, 1954, *Five Year Plan: Progress Report for 1953-54*, Delhi: Government of India Press.

Planning Commission, Government of India, 1956, *Second Five Year Plan: A Draft Outline*, Delhi: Government of India Press.

Planning Commission, Government of India, n.d., *Fourth Five Year Plan, 1969-74*, Delhi: Government of India Press.

Planning Commission, Government of India, 1971, *Fourth Five Year Plan, 1969-74: Summary*, Delhi: Ministry of Information and Broadcasting.

Planning Commission, Government of India, 1973, *Draft Fifth Five Year Plan, 1974-79*, Volume II, Delhi: Controller of Publications.

Planning Commission, Government of India, 1981, *Sixth Five Year Plan, 1980-85*, Delhi: Controller of Publications.

Planning Commission, Government of India, 1985, *Seventh Five Year Plan, 1985-90*, Volume II: Sectoral Programmes of Development, Delhi: Controller of Publications.

Planning Commission, Government of India, 1992, *Eighth Five Year Plan: 1992-97*, Volume I: Objectives, Perspective, Macro-Dimensions, Policy Framework and Resources, New Delhi: Controller of Publications.

Planning Commission, Government of India, 1992, *Eighth Five Year Plan: 1992-97*, Volume II: Sectoral Programmes of Development, New Delhi: Controller of Publications.

Pugh, C. 1989, "The World Bank and Urban Shelter in Calcutta", *Cities*, 6:103-18.

Ribeiro, E.F.N. 1985, *A Compendium on Indian Slums*, Ministry of Works and Housing,Town and Country Planning Organisation (TCPO), Government of India.

Roy, Maitreyi Bardhan, 1993, *Calcutta Slums: Public Policy in Retrospect*, Calcutta: Minerva Associates.

United Nations, 1989, *Population Growth and Policies of Mega Cities: Jakarta*, New York: United Nations.

Yadav, C.S. 1989, "Migration and Urbanization in India: A Case Study of Delhi", pp. 227-244 in Frank J. Costa, Ashok K. Dutt, Lawrence J.C. Ma and Allen G. Noble, (eds), *Urbanization in Asia*. Honolulu: University of Hawaii Press.

8 Health Planning and the Resurgence of Malaria in Urban India

RAIS AKHTAR, ASHOK K. DUTT, VANDANA WADHWA

Malaria has been recorded as far back as the Fourth Century B.C. by Hippocrates (Russell 1955), and ever since has been one of the greatest, if least spectacular, of killers. Even in this age of technological and medical advancement, it is responsible for two to three million deaths each year, and causes debilitating symptoms in millions more (Kaberia 1994). The war against this disease has been long, but nothing if not humbling. Despite nearly a hundred years of eradication attempts, there seems to be no certain method of total eradication, though prevention and cure are now possible in many areas of the world.

The 1960s saw a burst of optimism in this direction, and there was much talk of a malaria vaccine. In fact, by 1987, Manuel Pattaroyo and his colleagues successfully synthesized and tested a malaria vaccine. However, subjected to tests in Tanzania and Gambia, it did not prove to be very efficacious. Moreover, recent research (Nowak 1995) has revealed that the *Plasmodium falciparum*, the most dangerous malaria parasite, avoids the human immune system by changing its signature proteins frequently enough to outrun the host antibodies that attack it.

Warm, humid climates are optimal malaria breeding grounds, and this is the major factor responsible for the occurrence of this disease in India. The endemic areas of the forested tracts of central, eastern and southern India, and the marshy tracts of the west were ideal breeding grounds from which malaria spread to the rest of the country. Although for a long time it was believed to be a predominantly rural disease, the larger cities of Bombay (Mumbai) and Delhi have experienced it since the early part of this century. Urban malaria established its hold in the early 1960s, just when the disease was thought to be under control. However, the country experienced an overall resurgence in 1965 and urban malaria became a dreadful reality. Urban development has provided a more conducive environment for certain vectors of malaria, and over the last two decades, urban malaria has fluctuated through a series of combinations of varying intensities and geographical spreads.

Since the probability of total eradication through a vaccine does not seem a feasible course of action, the same end may be better achieved, or at least attempted, through prevention and control measures that complement and supplement health planning policies. Thus, in order to deal with this dreadful disease, it is imperative to study the past trends and patterns of its occurrence, so as to achieve a better understanding of its behavior. This is the first step toward any action that may be directed against it.

The ultimate purpose of this chapter therefore, is to study the changes in the intensity and spread of malaria in urban India, and to explain these particular patterns.

Malaria Vectors and Appropriate Ecology

Malaria is caused only by the female anopheles mosquito. Of the various anophelines found in India, the major malaria vectors are *Anopheles culicifacies*, *Anopheles fluviatilis*, and to a lesser extent, *Anopheles stephensi*. Of these, it is the last that has been of most concern to populations in the urban areas, as it shifted its ecology from rural environs, where it was mainly zoophilous, to that of the urban areas, where it is found breeding in areas of construction activity, cisterns and shallow wells, overhead tanks, and in the slums. This situation is aptly captured in the following quotation, "*A. Stephensi* is highly zoophilic and in the countryside (of Bombay State) conveys a highly unstable malaria or none at all, but within the city, it can find breeding places, and even a small prevalence causes severe (endemic) malaria" (Batra, et al 1979,112).

Urban Malaria: Past and Present

The history of urban malaria in India can be divided into a series of phases, based upon chronology and related technological advancement which reflects itself in the measures adopted to combat the disease.

The Pre-Independence Period (Beginning of the century to 1947)

This initial phase covers the period up to 1947. During the first decade of the century, the city of Bombay, was the focus of malaria incidence. According to

Turner (1910), Bombay was the world's first city to attempt to combat malaria utilizing modern scientific techniques. In Delhi malarial vectors were abundant as early as 1914. In fact, 13 percent of the total adult mosquitoes collected in the city represented *A. stephensi* and the breeding of this species was encountered in used and abandoned wells, both in the pre-monsoon and monsoon periods (Hodgson, 1914).

The basic thrust of the anti-malarial programs was to destroy mosquito larvae, both permanent and seasonal. Improved sanitation, extensive draining, and leveling, the filling in, and cleaning of wells and tanks, and extermination of mosquitoes was also suggested. Quinine was used as a curative measure.

The Period of Active Anti-Malarial Drives

This period (1948 to 1965) itself may be divided into two parts. The first decade was more of a stock-taking exercise, whereas the second decade was more action-oriented.

India received a fresh lease on life in 1947, at the time of independence. At this time, malaria was a major cause of mortality and morbidity, causing 800,000 deaths, and 75 million to suffer its ravages annually (Akhtar and Learmonth 1977). In 1948, the Malaria Survey of India prepared a map of healthy, partially endemic, and endemic areas. Endemic areas were those where malaria existed all year-round, due primarily to climatic conditions. Ideally, the temperature ranges between 65 and 90 degrees Fahrenheit, and rainfall typically exceeds 30 inches in such areas. The Ahmedabad region in the west, the swamps of the Kutch area, the southern peninsula, the lower Ganges plain, and Northeast India were identified as endemic areas. The forested tracts of the Ghats, the Madhya Pradesh and Orissa hills, and the Assam hills were termed as "hyper-endemic areas". Northwest India has been independently characterized as an epidemic area, and the interior parts of the Deccan peninsula and West India are known to be non-endemic (Dutt, et al. 1980).

The identification of these areas was followed by various anti-malaria drives beginning in 1953. By 1958, these efforts amounted to a veritable war against the disease in the form of the National Malaria Eradication Program. This program encompassed house and environs spraying, medical treatment of infected persons, and distribution of prophylactics. The success rate was a phenomenal 97.9 percent, declining from 75 million persons affected in 1952 to 0.1 million in 1965, which is now known as the "trough year" of malaria incidence.

However, the focus of this program was on rural areas, as malaria was thought at that time to be a rural disease. The first signs of urban malaria surfaced as early as 1962-63 in the four South Indian towns of Vishakhapatnam, Guntur, Salem, and Erode. By 1965, the number of towns had risen to ten, though malarial incidence in the country in general had declined. The major factors responsible for the spread of urban malaria were:

- The increase of developmental activities, both rural and urban, which resulted in conditions suitable for mosquito breeding, e.g., construction activities in urban areas, and increased farming activity coupled with irrigation in the rural areas, resulting in small areas of stagnant water.
- *Anopheles stephensi*, a malaria vector that until this time was not so effective, since it was frequently zoophilous, adapted to the new urban ecology and began to breed in cisterns, shallow wells, and overhead tanks abundantly found in urban areas.
- At the same time, *A. stephensi* and *A. culicifacies*, the two most common urban malarial vectors began to achieve immunity to DDT and HCH and DDT and BHC, respectively, which were the two most popularly used insecticides. (Akhtar and Learmonth 1977)

Period of Malaria Resurgence

In view of the increasing incidence of urban malaria, the government extended the urban malaria scheme to 28 towns by 1971. According to the National Malaria Eradication Program, this helped in reducing the incidence of urban malaria so that it accounted for no more than one percent of the total malaria cases in the country (Kondrashin and Rashid 1987). However, due to the Indo-Pakistan war (1971) which interrupted the program, disproportionate defense spending, overconfidence on the part of the government due to the recent success in controlling the disease, and several other factors as noted in the earlier section, malaria resurgence began from four major foci: the Kutch Salt Marsh, and the Madhya Pradesh, Orissa and Assam hill forests. These were all inaccessible areas where the full effect of anti-malaria programs had not yet been felt (Akhtar and Learmonth 1977).

There were many fluctuations in the pattern of malaria incidence over the 1965-76 time span, but by the end, the final scenario showed intensified rates in the northwest, west, and northeast parts of the country, and lower rates of malarial incidence in the traditionally endemic south and southeast. In these latter areas, the population had developed a certain level of resistance to the disease.

The resurgence of malaria rates from 0.1 million in 1966 to 6.4 million in 1976, led to an in-depth review of the eradication program, and a modified plan of operation was introduced in 1977 (Kondrashin and Rashid 1987). The basic objectives of the program were:

- Prevention of deaths due to malaria.
- Reduction of morbidity
- Maintenance of the status of those aspects of industrial development and the "green revolution" which promote and retain successes achieved in the war against malaria.
- Emphasis on bio-environmental factors, drug resistance in *Plasmodium falciparum*, and organizational and managerial aspects.

The Post-Resurgence Period

This program led to a significant reduction in malaria to 2 million cases (with negligible deaths) in 1984, which has more or less been maintained since then. However, there are shifts in malaria rates and geographical spread every few years, and these substantially influence the direction of the health planning measures within which anti-malaria efforts are carried out. The time span of this period (1978 to date) encompasses the time frame of the present study. Thus an update of current policy will be given below.

At present, the Malaria Action Program launched in January, 1995 has established new criteria for identifying urban areas that should fall under its purview and care. These are:

- The urban area must already have an existing malaria database (of at least a year).
- The population of the urban area must be a minimum of 50,000 (as opposed to 40,000 laid down in 1977).
- The Spleen Rates (SPR) defined as the percent of the population affected by spleen enlargement due to malaria infection, should be 5 percent or more and a 1:3 or greater ratio between clinical malaria cases and fever cases, as per hospital/dispensary statistics during the last calendar year (as opposed to an Annual Parasite Index (API), which is the number of positive blood samples in an area over the population of the area in thousands, of 2 per 1000 or more).

The Committee also recommended a more effective case detection mechanism, the opening of Fever Treatment Depots in slum areas at a rate of one per 2000 population, and the ready accessibility of anti-malarial drugs.

Research Design and Methodology

The research design employed for this study is comprised of four major sections: matrix preparation, data collection, and data analysis and summation.

Matrix preparation comprises the two core aspects of determination of the research problem, and the background discussion, which together form the grounding or matrix for the study. The former is necessary for determining the purpose of the study, whereas the latter provides a theoretical underpinning, the foundation on which the study rests.

The background and context of the study focuses on three aspects, malaria vectors and ecology, the incidence and control of malaria in urban India in the past and present, and the need for health planning derived through the debate regarding eradication by vaccination. This last is discussed after the results and findings of the study.

All effective research and planning are based on accurate data. The data required for this study are not comprised of a sample, but rather the universe itself, which is:

- Location of and APIs of all urban areas covered by the National Malaria Control Program in 1978.
- Location of and APIs of all urban areas covered by the National Malaria Control Program in 1993.

The above points are chosen to maintain continuity with earlier studies (1978), and to present the latest picture. This time span is also long enough to enable an adequate scenario of the shifts in malaria occurrence to be revealed.

APIs are used as a measure of malaria intensity, while location is employed to gauge its spread. The data source is the above mentioned National Malaria Control Program.

This empirical study reveals the differential rates of malaria spread and intensification over time, despite standard control programs. Toward this end, the analysis will employ simple statistical tools, and include the possible explanation for such spatio-temporal changes. It will also delve into the issues and implication related to such shifts as explained by the analysis.

The APIs for 1978 were categorized into areas of "high", "medium", and "low" occurrence of malaria, where "high" constituted any API value above the mean. These classes were mapped using the proportional representation technique. A similar exercise was conducted for the 1993 data base. A t-test was conducted to check for statistical significance in the difference between the mean APIs.

The rate of change in malaria intensity was computed, using a consolidated data base, where only those urban areas which figured in both data sets were included. This, too, was plotted on a map, depicting areas of increase, mild decline, and sharp decline. The result was a series of maps depicting the intensity and geographic spread of malaria for the years 1978 and 1993, and the spatial representation of the rate of change of this intensity.

From the background discussion, it is apparent that health planning has taken in hand the aspect of planning for malaria eradication and control. However, constant resurgence proves that such efforts are not adequate. Therein lies the significance of this study. The aim of this chapter is to trace the shifts in the pattern of malaria incidence in urban India, so that the findings may direct and strengthen future health planning measures.

The findings are presented in three stages, the situation in 1978, followed by that in 1993, and finally, the changes in incidence of malaria over this time span.

Pattern of Malaria Incidence in 1978

The number of urban areas included under the Malaria Eradication Program at this time were 55. For names and APIs of these urban areas, see Table 8-1. For location of intensive areas, see Figure 8-1. The mean API was 12.68.

The focus of malaria incidence was found to be in the drier, traditionally non-endemic area of northwest India, namely Delhi and the adjacent belt of newly developing towns in the state of Haryana, Chandigarh, and Amritsar. Malaria was present in the urban centers of the endemic areas of the south, but its intensity (i.e. rate of occurrence) was remarkably lower than that in the northwest.

The most probable reasons for this pattern of concentration in the northwest are:

- The population of the non-endemic areas had lower resistance to malaria and were thus more vulnerable to the disease. Resistance to the malarial parasite has a span of ten to twelve years in the human body (WHO, Expert Committee on Malaria 1979; Dutt, et al. 1980), and the southern population retained the resistance it had acquired when malaria was still endemic there in the early 1960s (Akhtar and Learmonth, 1977).
- The malaria vector *Anopheles stephensi* had by now adapted very well to its new urban surroundings, breeding in slums, cisterns, overhead water tanks, and other such urban features.
- This factor was exacerbated by the fact that the ongoing developmental activities in urban areas were creating ever more favorable conditions to promote vector breeding. For example, construction activity caused shallow puddles,

Table 8-1   India: Rate of Change in API, 1978 -1993

City	Rate Change	City	Rate Change
Jullandhar	-99.41	Bharatpur	-70.26
Ajmer	-99.20	Ratlam	-65.45
Sonepat	-99.15	Kota	-61.14
Meerut	-98.84	Jhansi	-59.33
Belgaum	-98.66	Bangalore	-58.77
Amritsar	-98.46	Ludhiana	-48.04
Delhi	-98.31	Hyderabad	-46.27
Bhiwani	-97.19	Aurangabad	-45.15
Panipat	-96.81	Erode	-11.65
Karnal	-96.58	Guntur	27.76
Patiala	-96.54	Agra	36.84
Pune	-95.46	Warrangal	58.99
Lukhnow	-94.31	Madras	71.56
Jaipur	-94.06	Bellary	73.23
Bokaro	-92.76	Dhulia	80.59
Rohtak	-92.68	Agartala	91.93
Jodhpur	-92.28	Vellore	187.67
Bhavnagar	-90.29	Vijaywada	304.07
Ahmedabad	-87.34	Bombay	527.59
Chandigarh	-86.43	Bhopal	937.89
Broach	-84.76	Tuticorin	1279.59
Bikaner	-79.03	Calcutta	2675.56
Salem	-71.87	Vishakapatnam	9505.66
Jalgaon	-70.83		

and an influx of rural labor force which not only carried the disease in with them, but also inhabited slums, which are ideal breeding grounds for mosquitoes that cause malaria.

• The Green Revolution that was a very strong movement in the northwest part of the country, also contributed to the intensification and spread of malaria in this part of the country because, even though it was rural in nature, the agricultural development activities such as canal irrigation caused the underground water table to rise, and in some areas it caused puddles, which were suitable for mosquitoes to breed. This was accompanied by dense vegetation in the fields, and waterlogging and fungi at the banks of the canals, which further aggravated the situation.

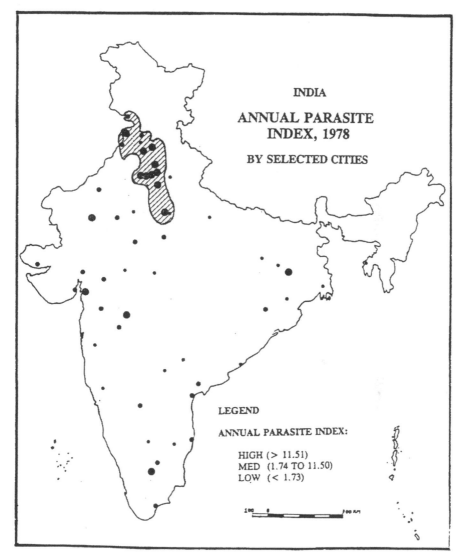

Figure 8-1   India: Annual Parasite Index, 1978, by Selected Cities

Pattern of Malaria Incidence in 1993

By 1993, the situation was considerably different as reflected in Figure 8-2.
The number of urban areas now under the wing of the Malaria Eradication
Program (MEP) grew to 86. For names and APIs of urban areas, see Tables

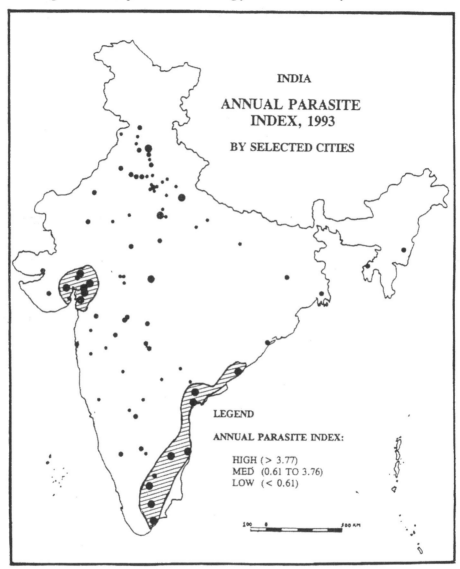

Figure 8-2   India: Annual Parasite Index, 1993, by Selected Cities

8-2 and 8-3, revealing the spread of the disease. However, the mean API plummeted to 3.67, reflecting a general decline in the intensity of malaria in urban India. This is also statistically significant, as shown by the t-value in Table 8-4. The regional polarity, however, remained as strong as ever, though there was a reversal of roles in the focus of occurrence.

The traditionally endemic areas were seen to have re-emerged as centers of urban malaria. These included the Ahmedabad region in the west, including the urban areas of Gandhinagar, Rajkot, Anand, and Broach, and Southeast India (Vijaywada, Guntur, Madras, Vellore). This shift was mainly due to:

- The ever conducive climate of this part of the country which renders malaria endemic and malaria is present throughout the year.
- The resistance of the population had passed the 12-year mark, and they were again prone to the disease. Also new members of the population, i.e. children under five years of age, were added who were most susceptible to malaria.

The northwest region displayed a decrease in rate of urban malaria, but an increase in number of cities affected by it. The major causes for this pattern were:

- The fall in malaria rates can be attributed to the fact that this area has received greater attention by virtue of its being close to a major administrative and economic center, the capital city of New Delhi.
- Another major factor for this decline is the building up of resistance against this disease due to the past epidemics of malaria.
- The spread of malaria is mainly due to the increase and spread of developmental activity, and the associated conditions that foster malaria vector breeding. In fact, the corridors of urban malaria spread are congruent to the corridors of urban development, radiating from the Delhi region.

The Change in Malaria Intensity from 1978 to 1993

Figure 8-4 showing the rates of change in urban malaria over the 1978-1993 period, vividly depicts the shifts in the intensity and spread of the disease.

The urban areas of the southeast region display an intensification of urban malaria, reestablishing their role of centers of diffusion, although the malaria rates are not as high as they were in 1978. The northwest region reveals a decline in malaria in individual centers, but a spread is apparent in geographical terms.

The reasons for these differential rates and spreads are mainly the aforementioned partiality toward the northwest, which is also related to the factor of increased developmental activity in this particular region. Also, the development of resistance against malaria in the northwest has led to a decline in malaria rates, though the disease has spread to a greater number of urban areas due to conditions made conducive for vector breeding due to developmental activities and the transmission of the disease (in a minor way) through

Table 8-2   India: API Rates, by city, 1978

City	API	City	API
Agra	0.08	Guntur	4.33
Calcutta	0.14	Bikaner	4.57
Vishakapatnam	0.16	Jhansi	4.66
Warrangal	0.22	Ajmer	5.51
Belgaum	0.30	Dahod	6.12
Bombay	0.32	Kota	6.33
Bangalore	0.33	Salem	6.90
Agartala	0.35	Madras	6.98
Bhopal	0.38	Ferozpur	8.52
Ludhiana	0.54	Bhavnagar	9.02
Raurkela	0.61	Jammu	9.38
Lukhnow	0.72	Ahmedabad	10.47
Hyderabad	1.01	Erode	11.63
Pune	1.04	Jalgaon	12.62
Daltonganj	1.23	Amritsar	13.31
Jaipur	1.28	Jodhpur	13.58
Ludhiana	0.28	Jalgaon	3.68
Ahmednagar	0.30	Calcutta	3.75
Sonepat	0.31	Nanded	3.76
Warrangal	0.35	Bhopal	3.94
Muzaffarnagar	0.37	Vellore	4.83
Gurgaon	0.42	Guntur	5.53
Hospet	0.45	Anand	5.62
Mathura	0.47	Broach	7.31
Hyderabad	0.54	Gandhinagar	7.61
Panipat	0.55	Badaun	8.89
Bhavnagar	0.58	Bharatpur	10.24
Nasik	0.59	Erode	10.28
Ratlam	0.60	Chandigarh	10.79
Agartala	0.67	Surendranagar	11.97
Hoshiarpur	0.75	Madras	11.98
Tumkur	0.76	Vijaywada	12.12
Berhampur	0.78	Nadiad	14.93
Imphal	0.79	Vishakapatnam	15.27
Hassan	0.88	Godhra	23.47

Table 8-2 continued

Ambala	0.92	Tuticorin	40.01
Bikaner	0.96	Dindigul	51.85
Meerut	1.38	Panipat	17.34
Hazaribagh	1.46	Bokaro	18.31
Vellore	1.68	Rohtak	24.75
Ratlam	1.73	Patiala	29.23
Dhulia	1.74	Bharatpur	34.43
Bellary	2.04	Sonepat	36.87
Aurangabad	2.16	Broach	47.94
Jullandhar	2.54	Karnal	54.19
Tuticorin	2.90	Delhi	58.07
Vijaywada	3.00	Bhiwani	62.38
Bhuj	3.13	Chandigarh	79.55
Sambalpur	3.73		

small scale migrations, mainly of the labor force. In the southeast and west, the cycle of malaria is played out again and again as the endemic area remains ever prone to this age-old disease.

Is Eradication Possible by Vaccine?

The fact remains that less developed, tropical and semi-tropical countries possess a suitable ecology for endemic malaria (Dutta and Dutt 1978). A majority of the population remains threatened by malaria because they cannot afford certain preventive measures such as mosquito nets and door screens, or facilities such as medical care, which are more accessible to the populations of the more developed countries. Immunity of the vector in tropical areas to DDT (used for mosquito control) and chloroquine (used to treat infected persons) are other factors that retard malaria eradication (Hudson 1995).

It is quite obvious that *total* eradication may only be achieved though the introduction of a vaccine. For a while there was a promising signal when the Columbian researcher, Manuel Pattaroyo, and his colleagues discovered a malaria vaccine known as SPF66, and which seemed to generate encouraging results (Glanville 1994). However, recent experiments in Africa have cast a shadow of doubt on that optimism. According to reports published in *Lancet*, British doctors have downgraded their optimism about the world's first malaria

Table 8-3   India: API Rates, by city, 1993

City	API	City	API
Varanasi	0.00	Karnal	1.86
Belgaum	0.00	Jhansi	1.90
Raichur	0.01	Salem	1.94
Jullandhar	0.02	Bombay	2.00
Meerut	0.02	Parbani	2.02
Khammam	0.03	Sirsa	2.27
Kanpur	0.03	Kota	2.46
Ghaziabad	0.04	Gandhidham	2.52
Lukhnow	0.04	Bhuswal	2.69
Ajmer	0.04	Dhulia	3.14
Pune	0.05	Rajkot	3.48
Ujjain	0.05	Bellary	3.54
Moradabad	0.06	Jalgaon	3.68
Yamunanagar	0.06	Calcutta	3.75
Bulandshahar	0.08	Nanded	3.76
Jaipur	0.08	Bhopal	3.94
Agra	0.10	Vellore	4.83
Bangalore	0.13	Guntur	5.53
Sholapur	0.19	Anand	5.62
Indore	0.20	Broach	7.31
Amritsar	0.21	Gandhinagar	7.61
Faridabad	0.21	Badaun	8.89
Delhi	0.98	Erode	10.28
Patiala	1.01	Chandigarh	10.79
Jodhpur	1.05	Surendranagar	11.97
Hissar	1.12	Madras	11.98
Aurangabad	1.19	Vijaywada	12.12
Akola	1.31	Nadiad	14.93
Ahmedabad	1.33	Vishakapatnam	15.27
Bokaro	1.33	Godhra	23.47
Bhiwani	1.75	Tuticorin	40.01
Rohtak	1.81	Dindigul	51.85

vaccine, saying it offered no significant protection in newly run trials on children in Gambia (Hudson 1995).

Initially this vaccine had shown encouraging results in Latin America, but a two-year test of the vaccine conducted by the WHO in Tanzania, concluded in late 1994 that it "could not cut the risk of clinical malaria in children by even one-third" (Hudson 1995).

Research on the vaccine now entails looking into a genetic base that triggers the parasite activity among the humans. Russell Howard of the Santa Clara Biotic Company Affymax Research Institute has found that *P. falciparum* EMP-1 genes under the conviction EMP-1 proteins, are "central to the malaria parasite's disease causing capabilities", and that the "parasite has a great capacity for variation to evade the immune system response" (Nowak 1995). Since the genes responsible for variation to immune response

Table 8-4 India: t-test, API 1978-1993

T-TEST /PAIRS 1978 1993.

Paired samples t-test: 1978
                        1993

Variable	Number of Cases	Mean	Standard Deviation	Standard Error
1978	47	12.6801	19.221	2.804
1993	47	3.6708	6.641	0.969

(Difference) Mean	Standard Deviation	Standard Error	Corr.	2-Tail Prob.	t Value	Degrees of Freedom	2-Tail Prob.
9.0093	20.020	2.920	0.050	0.738	3.09	46	0.003

have been isolated, there is a ray of hope for the development of a new, more effective vaccine, but not in the near future.

Summation

The preceding discussion reveals a differential rate of malaria intensity and spread over the country, and this is reflected in the northwest/southeast di-

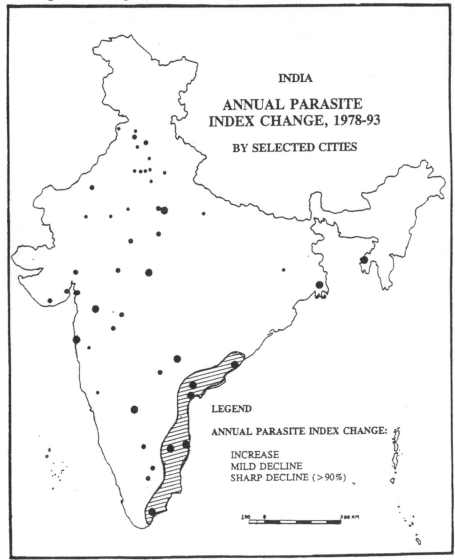

Figure 8-3   India: Annual Parasite Index Change, 1978-93, by Selected Cities

chotomy in the pattern of malaria incidence. The rate of malaria, i.e. its intensity, has declined considerably in the northwest region, but spatial diffusion has taken place. On the other hand, the southeast has experienced a sharp resurfacing of the disease, although in 1978 it had displayed depressed rates. This study has also shown that despite eradication and control measures taken by the Malaria Eradication Program in India, the disease still evades these efforts toward total eradication. This is due to several factors such the prevalent climatic conditions that are conducive to mosquito breeding, economic constraints, and technological constraints. Moreover, this disease is now widespread in urban India where construction activities, green belts in cities, and the slum areas (which simulate the rural environment), provide excellent breeding conditions for *Anopheles stephensi,* which is the major malaria vector in urban India. Alarmingly, this vector is fast acquiring immunity to the more popular preventive and curative substances used to contain malaria.

It is, however, not plausible to wait for a permanent measure such as a vaccine, such as that used for smallpox, because of the very nature of the malaria parasite, which has a capacity to "disguise" itself from the anti-bodies by changing its complex protein structure.

Looking to the Future

The resurgence of malaria has been an unpleasant experience, and the fact that urban India is the focus of the country's economy underlines the importance of studying this phenomenon. It is essential to study the pattern of malaria incidence in urban India, because the variations in malaria intensity over space and time have a profound effect on the thrust and direction of the strategies that need to be formulated.

The fact remains then, that until an effective vaccine is discovered, health planning measures consisting of meso- and micro-level control and prevention strategies will have to be formulated in order to further strengthen the ongoing programs.

"Everything about malaria is so moulded by local conditions that it becomes a thousand epidemiological puzzles. Like chess, it is played with a few pieces but is capable of an infinite variety of simulations" (Hackett 1937, as quoted in Gilles 1993).

References

Akhtar, R., and A.T.A. Learmonth, 1977, "The Resurgence of Malaria in India: 1965-1976", *GeoJournal*, 1:6:69-79.

Batra, C.P., R. Reuben, and P.K. Das, 1979, "Urban Malaria Vectors in Salem, Tamil Nadu: Biting Rates on Man and Cattle", *Indian Journal of Medical Research*, 70 (suppl, Dec):103-113.

Dutt, Ashok K., Rais Akhtar, and Hiran M. Dutta, 1980, "Malaria in India with Particular Reference to Two West-Central States", *Social Science & Medicine*, 14D:317-330.

Dutta, Hiran M., and Ashok K. Dutt, 1978, "Malarial Ecology: A Global Perspective", *Social Science & Medicine*, 12:69-84.

Gilles, Herbert M. 1993, "Epidemiology of Malaria", pp. 124-157 in Gilles, H.M. and David A. Warrell, (eds), *Bruce-Chwatt's Essential Malariology*, London: Edward Arnold.

Glanville, Hugh de. 1994, *The Lancet*, 343:8897 (March); 593.

Hodgson, E.C. 1914, "Malaria in the new province of Delhi", *Indian Journal of Medical Research*, 2:405-455.

Hudson, L. 1995, "First Malaria Vaccine Trial Results Prove Disappointing in Gambia", *India Abroad*, (September 15) 46.

Kaberia, K. 1994, "Super rice will feed a half billion more people", *Insight of the News*, 10 (Dec 12):32.

Kondrashin, A.V., and K.M. Rashid, (eds), 1987, *Epidemiological Considerations for Planning Malaria Control in South East Asia Region*, New Delhi, World Health Organization, SEARO.

Nowak, R. 1995, "How the Parasite Disguises Itself", *Science*, 269: (Aug) 755.

Russell, P.F. 1955, *Man's Mastery of Malaria*, London: Oxford University Press.

Turner, J.A. 1910, "Malaria in Bombay From 1901 to 1910", *Paludism*, 1:39-40.

WHO, Expert Committee on Malaria, 1979, *Seventeenth Report*, Technical Report Series, 640, Geneva: World Health Organization.

9 Residential Land Development in Highland Ecuador

BETTY E. SMITH

The last decade of the twentieth century has been a period of uncertainty for local and national level planners in Ecuador. To some extent this can be attributed to a changing international economic context. Ecuador has substantial oil resources and well-developed commercial agricultural production. However, national and local planning budgets have been indirectly affected by dramatic and unpredictable fluctuations in prices of the primary export products of Ecuador: petroleum, bananas, shrimp, coffee and cacao. To encourage competition on a global scale, foreign investment legislation has been enacted, foreign debt renegotiated, and free trade resolutions implemented. It is hoped these steps may improve planning processes by providing increased financial stability to national and local budgets.

Ecuador consists of three distinct geographic and climatic regions: 1) the humid tropical Pacific coast where the bulk of commercial agricultural production for export occurs; 2) the eastern and western ranges of the Cordillera between which lie the high inter-Andean valleys where the majority of cities are located; and 3) the humid tropical Ecuadorian Amazon that makes up the eastern one-third of the country and is the source of petroleum, piped to the Pacific coast for export. National planning and local development across these diverse environments has been uneven because of natural disasters, transportation problems associated with rugged topography, and fluctuations in global oil prices, as well as government policies designed to reduce the rate of inflation.

Healey and Barrett (1990) have suggested that understanding of planning and urban land development can be enhanced by improved understanding of the procedural dynamics produced by the action of public policy and land use regulation. However, development involves the administration of planning functions interfaced with the economics of market forces. The purpose of this chapter is to describe the planning and market processes of middle and upper income residential land development in two intermediate size cities in Ecuador. Following background, methodology, and description of the field investigation sites, the chapter proceeds with a discussion of key participants in urban land development; residential subdivisions; home buyer preferences; site preferences of developers and lot purchasers; and a few problems related to development.

The Ecuadorian Background

Cities in Ecuador derive their local authority from the *Ley de Régimen Municipal No. 680*, January 31, 1966. However in practice, local government remains subordinate to a strong centralized government, in spite of extensive decentralization rhetoric (Nickson 1995). Since the founding of the Republic of Ecuador in 1830, the country has been governed by a series of national leaders who have maintained strong centralized control by military authority. Ecuador has a political history of military coups and juntas deriving from diverse and often conflicting economic and social interests. Between 1830 and 1985, there were only two periods (1912 to 1925 and 1948 to 1961) when presidents were elected, completed their terms, and passed power to newly elected chief executives (Rodriguez 1985). More recently the national government has passed peacefully by democratic election from President Rodrígo Borja (1988-1992) to President Sixto Durán Ballén (1992-1996), and once again by democratic election in July 1996 to President Nebot, whose term, however, was short-lived. Public pressure forced him to stand down in early 1997, placing the country's democracy on a less steady footing.

Ecuador is a unitary state divided administratively into twenty-one provinces (*provincias*) and 193 districts (*cantones*). The estimated 1994 population of Ecuador was 10.6 million. Each *cantón* has one municipality, its capital city, which has jurisdiction over both the rural and urban parishes of the *cantón*. Municipal mayors and council members are chosen by direct popular vote for four-year terms, simultaneously with national and provincial elections. During the period of the 1970s oil boom, fiscal resources were transferred from central government to municipal governments to support infra-structure and services, leading to economic dependence on central government and the erosion of municipal administrative autonomy. Not until the economic restructuring during the Borja and Durán administrations (1988-1996), involving the World Bank and the International Monetary Fund, were these trends reversed (Carr 1992, 84).

During the 1980s the national municipal association, *Asociación Municipalidades Ecuatorianas (AME)* provided training in land use planning and cadastral surveying support services to local government. In the 1990s the function of this national agency was superseded by two international and regional organizations called the International Union of Municipalities and Local Powers (IULA) and the Latin American Center for Training and Development of Local Governments (CELCADEL). These organizations provide support services such as local land use planning and local public policy training seminars for municipal public officials.

While much has been published about planning and growth in the largest Latin American cities (see for example, Greenfield 1994; Lowder 1986; Violich 1987), including the largest Ecuadorian cities of Quito (Carr 1992; Carrion and Vallejo 1992) and Guayaquil, less has been written about planning and development in the intermediate size cities of Ecuador. A very broad discussion about intermediate size cities across all of Latin America was published by the United Nations Centre for Human Settlements, Habitat (CNUAH 1993), including six intermediate size city case studies in four Latin American countries (Mexico, Venezuela, Peru, and Brazil). A similarly broad approach by the Federal Republic of Germany in 1991 suggested strategies for development cooperation in urban development management with intermediate size cities in Third World settings (Drewski, Kunzmann and Platz 1991). The decentralization of political power, planning responsibilities, and resources in Latin America continues to be studied extensively by the Urban Management Programme (see Gangotena 1995 for a discussion of Ecuador), among others (Coraggio 1991; Melo 1988). Also at the national level of analysis, the Institute of Social Studies and the Netherlands Ministry for Development Cooperation (Teekens 1988) have reviewed theory and policy planning in Ecuador.

However, a need still exists for field work and analysis of planning and development processes in intermediate size cities in Ecuador, as well as in other parts of Latin America. A World Bank policy paper notes that, "In contrast to the extensive investment made in urban research during the 1970s, the decade of the 1980s has seen a decline in the quantity of urban research in both developed and developing countries"(World Bank 1991, 75). Several areas in need of future research are identified, including the role of government in the urban development process (World Bank 1991, 77).

Some Ecuadorian local level planning documents on the role of government in the urban development process have been published by the *Asociación de Municipalidades del Ecuador* (*AME*) and the *Unión Internacional de Municipios y Poderes Locales* (*IULA*) such as Darquea, García and Gallegos (1994) and Vega (1994); *Instituto Francés de Estudios Andinos* (*IFEA*) and *Centro de Investigaciones CIUDAD* (Carrion 1988), and; *Comisión de Desarrollo Urbana y Regional* (*CLACSO*) (Carrion 1986; Larrea 1986; Verdesoto 1986), to name a few. Monographs about the cities examined in this research, Ibarra and Riobamba, are descriptive and informative, but shed little light on the investigative questions of this study (Tobar 1985; Heredia 1992).

Research Questions and Methodology

The organizing principles of the field investigation posed five research questions: 1) Who are the key parties and financial institutions involved in urban residential land development? 2) What procedural steps are involved? 3) What are the housing preferences of home buyers?; 4) What are the site preferences of developers and lot purchasers? and 5) What local urban land use issues and solutions emerge from the interview data? Answers to these questions should clarify planning and land development processes and enhance understanding of the procedural dynamics produced by the action of public policy and land use regulation as suggested by Healey and Barrett (1990). The provision of housing is not simply a technical planning problem, but a political process as well. Ideally, by answering these questions, deficiencies can be identified in terms of conceptualization of planning problems, issues of coordination among local and central government agencies and between government agencies and the private sector. It is critical to note, however, that the study focuses on the issues of urban and residential land development in the formal sector. Thus, the research is framed principally on the actions of officials, white collar businesses, and middle and upper income home owners.

Municipalities in Ecuador find themselves embedded in a network of global influences, changing national and local institutional relationships, and dynamic social and economic contexts. The open-ended, non-standardized interview format offers the opportunity to discover unanticipated responses which explain institutional, developer, and home buyer behavior. This technique is particularly appropriate in times of "economic and social change that challenge traditional analytical categories and theoretical principles" (Schoenberger 1991, 180). It has been argued that open-ended interview questions are more sensitive to historical and institutional complexity than more structured survey instruments. In this research, interviews with key personnel provide insight about the procedural dynamics of planning and housing subdivision development.

Two survey instruments were designed in Spanish, one for public officials and one for architects and other professionals involved in development. The questionnaire for municipal officials emphasized questions pertaining to planning procedures, whereas the questionnaire for development professionals addressed planning procedures and clientele housing preferences. Approximately half of the questions were the same for both groups of interviewees.

The interviews ranged in time between one and three hours and took the form of an interactive dialogue as the interviewer and the interviewee worked through and discussed the research problems presented in the questionnaire. Those interviewed explained their role in the overall residential land and housing development process. In Ibarra, key personnel in those offices participating directly in the urban land development process were interviewed: 1) Municipal Planning Department; 2) Municipal Cadastral Department; 3) Ecuador Bank of Housing (*Banco Ecuatoriana de Vivienda- BEV Ibarra*); 4) Imbabura Mutual Bank (*Mutualista Imbabura*), and; 5) Ibarra Branch of Social Security (*Instituto Ecuatoriano de Seguridad Social- IESS Ibarra*). Nine interviews in Riobamba provided additional insight about urban land development in Ecuador. Key personnel in Riobamba included those from: 1) Municipal Planning Department; 2) Municipal Cadastral Department; 3) Ecuador Bank of Housing (*Banco Ecuatoriana de Vivienda- BEV Riobamba*); 4) Chimborazo Mutual Bank (*Mutualista Chimborazo*); 5) Riobamba Branch of Social Security (*Instituto Ecuatoriano de Seguridad Social- IESS Riobamba*); , and; 6) Office of Electricity (*Empresa Electrica*). Those interviewed were predominantly architects by training and in charge of housing programs in their respective agency.

Field Investigation Sites

Ibarra and Riobamba represent two intermediate-sized cities located in the *sierra* region of Ecuador (Table 9-1), playing secondary roles to the million-plus populations and dual primacy of the two largest Ecuadorian cities, Quito and Guayaquil. New residential subdivisions in Ibarra and Riobamba, while usually peripheral in location, present an impression of orderly urban growth, not unlike many North American or European cities of similar size. According to the 1990 Census, Ibarra is smaller, population 80,991 (INEC 1992a), than Riobamba, population of 94,505 (INEC 1992b). However, Ibarra has grown more rapidly over the past decade than Riobamba. Ibarra has grown at a rate of 52 percent compared with the 25 percent population growth rate of Riobamba. According to the 1982 census, Ibarra's population was 53,428 (INEC 1984a) and Riobamba's 75,455 (INEC 1984b).

Ibarra lies on a southwesterly sloping volcanic and sedimentary plain, at an altitude of 2,204 meters, about two hours driving time by paved road north of the capital city of Quito, encouraging a weekend commute for some Ibarra residents. The city exhibits a north-south grid pattern core surrounded by new residential development extending outward in a radial and tentacular pattern,

Table 9-1   Population ('000s) of Primary and Intermediate Size Cities in Ecuador

(after Lowder 1990:85 and Fernandez 1994:217)

	1950	1962	1974	1982	1990	Region
Dual primate cities						
Guayaquil	258.9	510.8	823.2	1199.3	1508.4	coast
Quito	224.3	365.7	599.9	866.5	1100.8	sierra
Secondary cities						
Cuenca	40.0	60.4	104.5	152.4	195.0	sierra
Machala	7.5	29.0	69.2	105.5	144.2	coast
Portoviejo	29.1	48.3	80.2	102.6	132.9	coast
Manta	19.0	33.6	64.5	100.3	125.5	coast
Ambato	34.0	53.4	78.8	100.5	124.2	sierra
Santo Domingo	1.5	7.0	30.5	69.2	114.4	coast
Esmeraldas	13.2	33.4	60.3	90.4	98.6	coast
Milagro	13.7	28.1	53.1	77.0	93.6	coast
Riobamba	37.5	41.6	58.1	75.5	94.5	sierra
Loja	21.1	26.8	47.7	71.7	94.3	sierra
Quevedo	4.2	20.6	43.1	67.0	86.9	coast
Ibarra	18.1	25.8	41.3	53.4	81.0	sierra

often coinciding with major and secondary transportation arteries. The city of Ibarra is known as the *ciudad blanca*, the white city, because all structures in the city are required to be painted white. The permit requirement for painting is not unlike those imposed in special architectural review zones of historically significant city areas, including Quito, as well as many cities in North America and Europe.

Today Ibarra, located on the Pan-American Highway between Quito and the Colombian frontier (Figure 9-1), is the largest Ecuadorian city north of Quito, and serves as an economic, administrative and transportation hub for the northern highland region. Government offices, service, manufacturing and commercial sectors are all economically important. However, there is no single predominant employer and no industrial base. Although the city has an industrial park (40 hectares) capable of providing services to 114 industries, the infra-structure is incomplete and the park remains unoccupied.

The direction of city growth in Ibarra is southerly and westerly, with a greater emphasis in the direction to the south. Reasons for this tendency for

Figure 9-1   Primary and Secondary Cities in Ecuador

growth in the southerly direction include: 1) availability of potable water to the south; 2) lack of infra-structure in other directions; 3) proximity to the Pan-American Highway which extends southwesterly to Quito; 4) availability of a large number of level lots to the south; 5) marketability of lots in the direction of Quito, and; 6) limitations of steep topography to the north and east. Construction jobs related to urban growth, and government office jobs are the main sources of employment in Ibarra. Infra-structure development in large subdivisions offers employment opportunities, as does the service based economy, that provides jobs in restaurants, hotels and family-operated businesses. Job sites tend to be dispersed, since there is no single dominant employer.

The second study site, Riobamba, is located three hours driving time by paved road, south of the capital city of Quito, at a point on the highway approximately halfway between the national capital and the large coastal city of Guayaquil. The city infra-structure is organized in a rectilinear pattern, exhibiting an extensive grid of streets surrounded by a broad, elliptical-shaped avenue called the *Circumvalación* that represents the extent of full urban infrastructure and services. The agricultural lands surrounding the city are being subdivided as the city continues to grow. However, the Riobamba area is considered economically less dynamic than the region north of Quito where Ibarra is located. This impression is supported by a study of economic location quotients (Vega 1995), in which Ibarra exceeds the 1990 national average in seven of the eleven economic sectors tested, while Riobamba exceeds the national average in only three sectors.

Riobamba is growing to the north where the two main transportation routes enter the city, and to a lesser extent to the south. In spite of better infrastructure availability to the south, the demand is stronger for residential locations to the north, which are more marketable because of the demand for accessibility to Quito and Guayaquil. Job sites in Riobamba are scattered; for example, *BEV Riobamba* provides a major source of employment issuing contracts for housing infra-structure and home construction at various locations. Other sources of employment include *Cemento Chimborazo* (about 300-400 employees) located to the south well outside the city limits, and *La Cerámica* and *Tubasec* asbestos tubing manufacturing companies which employ about 300 persons and are located in the industrial park within the city. In addition, the military employs about 1600 persons and municipal offices employ another 900. The largest employers, as in Ibarra, are government offices.

Roles of Key Participants in Urban Land Development

Field work in the city of Ibarra revealed several important participants in the urban land and housing development process including the Municipal Planning Department, the Municipal Cadastral Department, the national housing institution, a regional financial institution, and the social security administration. The Ibarra Development Plan guides the Planning Department, whose responsibilities include the review of subdivision requests and the issuance of building permits for all new construction and remodeling, as well as weekly inspection tours of building projects in progress. The Municipal Cadastral Department oversees property valuation and record keeping for the purpose of title transfer and real estate taxation. The functions of the Planning Department and the Cadastral Department closely parallel those of planning departments and assessors offices in cities and counties of the United States.

An important participant in the provision of housing is the *Ministerio de Desarrollo Urbano y Vivienda*, a national housing institution containing two branches: 1) *Banco Ecuatoriana Vivienda (BEV)*, the financial branch and 2) *Junta Nacional de Vivienda (JNV)*, the technical branch. The institution develops and sells lots and homes, offering one hundred percent variable rate mortgages with an interest rate less than the prevailing commercial interest rate. In September 1992, the interest rate for the purchase of a lot or home was 37 percent per annum, while the prevailing commercial interest rate ranged between 55 percent and 60 percent per annum. In September 1992, it was possible to rent a home in Ibarra for the equivalent of fifteen to twenty dollars per month, whereas the cost to purchase a minimum home through *BEV Ibarra* was about twenty five dollars per month. This national housing agency has functioned since 1976 and been involved in four large residential land and housing development projects in Ibarra.

Also involved in housing is a quasi-public institution, *Mutualista Imbabura*, primarily a regional financial institution, that in 1992 was marketing a forty-eight lot subdivision called *La Cuadra*, developed with full services in the central city area. The financial institution makes ten year adjustable rate residential loans to individuals for the purchase of lots, purchase of existing homes, construction of new homes, or for the enlargement of existing homes. Construction of a bank-financed home usually takes six months and involves five or six progress inspections during construction.

Of the institutions interviewed in Ibarra, the *IESS Ibarra* offered the most favorable home financing terms, although these are limited to persons who have worked and paid into the social security system for a minimum of four years, are retired, or who are married with at least one child. The interest rate on the twenty-year loan is adjusted to the income of the borrower and the loan may be used to purchase a lot for construction of a home or for the purchase of an existing home. In September 1992, the maximum loan for an individual was 12,000,000 Sucres ($6,666 US) or for a family, 18,000,000 Sucres ($10,000 US). The institution has been involved for more than fourteen years in its first and only subdivision in Ibarra, *Pilanqui IESS*, located near the airport southwest of city center.

Key participants in Riobamba are quite similar to those in Ibarra, in terms of their planning and development functions. The Municipal Planning Department of Riobamba reviews requests to subdivide land for residential development and reports to the Planning Commission (*Comisión de Planemiento*), who recommends approval or disapproval to the City Council (*Consejo Cantonal*), the elected body that makes the final decision to approve or disapprove the proposed subdivision. The Planning Commission consists of three publicly elected commissioners, three municipal architects usually from the planning department, and one secretary. The City Council consists of three elected council members and the elected *alcalde* (mayor). The Planning Commission provides a planning advisory function to the City Council, while the City Council makes the final decision for approval of a given land development. The Cadastral Department establishes land values for the purpose of taxation and maintains ownership records.

The *BEV Riobamba* and *IESS Riobamba* each serve similar residential land development functions as those described for Ibarra. *BEV Riobamba* has been active in the development of subdivisions and marketing of homes including two multi-family developments in which apartments were sold, as well as two large residential subdivisions which have been built out in five phases. In the case of Riobamba the *IESS Riobamba* has developed three subdivisions, but to date the institution has not marketed homes in these subdivisions, pending final approvals. *IESS* offers extremely favorable financing terms for the purchase of used homes as well as for new construction.

Mutualista Chimborazo is a regional bank which provides banking and borrowing services to the province of Chimborazo where Riobamba is located. It makes residential loans for the purchase of lots, the purchase of existing homes or for the construction of new homes. *Mutualista Chimborazo* has developed six subdivisions. However, homes have been constructed in only one subdivision. The other residential developments only have single family lots for sale.

Residential Subdivisions

During the period 1990 to 1992 there were twelve subdivisions approved in Ibarra, each relatively small, containing fifty to one hundred lots. The subdivisions remain vacant, in most cases pending completion of infra-structure. Of the twelve approved subdivisions, five are located to the northwest of the Tahuando River near the recently constructed *Universidad Técnica del Norte* and five are located to the southwest of city center. Two large on-going projects were approved during the 1980s. While theoretically it is possible to conclude all steps in the housing subdivision process in one year, it frequently takes five years, and there is one case of a subdivision development which has been underway for twenty years.

Subdivisions approved between 1990 and 1992 in Riobamba were located primarily to the northwest of city center, and to a lesser extent, to the south of city center. In Riobamba most proposed subdivisions are approved after modifications suggested by planners are incorporated in the design. Projects that are discouraged at an early pre-hearing stage usually lack sufficient available infra-structure. The time required to process a subdivision in Riobamba varies, from one to two months to create five or fewer lots, to one or two years to design and secure approval of a full subdivision including infra-structure improvements. Depending upon cost and demand for completed lots, the project may require longer to complete.

Residential developments are often approved in the preliminary stages but fail to complete the studies and infra-structure requirements for the sale of lots. Consequently, many developments in Ibarra and Riobamba remain in the preliminary stages for years. Since preliminary approvals are effective for two years and can always be extended, there is no urgency to complete the development. Just as institutions may be slow to complete developments, so too, do individuals often require years to complete construction of a home. In some cases the home is constructed in stages over a period of years while the owners live at another location in the city; in still other cases the lots are held for future resale. It is estimated by the Ibarra Cadastral Department that there are approximately 18,500 residential lots in the city of which 13,000 have homes. The remaining 5,500 approved lots remain vacant. In the city of Riobamba in 1992 there were 26,173 lots with full services available, of which 7,437 were vacant. Within the approved subdivisions approximately twenty to twenty-five percent of lots remain vacant.

Size of lots in Ibarra varies by zone. However, exceptions to lot size regulations are often made for the creation of small residential lots (6 m by 18 m) in the large subdivisions developed by *BEV Ibarra*. Lots created in times past, near the city center, tend to be larger than lots created since 1980, that are substantially smaller and more distant from the city center. This is a response to increasing development costs and ongoing demand for affordable lots. Groups of ready-made lots are not available for purchase to large institutional developers, hence, the same institutional entity will function as land speculator, subdivider, developer, builder and financier.

Size of lots in Riobamba has decreased over the years as well, as the demand for economically priced building sites has intensified. Historically, Riobamba lots have averaged 10 meters by 24 meters. However, increased costs of development have forced developers to create smaller lots (6 m by 15

m) in order to maintain profit margins. *BEV Riobamba*, like *BEV Ibarra*, has had difficulty securing municipal approval because lots are less than the minimum size required by local land use planning regulations. However, exceptions to lot size regulations have been justified on the basis of economic need for affordable housing among lower and middle income citizens.

Home Buyer Preferences

Buyers with sufficient financial resources generally prefer to purchase a lot and construct a home incrementally to their own specifications, whereas, those with more limited financial resources purchase a pre-constructed home. Thus, a fairly typical sequence of home ownership includes the early years of family formation in a pre-constructed home, followed by a period of investment in a building lot and the gradual construction of a new residence. A lot in most cases must be purchased for cash, since private sellers of lots hesitate to provide owner financing because of high rates of inflation and uncertain economic conditions.

In Ibarra, preferred residential areas are located predominantly in the southside and westside neighborhoods, as well as city center neighborhoods. These areas provide total services and a tranquil neighborhood atmosphere. However, urban amenities are not necessarily proximate to areas of preferred living. Examples of urban amenities in Ibarra include basketball and tennis courts, swimming pools, the motorway which encircles Lake Yahuarcocha (located within the city limits), city parks, universities, municipal stadium, municipal coliseum, municipal gallery, and nearby lakes located outside the city. Those interviewed indicated that the least preference was for neighborhoods judged to lack full infra-structure and to have a history of poor public safety.

Likewise, preferred places to live in Riobamba are those with the best public infra-structure. These neighborhoods are located in the center of the city and to the north. Principal public amenities of the city are dispersed and include outdoor and covered markets, a stadium, several public swimming pools, a large children's park, and various small parks in residential areas. *Cemento Chimborazo* functions as a dis-amenity to the south of the city, creating severe air pollution problems by emitting heavy dust.

Middle class home buyers buy pre-constructed homes. They prefer a three bedroom home with two baths, living room, dining area, separate kitchen, space for laundry services and, if possible, a garage and living quarters for a domestic employee. An average home size is two hundred square meters. These purchasers of pre-constructed homes are: 1) generally between the ages of thirty and forty, 2) have two children, about ten or twelve years of age, in primary or secondary school, and 3) tend to be shopkeepers or workers without a university level education. Although much of the housing constructed by *BEV Ibarra* is two story attached dwellings, purchasers prefer to purchase detached single family dwellings, if financially feasible.

In selecting a pre-constructed home, important buyer considerations include lighting, ventilation, a third bedroom, living room, dining area, and separate kitchen. A garage and domestic employee quarters are also factors to be considered, depending upon the socio-economic level of the purchaser. Other factors include access to shopping and proximity to public bus transportation. A detached single family dwelling is usually preferred over multi-family, common-wall construction.

Although it is more economical to purchase a pre-constructed home, the majority of buyers prefer, when feasible, to purchase a lot, in order to incorporate personal tastes in the construction of the home. In a typical Riobamba scenario, a buyer first owns an economically priced pre-constructed home for five to ten years, and then purchases a lot in an area to the north of the city for construction of the next home. In general, buyers of homes and lots range in age between thirty and fifty years of age, and are professional employees, store owners, professors or other middle level professionals. Small families may average three or four members, while a large family may have between six and ten members.

Site Preferences of Developers and Lot Purchasers

A number of physical site characteristics are important considerations when developing a large tract of land or selecting an individual lot. Shape, slope, topography, soil suitability, drainage, quality of adjoining neighborhoods, access to shopping, quality of street access, proximity to schools, access to public transportation, and availability of services such as water, electricity, telephone, sewer, and waste disposal service, are all important considerations. When one or more of these features is lacking or of poor quality, a potential housing subdivision site or building lot is less than prime for development.

Typically, in Ibarra and Riobamba, preferred locations for housing subdivision development are adjacent to already developed sites, rectilinear in shape, and near available service infra-structure. Those interested in purchasing individual lots will usually find them within the urban city limit, but outside the central downtown area, often clearly delineated by walls and fences, and often part of a subdivision that may or may not have completed infra-structure. Lots are usually level and treeless in these two cities, requiring only minimum clearing and grading prior to home construction. Since real estate sales occur by owner rather than through agencies, lots for sale often have large "for sale" announcements painted on nearby brick walls. Individual lot purchasers prefer a lot in a residential area, with full service infrastructure, saleable if necessary, and in an appropriate socio-economic setting. Less important, for an individual lot purchase within a developed subdivision, is proximity to schools and public transportation, since schools and public transportation are generally accessible in these intermediate size cities.

Much urban residential land development occurs at the edge of the urban area. The transition from urban to non-urban land use is sudden, determined by the extent of urban infra-structure, and is further characterized by the interspersing of residential streets stubbed into adjoining agricultural land in anticipation of future development. Some low density continuous development and ribbon development can be observed along major transportation routes entering the cities. The following are the locational site selection criteria for land development: 1) physical suitability; 2) infra-structure and service availability; 3) proximity to recently constructed institutional buildings (such as a university or hospital); 4) proximity to previously developed subdivisions, and; 5) proximity to predominant direction of city expansion.

A Few Problems Related to Development

Much lot speculation exists in Ibarra and Riobamba because the value of a lot increases at a rate faster than inflation, hence, the purchase of a residential lot is an excellent means of preserving capital in an economy with variable and high rates of inflation. Speculators withhold land from the market, anticipating a higher price in the future and in some cases create shortages of available land for subdivision and building lots. Many vacant lots are interspersed among the developed residential sites, perhaps because many persons have the capacity to purchase a lot, but for economic reasons are waiting to construct a home.

Lot speculation is problematic for planning administration because extension of services is more expansive, and hence more expensive, when lots remain vacant with unused services available in the street. Lot speculation is also resented by lot buyers and home buyers who see the practice as inflationary. In Riobamba, a regulatory solution was implemented, in September 1992, to encourage the infilling of vacant lots and, thereby, to improve infra-structure efficiency. A development moratorium was implemented and no new subdivision approvals were allowed.

Several other planning and development problems have been encountered in Ibarra and Riobamba, such as increased commercial uses along the main transportation corridors, land uses inconsistent with the city development plans, and issues of solid waste disposal. In Ibarra, a revised general plan provided for additional two story zoning in areas that already had two story construction, an apparent response to prevailing conditions. In Riobamba, a city development plan is being revised and a special study of the relationship between rent and property values is underway. Several studies are underway in both cities to better understand and resolve specific urban land use problems. These are: 1) traffic congestion at the city markets, 2) rehabilitation of city-owned urban amenities, and 3) need for new solid waste disposal sites.

Summary

Field work in Ibarra and Riobamba demonstrated the importance of institutional builders, the function of planners, and the nature of financial arrangements with home buyers. It also has suggested a number of conclusions with regard to the dynamics of city planning and housing development in intermediate size Ecuadorian cities. City planners, elected officials, national housing development agencies, lending agencies, and individual owner-builders are key participants in the urban land development process. City planning exerts a profound impact upon middle and upper class residential development, applying institutional mechanisms such as zoning ordinances, subdivision regulations and the provision of public services. Equally important is the role of local elected officials in controlling the size, design, and timing of residential development. National level institutions such as *BEV* and *IESS* construct the majority of subdivision houses and provide favorable financing, thereby increasing marketability of new and used homes in Ibarra and Riobamba. Generally, homebuilders in Ibarra and Riobamba are either large institutional builders who provide their own financing, or individual owner-builders who have purchased a single lot to construct their next family home. There is essentially no incidence of the small or medium-sized private corporate builder.

Residential subdivisions generally take one or more years to design and secure local municipal approval; often revisions to initial proposals are necessary, adding time to the development process. Over the years the size of individual lots has declined and the prevalence of vacant lots within developed subdivisions has become problematic. Properties are marketed by individual owners or by institutional developers rather than by real estate professionals. Real estate brokers are not involved in the marketing of homes in Ibarra and Riobamba, although they are often involved in the larger cities of Quito and Guayaquil. Nor is there a real estate multiple listing service or public record of sale prices. The result is a lack of public knowledge about pricing, a wide range of sale prices, and a variety of institutional financing terms available. Mortgages, adjusted to income affordability, reduce the cost of home purchase for many buyers, most notably mortgages made by the *BEV* and *IESS*. Thus, a large variance in long term costs of home ownership in Ibarra and Riobamba can be inferred.

Problems of local land development identified include traffic congestion, rehabilitation of city amenities, need for solid waste disposal sites, and the vexing issue of vacant lots within developed subdivisions, resulting in increased overall cost of infra-structure and service delivery. To encourage the infilling of the vacant lots within developed subdivisions, city planners occasionally place a moratorium on the approval of new subdivisions. Overall, lot size has declined over the years, due to increasing costs of development and demand for affordable lots, consequently, smaller than usual lot size exceptions have been made (on the basis of social equity and public need) for large institutional developers such as *BEV* and *IES*.

Much work remains to be done in the study of planning and residential development in highland Ecuador. A similar study of other medium size highland *sierra* cities such as Cuenca, Ambato, Loja, and Latacunga, would be useful and serve to confirm and expand the findings of the present study.

Acknowledgement

The author wishes to thank Fausto Yepez, Giocanda Benavides, Carlos Vasquez, Gonzalo Montes, and the Celso Santacruz family. The author gratefully acknowledges the kind help of *Alcalde* Carlos Castro, Jose Ruiz, and Rosendo Moreno, and the Inez Rodriguez family. The invaluable assistance of Nestor Vega Jiménez, Gonzalo Darquea Sevilla, and the support of Mark Diamond Research Fund is gratefully acknowledged.

References

Carr, M. 1992, Metropolitan Planning and Management in the Third World: A Case Study of Quito, Ecuador, pp. 83-137 in A. Ramachandran, (ed.), *Metropolitan Plan-*

ning and Management in the Developing World: Abidjan and Quito, Nairobi: United Nations Center for Human Settlements (Habitat).

Carrión, F. 1986, Ciudades Intermedias y Poder Local en el Ecuador: Una Aproximación Analítica, pp. 67-88 in D. Carrion, J. Hardoy, H. Herzer, and A. Garcia, (eds), *Ciudades en Conflicto: Poder Local, Participación Popular y Planificación en las Ciudades Intermedias de América Latina*, Ecuador: Centro de Investigaciones (CIUDAD).

Carrión, F. 1988, La Investigación Urbana en el Ecuador, pp. 85-111 in F. Carrión, (ed.), *Investigación Urbana en el Area Andina*, Quito: Centro de Investigaciones (CIUDAD).

Carrión, F. and R. Vallejo, 1992, La Planificación de Quito: Del Plan Director a La Ciudad Democratica, pp. 143-170 in F. Carrion, (ed.), *Ciudades y Políticas Urbanas en América Latina*, Quito: (CODEL).

Centro del las Naciones Unidas para los Asentamientos Humanos (CNUAH), 1993, *Gestión Urbana en Ciudades Intermedias de America Latina*, Nairobi: United Nations Center for Human Settlements (Habitat).

Centro Ecuatoriano de Investigaciones Geographic, 1987, *El Espacio Urbano en el Ecuador: Red Urbano, Region y Crecimiento*, Quito: Instituto Geografico Militar.

Coraggio, J. 1991, *Ciudades Sin Rumbo: Investigación Urbana y Proyecto Popular*, Quito: Centro de Investigaciones (CIUDAD).

Darquea, G.; G. García and F. Gallegos, 1994, *Marco General de la Planificación Local Participativa*, Quito: Unión Internacional de Municipios y Poderes Locales (IULA) and Asociación de Municipalidades del Ecuador (AME).

Drewski, L.; K. Kunzmann and H. Platz, 1991, *Promotion of Secondary Cities: A Strategy for Development Cooperation*, Eschborn, Germany: Deutsche Gesellschaft für Technische Zusammenarbeit (GTZ).

Fernández, M. 1994, Ecuador, pp. 215-251 in Greenfield, Gerald, (ed.), *Latin American Urbanization: Historical Profiles of Major Cities*, Westport, CN: Greenwood Press.

Gangotena, R. 1995, El Proceso de Descentralización en el Ecuador, pp. 131-193 in P. Trivelli, (ed.), *Municipalidades y Descentralización: Presente y Futuro*, Quito: Sociedad Alemana de Cooperación Técnica y Programa de Gestión Urbana.

Greenfield, G., (ed.), 1994, *Latin American Urbanization: Historical Profiles of Major Cities*, Westport, CN: Greenwood Press.

Healey, P. and S. Barrett, 1990, "Structure and agency in land and property development processes: Some ideas for research", *Urban Studies*, 27:89-103.

Heredia, C., (ed.), 1992, *Riobamba en el Siglo XX*, Riobamba, Ecuador: Ilustre Municipalidad de Riobamba.

Instituto Nacional de Estadística y Censos (INEC), 1992a, *V Censo de Población y IV de Vivienda 1990 Provincia de Imbabura, Ciudad de Ibarra*, Quito: Dirección de Informática.

Instituto Nacional de Estadística y Censos (INEC), 1992b, *V Censo de Población y IV de Vivienda 1990 Provincia de Chimborazo, Ciudad de Riobamba*, Quito: Dirección de Informática.

Instituto Nacional de Estadística y Censos (INEC), 1984a, *V Censo de Población y IV de Vivienda 1982 Provincia de Imbabura, Ciudad de Ibarra*, Quito: Dirección de Informática.

Instituto Nacional de Estadística y Censos (INEC), 1984b, *V Censo de Población y IV de Vivienda 1982 Provincia de Chimborazo, Ciudad de Riobamba*, Quito: Dirección de Informática.

Larrea, C. 1986, Crecimiento Urbano y Dinámica de las Ciudades Intermedias en el Ecuador (1950-1982), pp. 89-126 in D. Carrion, J. Hardoy, H. Herzer, and A. Garcia, (eds), *Ciudades en Conflicto: Poder Local, Participación Popular y Planificación* en las Ciudades Intermedias de América Latina, Quito: Centro de Investigaciones (CIUDAD).

Lowder, S. 1986, *The Geography of Third World Cities*, Savage, Maryland: Barnes & Noble Books.

Lowder, S. 1990, "Development Policy and Its Effect on Regional Inequality: The Case of Ecuador", pp. 73-93 in D. Simon, (ed.), *Third World Regional Development*, London: Paul Chapman.

Melo, F. 1988, *El Municipio: Gobierno de la Comunidad Local*, Quito: Talleres Gráficos.

Nickson, R. 1995, *Local Government in Latin America*, Boulder, CO: Lynne Rienner Publishers.

Rodríguez, L. 1985, *The Search for Public Policy: Regional Politics and Government Finances in Ecuador, 1830-1940*, Berkeley: University of California Press.

Schoenberger, E. 1991, "The Corporate Interview As a Research Method in Economic Geography", *Professional Geographer*, 43:180-192.

Teekens, R. 1988, *Theory and Policy Design for Basic Needs Planning: A Case Study of Ecuador*, Aldershot, Netherlands: Gower.

Tobar, C. 1985, *Monografía de Ibarra*, Ibarra, Ecuador: Centro de Ediciones Culturales de Imbabura.

Vega, N. 1994, "Los Municipios del Grupo Andino: Ejemplo de un Estudio Comparativo", *Revista Interamericana de Planificación*, 27:107-108:216-224.

Vega, N. 1995, ¿Como Conocer la Base Económica de una Región?, *Poder Municipal* 25:22-25.

Verdesoto, L. 1986, Resultados Electorales en las Ciudades Intermedias: Ecuador (1978-1979), pp. 259-268 in D. Carrion, J. Hardoy, H. Herzer, and A. Garcia, (eds), *Ciudades en Conflicto: Poder Local, Participación Popular y Planificación en las Ciudades Intermedias de América Latina*, Quito: Centro de Investigaciones (CIUDAD).

Violich, F. 1987, *Urban Planning for Latin America: The Challenge of Metropolitan Growth*, Boston: Oelgeschlager, Gunn & Hain.

World Bank, 1991, *Urban Policy and Economic Development: An Agenda for the 1990s*, Washington: The World Bank.

10 Sustainable Development in China: Reconciling Modernity with Tradition

PATRICK H. WIRTZ, ERIC J. HEIKKILA

As China undergoes perhaps its most far-reaching transformation in recent history, new approaches are arising to relieve the accompanying political, economic, and cultural stresses. One approach is sustainability, formally initiated by the central government in a program to develop "sustainable cities" throughout the nation. The appeal of sustainability in China is its ability to steer the often contentious course between modernism and the socially-affirming precepts of tradition. As a rhetorical device, sustainability offers the government a chance to maintain two primary commitments, the first, a commitment to the political stability of the state and the second, a commitment to pursue economic reforms which, some critics have argued, threatens commitment to the first. While sustainability is rhetorically popular, it is still largely untested as instrumental policy. It has yet to prove that its ideals can accommodate both commitments without straining under the tensions pursuant to each.

A notion of dual conflicts underlies this discussion. In the first case, the phrase "sustainable development" itself suggests an oxymoron, a clash between the act of preserving and the act of transforming. Resolving how something can be preserved over time given a constantly changing world, is the challenge facing sustainable development. In the second case, the PRC leadership must deal with reconciling market capitalism with the socialist precepts upon which it constructed its political legitimacy. Perhaps best exemplified by a characteristic use of euphemistic terminology, China has been experiencing socialism with a Chinese face, for over a decade and a half. The Chinese record in shifting towards a market-socialist hybrid, increasingly geared toward the former, is a testament to political authority overriding ideological tension. This remarkable feat has required deft handling, particularly the ability to interpret ideals and principles as broad in scope as possible. During the current transformation, Chinese authorities can use the concept of sustainable development to achieve an important goal, namely to alleviate the tensions inherent in their version of market-socialism and thus sustain social cohesion and the party's political authority.

This chapter asks whether sustainable development in China is in truth a viable concept. This is as much a test of the authority of sustainable development as a universal principle as it is an examination of the capacities of inter-cultural interpretation and the assimilative capacities of Chinese cultural traditions. Sustainable development persists as a contested concept even within the primarily Western environmental ethics and international development policy-making fields from which it derived. The applicability of sustainable development amid the transformative changes occurring in contemporary China begs the issue of whether a concept that has evolved under different cultural and historical contexts is compatible with China's persisting cultural traits and the dominant political rhetoric. We suggest that sustainable development is a viable concept. Cultural threads running through Chinese traditions lend support for a nascent concept of sustainability in China.

Any concept, particularly claiming a comprehensive scope, requires the authority to legitimize itself. Authority can rest in many sources. However, given a reactionary climate during periods of rapid and far-reaching structural changes, nativistic sentiment often assumes an enlarging role in claims to cultural authority. This type of sentiment is often supported rhetorically through a reassessment, and consequent preservation, of customary traditions; thus, in times of stressful transformation the search for authority may serve to reinforce the persistence of certain cultural patterns. China exemplifies this issue, as there still flourishes there a powerfully influential tradition based upon recognizing continuities in social patterns. That continuing patterns support the rhetoric only serves to reinforce the appearance of Chinese cultural immutability. Despite a historical record of revolutionary changes, a certain set of cultural attributes have preserved themselves over millennia.

The relevancy of this issue extends beyond acknowledging the continuity of Chinese cultural traditions; a concept of Chinese sustainability should inform any evaluation of Chinese development policy and its effects upon social traditions. The issue here is whether the transformation of China by the market economy remains compatible with these traditions, or does it require a subsequent transformation in social structure and individual identity, such as occurred in the West. Sustainable development in the Chinese context suggests a means to test this issue. Moreover, sustainability in the Chinese context also suggests a means to test the embodying principles underlying the concept of sustainability in general: Is it possible to maintain an overarching societal structure given change over time? In this chapter we first review the principles of sustainability as they have evolved since the coining of the terms in Western development literature. This follows with a highlight of Chinese cultural traditions that embody or complement these principles. This exposition grounds the often abstract concept of sustainability within a specific cultural context, suggesting a means to judge its instrumental worth and also its limitations.

Contemporary China: The Context for Sustainability

Visitors to contemporary China, particularly the major population centers on its eastern seaboard, often remark on the tremendous transformation they are witnessing, a landscape reconfiguring itself amid persisting symbols of an older, traditional order. An economy growing at an annual rate of close to 10 percent since 1988 appears manifest in massive new developments, signs of surging near-term construction, and expanding consumer culture. These and other signs suggest that there is truth behind the rhetoric, touted in international business and popular journals, of a New China rising phoenix-like from an extended period of malaise and missed potential.

Under this rubric of a New China, tensions within the social structure have arisen, exacerbated by the rush for new development. Many who are following the progress of China see potential disruptions in social stability as a major obstacle to continued, steady development. Two significant tensions exist in contemporary China. One involves the ideological shift from socialist-Maoist principles underlying the 1949 communist revolution toward free-market capitalism since the Open Door policy of the late 1970s. Unlike the experience of Eastern Europe and the former Soviet Union, this transition has occurred in China without a revolutionary change in the state's political structure. We should not conclude from this apparent success that the current administration has discovered a formula for assuaging the inherent social tensions (Chen 1995; Naughton 1995; Nolan 1995). Indeed, the contentious political events of the last few years suggest that this issue remains a top concern among government officials. This issue is not merely the political struggle over who shall lead the nation after Deng Xiaoping. At stake is not only the level of commitment to the reforms, but also the regions within China that will benefit most from them.

The second tension delves into deeper historical and cultural roots. It is the central dilemma arising between modernism and traditionalism, expressed as a conflict between traditional ethics and wealth and power. In this conflict modernism, embodied by the recent reforms and the consequent changes, represents the accumulation of material wealth and power. However, in this process modernism perverts the social order by systematically negating ethically-proscribed relationships between individuals (Madsen 1992). According to this view, personal loyalties shift from culturally-sustaining traditions based upon rituals and familial obligations toward the appeasement of individual satisfaction levels, or in other words, the individual at the expense of a larger collective welfare. The Chinese frame this conflict as a struggle to maintain their national and cultural identity, the players being Chinese tradition on one side, the "West", particularly the United States, being the agent of modernism, on the other, and the contested playing field being the contemporary Chinese population.

As implied above, there are multiple ways to contextualize the current trends within Chinese society, with perhaps significant variations between non-Chinese and Chinese interpretations. To many non-Chinese, the recent changes represent the opening toward the world economy of a latent market formerly encumbered by misguided egalitarian policies, which fostered mis-management and inefficient use of resources rather than the wealth of the nation. In this view, the transformation of Chinese society remains largely couched within economic references. While the leadership of the People's Republic must indeed struggle to justify their seemingly abrupt reversal in economic policy, there are also other ways to view the recent transformations. Within China an older tradition exists of drawing upon the rhetoric of China's immutability. Through this lens, the acknowledgment of persisting cultural traits, however vaguely defined as "Chinese", still informs discussion about the legitimacy of political structure and its accompanying policy. Indeed, the continuation of this older rhetorical tradition suggests an important component in understanding how the current administration has accommodated the monumental social and economic transformations of the last two decades and, despite these developments, how certain cultural threads have sustained themselves in daily life. The conflict between modernism and tradition draws heavily from this rhetoric and persists in the form of Chinese sustainability.

In 1992, the People's Republic of China announced *China Agenda 21*, a development program adapted from the United Nations document Agenda 21, in turn derived from the proceedings of the 1992 Earth Summit in Rio de Janeiro, Brazil. *China Agenda 21* proposes the implementation of a sustainable development program in Chinese urban centers, with a model city, Changzhou, in the eastern province of Jiangsu. With the prominent use of the term, sustainable development, the Chinese administration reinforces the increasing popularity of a concept originally rooted in the environmental discourse of the 1970s (Kidd 1992). The spread of the term into the development literature in the subsequent period attests to the general appeal of its call for ecological and socially-responsible measures to rectify recognized insensitivities in modern development policies. Its use by Chinese authorities reflects in part their growing recognition of the major ecological issues facing China. Moreover, their use of the term also underscores an ideological tension within a government increasingly severing itself from the rhetorical roots of its legitimacy, i.e. as a vanguard for socialist utopianism. For amid the radical shifts within Chinese society, rhetorical devices such as sustainable development offer a means to deal with this tension. At one level, using sustainable development as a rhetorical device informs official policy. At another level, the process of highlighting persistent cultural traits, held as traditional links to the larger national identity, prepares the basis for a Chinese variant of sustainability.

Sustainability and Sustainable Development

The explicit purpose of *Agenda 21* was to take the much publicized concept of sustainable development and prepare the foundation for its implementation as policy. This addresses accusations of the term's vapidity, namely that the term remains largely an appeal for undefined notions of comprehensiveness in development policy instead of a practical litmus test to judge policy measures (Beckerman 1993; Wildavsky 1994). As one critic of the term says "... if the Emperor of Sustainable Development has any clothes at all, they are pretty threadbare" (Beckerman 1993, 191). Sustainable development in China remains fraught with trying to instrumentalize ambiguous phrases.

Perhaps such criticism stems partly from the manner by which many authors use the terms sustainability and sustainable development. Indeed, they are often used without definition and often interchangeably, ignoring the subtle distinctions. Sustainability refers to the ontological associations between agents regarding the maintenance of the human and biological systems we inhabit over time. Sustainable development refers to the development processes which have the capacity to sustain themselves over time. Despite charges of its ambiguity, consistent themes can be derived from the sustainable development literature. In general, there are two broad inter-related principles: 1) intergenerational equity and 2) preservation of available capital stock.

The first of these two general principles of sustainable development, intergenerational equity, means that a sustainable path of development allows every future generation the option of being as well-off as its predecessors. Personal responsibility toward others, even those not yet born, is extended, which translates instrumentally into a penchant for conservation, i.e. savings over present consumption (Solow 1993). It also recognizes that each generation exercises power over its successors; we partly determine the type of environment our descendants inherit and are thus ethically responsible to them. Conversely, humans are also subjects to, as well as possessors of, power since we are dependent upon what our ancestors endowed to us (C.S. Lewis as quoted in Daly and Townsend 1989, 230). Although many cultures condemn gratuitous waste, sustainable development frames resource use as a matter of personal, social and cultural responsibility.

Interrelated with the first principle is that of preservation of the available capital stock. Capital stock in this context is not solely natural resources, instead it may be thought of as a bundle of endowments left for future generations. According to economist Robert Solow, this bundle of endowments includes nonrenewable resources, stock of plant and equipment, inventory of technological knowledge, general level of education, and supply of skills (Solow 1993). This is not about maintaining a status quo in perpetuity; environmental change is natural. Instead the rate of change within the system becomes the emphasis. Preservation thus involves whether current practices outpace the ability of a system to adapt to the rate of transformation.

Recognizing that forms of capital stock change over the long-term and cannot be preserved forever, sustainability offers the notion of compensation. Many argue that sustainability does not necessarily conserve everything, nor should it (Jonas 1984). According to Solow (1993), what matters is that the inherited endowment is adequately compensated. The important point is not the particular form of replacement, but the capacity to fulfill this goal. The primary focus of sustainability hence shifts to depletion and investment decisions related to the available capital stock, broadly defined.

Sustainable development does not solely preserve contemporary artifacts but preserves the processes, and the relationship between agents creating those processes, that transform societal values into implementable actions. Thus sustainability accepts the necessity of adaptation to different contexts. Moreover, a sustainable system must be flexible since communities evolve over time. This does not imply that everything is appropriate and should be accommodated. Rather, some processes are acknowledged as being more beneficial to the long-term health of a society than others. Sustainable development is, therefore, development working within the boundaries established by these recognized processes; it consciously promotes system maintenance over time through investment decision-making while adjusting for changes in the environment.

China today faces this issue. The treatment of capital stock thus becomes the centerpiece of political rhetoric. In practice Chinese authorities have focused the political tensions pursuant to the shift from socialism to market capitalism on the issue of material well-being, the utility value of available capital judged in regard to raising the material living standard for the average citizen. In the 1980s the most ardent supporters of the economic reforms in the PRC leadership interpreted socialism as broadly as possible using this rhetoric. Premier Deng argued that the ultimate goal of socialism was raising living standards, any process leading towards this goal conformed to the socialist mandate. To Deng, political legitimacy rested in raising economic productivity. If socialism could not surpass capitalism in speed of growth and economic efficiency, then socialism was nothing but a boast (Chen 1995, 58-59). In this line of reasoning, the socialist goal would remain rooted in fairness (*gongping*), but the means to achieve it did not have to be limited to a narrow range of Maoist or Leninist policies. Simply put, Deng expressed in the midst of implementing market reforms to the economy:

> Socialism means eliminating poverty. Pauperism is not socialism, still less communism. The superiority of the socialist system lies above all in its ability to increasingly develop the productive forces and to improve the people's material and cultural life. The socialist system can enable all the people to become well off. This is why we want to uphold socialism, (Nolan 1995, 162).

Obviously not everyone is pleased with this reinterpretation of socialism, narrowing its standard of instrumental worth to economic productivity to create an ambiguous umbrella. Currently, among systematic ideological objections, the rhetoric of equitable distribution continues to have strong appeal, particularly when regional disparities consequent to the market reforms have exacerbated interregional rivalries. These in turn, support a contentious political discourse tied to leveling regional disparities. Which regions benefit from the transformation of resources into material goods under this new policy of socialism, focuses attention toward the old stand-by of central planning, national policies regarding investment and redistribution.

Within this context, sustainability provides Chinese authorities a means to deal with the tensions arising from a broadly-defined socialism. The concept of sustainability offers the potential to redirect attention away from regional disparity and refocus it toward investment decision-making aimed at comprehensive development. Another ambiguous phrase, comprehensive development translates in this instance into redirecting ideological principles away from explicit material motives. This is done by reemphasizing the quality of the social environment and standards of fairness not necessarily tied to economic productivity. It is here that sustainability in the contemporary Chinese context differs from the West. In its conceptual evolution in the West, sustainability remains centered around its ideological roots in environmental discourse. In

China, sustainability provides the administration with a policy tool to reinforce its legitimacy during this period of rapid economic and social change. The ontological principle behind *China Agenda 21* is not that of an ecologically-sound nation, but a stable society sustaining itself over generations.

We do not suggest that, in China, sustainability is in principle discordant with ecological principles, but that the transposition of the phrase between general Western and Chinese contexts must be accompanied by recognizing the different discourses it draws upon. In China, the issue focuses on alleviating social tension. In the West, the issue remains largely one of interpreting and redirecting development processes through the lens of ecological principles. Despite these differences, an interdependent relationship exists between these two focuses. If Chinese authorities must deal with the environmental consequences of their development policies, the effectiveness of a set of ecological principles as formal action depends upon their authority to motivate people and to influence social relationships. Thus we have to recognize that the two contexts, Western and Chinese, have autonomous and mutually-informing characteristics.

Reflecting its origins in Western environmental discourse, sustainable development affirms a tenet of ecological thinking, that relationships between agents in a system are inter-connected; none exists in isolation within discrete boundaries. For example, a change in the hydrologic cycle creates multiple effects upon other agents in a system through a series of networks, be it the water quality downstream or agricultural output due to changes in flow. In such a case, those involved in the formal decision-making processes are accountable not only for the factors they included in their decision-making, but also for justifying the exclusion of factors, why something in the hydrologic system does not belong or is not encompassed within the decision-making purview.

However, determining every agent affected within a process is impossible. Limits do exist. An inherent liability when discussing sustainability is this notion of comprehensiveness. It is that we cannot recognize every factor that influences a system or every relationship within one. We also cannot forecast *a priori* all the consequences of an action or even the magnitude of the ones we can forecast reliably. The quest to be comprehensive is theoretically endless. Thus the issue becomes one of determining the ethical point where action should occur (Ulrich 1993).

Dominating the two general principles of sustainability is the idea of carrying capacities, presenting a basis for action. Proponents of sustainable development often uphold the belief that every system maintains a maximum capacity for use, additional accommodation leads toward malfunction. Given the supposition, there develops a point of contention about sustainability, namely, what are the inherent limits to human-based resource use? Roughly, there are two opposing perspectives. On one side modern ecological thought holds that natural ecosystems are relatively nonresistant when confronted by large-scale contemporary patterns of human resource use. On the other side are the principles of free-market economics that technology and human ingenuity work within the free market system to duly compensate resource loss.

Sustainable development proponents adhere to the former perspective. They argue that people ignore limits to natural ecosystems because of ignorance of biologic processes and faith in human-created substitutability. Sustainability proponents emphasize that vague assumptions about the capacities of environmental cycles distort decision-making; the capacity of ecological sinks to regenerate is still undefined. Historically, refuse tends to accumulate rather than recycle (Daly and Cobb 1989, 9). They point out that the faith in technology creating substitutes to offset depletion is primarily supported not by physical or biological scientists, but by politicians and business people, both of whom have vested interest espousing continuous growth through technological innovations (Bartlett 1994). They hold that even faith in human-created substitutes does not answer whether new substitutes are preferred, or whether we are simply making the best out of a worsening situation.

The other perspective holds that agent autonomy within market economics contains a self-regulating component that maximizes efficient use of existing resources which ultimately averts resource depletion. Accordingly, if a resource becomes scarce, its price subsequently rises, stimulating the search for new supplies or the creation of natural or synthetic substitutes. Proponents of this view point to numerous examples where the market process has maintained certain levels of a commodity rather than leading toward the commodity's depletion. They also point to the largely unsuccessful track record of future catastrophes held by environmental doomsayers adhering to their narrower conception of resource limits (When the Boomster... 1995).

Ultimately, arguments over what actions are sustainable are rooted in value-judgments. This is an issue of whether some states of existence are preferable to others; at its very least, sustainability is not about preserving states of misery. Hence, cultural values are an intrinsic part of this issue. System boundaries must exist, but where to set them and the distribution of activity within them are all subject to social values and physical capacities. In this light, the recognition of system linkages becomes less about scientific rules of causation, as in the hydrologic system example, and more about how community values facilitate or hinder recognition of these linkages. As systems evolve, community values accept or reject these changes. Thus, while sustainability affirms that human systems are contingent upon natural systems, sustainable development is a form of development; it is not about maintaining a status quo or returning to past ecological conditions.

Chinese Sustainability

Reflecting its slightly different focus, a case for Chinese sustainability places greater emphasis on cultural recognition of sustainable processes over empirical evidence derived from rigorous experiments upon ecosystems (*China Agenda 21* 1992). This is not a denial of the severe environmental problems facing the nation. Instead, this underlies the strong belief that the process of cultural transmission, whereby certain culturally-defining attributes maintain themselves over generations, contains the ability to accommodate changes in the environment while still sustaining a core of essential cultural values. Although this emphasis does not explicitly incorporate such concepts as carrying capacities, it affirms the role of community values in establishing what is sustainable.

This emphasis is important in comprehending the context of Chinese sustainability because it highlights a divergence from the common use of the phrase in Western development literature. In the West, belief in ecological processes, the understanding of which have evolved over time, intimates an existence above and beyond human endeavors. Their structure, and ultimately their worth, is not dependent upon human recognition; it is the responsibility of persons in a sustainable society to recognize their autonomy and adapt their lifestyles accordingly. The emphasis in China is more anthropocentric, ecological processes become relevant only as they affect human endeavors; a sustainable society may legitimately alter the physical conditions influencing these ecological processes. These two perspectives need not be incompatible. Ideally cultural adaptations affirm existent ecological processes, but the point is that they need not do so. In the Chinese context the relationship of humanity to the natural environment remains an important issue. The emphasis, however, becomes the capacity within the belief system of the sustainable society to incorporate and assimilate in cultural terms changes in the natural environment.

There are ample reasons to be pessimistic about sustainable development in China. From a Western perspective, perhaps the most arresting reason is the extensive and accumulating evidence of environmental degradation present in the Chinese landscape. Vaclav Smil (1984 and 1993) has well-documented the environmental crisis within the People's Republic. According to his analysis the critical problems are not so much general pollution and reduction of biomass, but the gradual degradation and loss of invaluable environmental services. These problems accrue from wind and water erosion, nutrient decline in agricultural soils, salinization and alkalization of irrigated farmland, overdrawing of groundwater, and extensive deforestation and desertification (Smil 1993, 10). Unfortunately, given the current trend in Chinese development, his diagnosis holds that continued degradation is inevitable.

There are other important reasons for a pessimistic view of China's future. They include an extensive level of general poverty which may diminish official and individual commitment to non-economic goals, the possibility of less stability in social relations during times of rapid transformation, and persistent population pressure on a limited resource base. Part of the problem with dealing with these issues is the relative lack of precedence to draw upon. Although other regions in the world face similar pressures, the scale of the Chinese context, particularly projected consumption patterns, renders comparisons problematic. Even comparing the current Chinese growth pressures with the rapidly expanding nineteenth century European economies and societies is also problematic. A pivotal difference is that nineteenth century European states were much more successful in obtaining from beyond their borders the natural resources to fuel their growth (Pomeranz 1993). Very few, if any, contemporaries suggest that China use the imperialism model to alleviate its resource limitations.

Given these serious issues, sustainable development appears to offer the framework to successfully confront and mediate these problems. However, issues of transferability and interpretation remain. The environmental ethos which influenced the development of the concept in the West does not have a comparable expression in contemporary China. Moreover, while there exists a body of traditional Chinese thought concerning humanity and its relationship to the natural environment, the philosophical precepts underlying these ideas often contrast with the precepts underlying the environmental ethics which have developed in the West. Granted these differences, there are also many comparable traits found in Chinese philosophical traditions and the concept of sustainable development. In particular, the philosophical traditions that have contributed a major role in maintaining customary procedures and cultural identity over centuries provide a foundation for sustainable development. In one sense, the very persistence of customary procedures suggests that certain processes have proven commensurable with sustainability. In support of sustainable development, there remains a persistent popular appeal in China of a philosophic tradition that affirms some tenets of sustainability, as outlined within environmental and development literature. However, attitudes regarding the relationship of humanity within the natural world are never monostatic. Even within Chinese tradition there have been different, often opposing, beliefs that have changed over time. More often that not, traditions become lumped with other traditions and customary rites under the rubric of traditional Chinese culture. This can be misleading as each of these traditions harbors a different perspective on humanity's relationship toward the natural environment.

The leading tradition, the one that has provided the dominant overarching structure for Chinese cultural development over the past two millennia, is Confucianism (Rubin 1976). Originating in the fifth century BC from the teachings of Kong Fu-zi (latinized as Confucius), the influence of Confucianism has ebbed and flowed in different periods of Chinese history. While there has not been unilateral acceptance of Confucian principles in Chinese history, for most of the past two thousand years, particularly since the Neo-Confucian revival during the Sung dynasty (960-1279 AD), Confucianism has received official state recognition and has been accepted as the orthodox way of life by the majority of scholars and officials throughout China (Smith 1973, 15). It is largely because of the influence of Confucianism that the rhetoric of Chinese cultural immutability, while debated in academic arenas, maintains a strong presence in modern Chinese identity (Weiming, et al 1992). In describing this phenomenon, people point to the continuance in Chinese society of Confucian values. A difficulty exists, however, in defining Confucianism, as it is not a religion in the sense that Christianity or Islam is defined. There is no divinely proscribed dogma in which adherents are required to believe. Instead, Confucianism is more akin to a set of ethics and customary behavioral patterns than a set of ideas. There also exists difficulty in defining its directly attributable influences on Chinese society.

Although Confucian and Chinese thought are not synonymous, their strong inter-relationship, evolving over two millennia of mutual influence, has made the issue of distinguishing each without reference to the other problematic at best. What makes the debates difficult is this very fact of close historical interdependence. Indeed, to many Chinese, Confucian ethics still inform their sense of Chinese identity despite several traditions which have threatened to prevent its transmission. Others reaffirm that the contemporary landscape of East Asia can only be properly interpreted in the context of the Confucian-influenced culture within which it is embedded (Heikkila and Griffin 1995).

Given the persistent legacy of this tradition, the issue arises whether there are component factors compatible with sustainability. Which precepts underlying the cultural traditions affirm the precepts underlying the construction of sustainable development? There is a customary belief within Western observations about China that a different set of relationships governs the treatment of nature in Western and Chinese social traditions.

On the one hand, in Western societies the historical legacy of massive capital accumulation derived from intensive resource use and over two centuries of industrialization have substantially transformed the natural environment. Particularly since the nineteenth century there has existed a significant discourse built upon criticism of this widespread transformation of the landscape. Critical targets include the economic processes underlying this transformation and the ideological structures and social customs supporting these processes.

On the other hand, despite clear evidence of large-scale contemporary and historical environmental degradation and intensive change to the natural landscape by human endeavors in China, observers of Chinese society remark on humanity's conceptual position in life, the view that humanity is essentially at home in, not alienated from nature (Allison 1989, 15). This is part and parcel of a belief found also in the modern environmental ethos, namely that of holistic unity, that the spheres of human activity and the natural world are not distinct, autonomous realms, but are part of a transcendent whole, each contributing to the existence of the other.

A few convergences do not imply that the traditions of Chinese philosophy affirm the modern environmental ethos. An investigation of the Chinese track record in sustaining natural landscapes, according to the standards of modern preservation aesthetics, would probably not elicit any special Earth Day commemorations. Despite this, certain features of this philosophic tradition suggest a nascent Chinese sustainability ethos. In particular, two aspects prepare the foundation from which to further articulate the construction: the belief in universal holism and a distinctively pragmatic approach in conceptualizing human endeavors.

The first aspect, the belief in universal holism, describes a form of conceptual association among the component parts of existence. Within this rich, centuries-old, philosophic tradition, Chinese cultural and political authorities have investigated the need to harmonize the relationship among Tian (heaven or universe), Di (earth or resources) and Ren (people or society) (Wang and Ouyang n.d.). Unlike the Western philosophic tradition, these three primary elements of the Confucian cosmos are not conceived in terms of transcendence, whereby one element assumes a hierarchically superior position to another. Instead these elements exist in conceptual polarity: they are symmetrically related, each requiring the other for adequate articulation. In sum, all elements are correlative; every element in the world is relative to every other (Hall and Ames 1987, 17-18). Conceptually, this reiterates the beliefs underlying the relationships between systems in Western sustainable development literature.

In the second aspect, scholars of Chinese philosophy stress the practical, as opposed to the theoretical, character of Chinese philosophy. Chinese philosophic tradition places greater emphasis upon, and consequent development of, the practical. This difference from the Western tradition is not one of kind but of degree of emphasis (Allison 1989, 10-11). An emphasis on the practical denotes that concepts derive their meaning more from exigent factors than universal principles. This means that ethical actions cannot be suitably defined without regard to the current needs and conditions of society. An altered context has the potential to shift the meaningful consequences of actions. In such a system, this increases the capacity for adaptation to changing circumstances. Universal principles become broadly-defined, their systematic descriptions conditional. Arriving at the ethical truth of a situation is based on reasoned argumentation, in turn built upon the learned experience of past situations. In an important sense, these traits are prerequisites for the foundation of a sustainable system: different contexts, the same overarching structure.

Both of the above aspects are centrally concerned with the idea of harmonizing relationships, among human agents and between humanity and the rest of matter. In contemporary Chinese society, a defining tension is the struggle to harmonize two sets of associations: the historical, culturally-affirming ones defined by tradition, and the newer, potentially-disruptive associations accompanying modernism. Central to the distinction between these two sets of associations is the role of the autonomous individual within a community of fellow individuals. Namely what are the responsibilities of individuals in supporting the larger community and what is the collective responsibility in influencing individual action?

The Chinese philosophic tradition offers some divergent views. Legalism holds that the institution of the state is the highest responsibility to uphold. Ideal humans are simply cogs in the state apparatus. They have no right to exercise free will over actions. In the legalist system, rewards and punishments are directed toward the highest goal: unquestioning obedience to the omnipotent and omniscient state. In the Taoist tradition, humanity is viewed as an exception to the natural order. Humans are feeble degenerates caught up in their own senseless affairs and thus the larger construct built upon these affairs—human community—demands no ethical responsibility among individuals to support it. Society has no right demanding sacrifices from individuals. Instead, individuals are encouraged to seek the Tao, the embodiment of the universal integrity, and to reject human civilization.

In contrast to Legalism, Confucianism places value on the cultivation of perfection in every human and the capacity within each individual to use reasoned argumentation in developing an ethical system. The problem of the personality occupies the central position in the meditations of Confucius. Humans may be imperfect, but they have an infinite capacity for perfecting themselves. The means to do so lie in the continual assimilation of ancient traditions, within which are embodied moral and cultural values. The chief aim is perfection of personality. Confucian teaching places great emphasis on manners and restraints of propriety informing a person's disposition and guiding their conduct regarding their relationship to others and to their own cultural standing (Smith 1973, 73). The ethical correctness of these manners and restraints evolve from the mediating role of customs in an individual's life: individuals arrive at truth through reflecting upon historical antecedence, informed argumentation with others, and the exigent context. The ideal character is *chun-tzu*, which inculcates the principles of humanity and justice in life, with honors, proscribed rituals and ethical relationships, contingent upon the changing state of affairs.

In contrast to Taoism, the Confucian tradition holds that the central goal in human life is to create a cultural order in the world. This is a transformation of the environment accruing from an ideal that human efforts should focus on harmonizing relationships between persons, cultivating individual moral capacity, and emphasizing duty (Lao 1989). Confucian tradition firmly roots ethical responsibility for agents in duty toward others and the developing sense of rightness (*yi*) which accrues from this attempt. Furthermore, this ideal occurs in a universe of perpetual change and unremitting activity; there is no complacency for agents, the responsibility to cultivate moral capacities continues throughout time.

As disparate as these three traditions are theoretically, through the course of history each has to some degree influenced the others. Confucianism has been the predominant orthodox tradition, particularly through its influential role in shaping Chinese education. It is the Confucian tradition that provides the prime resources to develop the framework for sustainability. In some sense, the current discourse over sustainability is a contemporary version of acknowledging a state of conceptual polarity. The struggle for a harmonious relationship between Tian, Di and Ren lies at the heart of the central dilemma in contemporary Chinese society: the conflict between modernism, represented as a discourse on wealth and power and the means to achieve them, and tradition, represented in the rhetoric as Confucian values. The conflict concerns whether the means to achieve modern material wealth is antithetical to the social and political order (Weiming et al 1992, 34). This view holds that the current changes in Chinese society due to modernism are incompatible with maintaining culturally-sustaining traditions and, in another vein, the continuation of the existing political structure.

Proponents of sustainable development contend that community-supported social structures are the roots of any continuing ethical system which has the ability to maintain a sustainable society. In the context of Chinese sustainability, are there traits of the Confucian cultural ethos that favor sustainability? From one perspective, a forceful one which values the autonomy of individual choice-making over collective judgments, this issue is paramount, as critics against sustainable development charge that the concept biases against individual preference autonomy. In a sustainable system, individual preferences would need to be subordinated to support the maintenance of the system. As such, the legitimacy of the authority prepares the basis for the legitimacy of sustainability, and vice-versa.

Indeed for sustainable development to be effective, proponents insist that it must not remain a model imposed by illegitimate authority. Without broad-based and in-grained social acceptance, the effects of sustainable development remain limited. Even authority broadly supported by the public is not the primary foundation in which to establish the concept. Sustainable development cannot rely on official government pronouncements and expect positive results. The foundation is more immediate than government. People must first internalize certain values which lead toward accepting lifestyle trade-offs. In doing so they become participants in the development process. In effect, sustainability depends upon what Leonard Doob, (1991) calls 'sustainable people' who possess the conviction that they are able to control many aspects of their existence including themselves and their physical environment. As with the *chun-tzu*, sustainable people must hold the conviction that they have the capacity to improve both the self and the world outside it.

While people need to accept the notion, sustainable development, in turn, takes responsibility for the maintenance and well-being of the societal structure. The goal is to engender a mutually supportive relationship between individuals, their community, and policies. Sustainable development depends upon strong individual identification with communities, the underlying assertion being that humans are social creatures who live in specific communities. Further, though we have individual identities and responsibilities, our membership in societies influences these identities. Persons make individual decisions, but they depend upon the community to support their decisions (Daly and Cobb 1989, 161). Thus 'sustainable people' create, and are products of, the social structure. The construct of community implies other-regarding behavior. It implies a mutual sense of obligation and responsibility toward the members, despite autonomous desires that might conflict. Indeed, community requires that individuals refrain from fulfilling some autonomous desire in order to preserve the larger construct. In this case, mutual interest is self-interest.

Within the Confucian tradition, the basis for self-interest rests in the context of responsibility and duty toward other individuals and institutions. It is emphatically other-regarding. Individual actions, to be judged ethical, must conform to customary duties underlying the relationship between agents. Stress lies on individual duty not individual rights. Confucian scholar Tu Weiming cites three hierarchical moral bonds (*sangsang*) father-son, husband-wife, and ruler-ruled (Weiming et al 1992, 7). These bonds develop into a system of moral authority within two primary institutions, the family and the state. Individual responsibility towards larger institutions is not necessarily innate, but requires cultivation.

Confucian ethics necessitates transforming the world in the name of self-improvement. While a modernist tenet upholds the value of individual autonomy, a moral universe in traditional Confucian thought is one in which humans have developed their innate capacity according to the prescribed relationships among one another and between humanity and the rest of matter. According to Mengzi, *xing*, the moral capacity for self-transformation, is a universal trait within all humans. However, if *xing* is innate, this does not mean that it expresses itself effortlessly. On the contrary, this set of ethics holds individuals and society jointly responsible for cultivating moral capacities, persons must cultivate their own and society must likewise provide the social and political support structures enabling its pursuit.

Conclusion

In terms of the tension between socialist-Maoist principles and the recent economic reforms, sustainability provides a convenient form of ideological continuity amid the transformations within Chinese society. Internal struggles within the PRC government over the change in direction in the political economy are far from complete. Indeed not long after the new reforms of the Open Door policy criticism did arise.

One method by which the conservative factions in Beijing control the areas most accommodating to foreign investment is through political appointments and dismissals. In the later half of the 1980s, political dismissals of high-ranking officials in open investment areas (Guangdong, Fujian, Hainan) on corruption charges signaled a reassertion of centralized authority over these regions. By the 1990s, although government officials have not taken steps to change the reforms, the PRC leadership was again reemphasizing socialist goals and links to Marxism. Although speeches reemphasizing socialist goals may in practice be largely lip-service, this phenomenon reveals the ideological and political tension in contemporary China. For all its recent economic affluence and nascent modernization, China's population is still 60 to 70 percent rural and the communist government is still the legitimate authority (Linge and Forbes 1990, 24).

Sustainable development offers a sense of accounting for the welfare of the entire population as social criteria play a dominant role in assessing development decisions. In his analysis of East Asian authoritarian states, Chalmers Johnson argues that state legitimization rests ultimately not on ideological pretensions but on the results achieved in pursuing modernization. But no matter how true this might be, the importance of ideology in interpreting the means toward modernization should not be downplayed. Even Premier Deng recognized the need to publicly define the socialist mandate while the consequences of the reforms were being felt.

Chinese sustainable development reflects what John Page (1994) refers to as the model of recent East Asian development success. According to this theory, East Asian economic miracles such as Korea and Taiwan achieved their development gains by successfully allocating capital to high-yield investments and technological progress. The governments of East Asian development success stories were all committed to export push strategies and realized that firms that export have better access to the latest technology. Indeed, the PRC government stresses that the foremost task of the coastal region is to develop an export-oriented economy. Additionally, China is similar to other East Asian countries in its emphasis on diffusing urbanism from large, megacities by strengthening the economies of secondary and intermediate urban centers with growth potential (Rondinelli 1994).

According to Page's model, the governments then avoided serious social and political unrest through the Principle of Shared Growth. In this principle, government programs visibly demonstrate through prominent public spending in education and other social services that all are to benefit from the economic growth. This suggests one method of accommodating the tensions pursuant to rapid development; Chinese authorities can pursue economic growth while still publicizing egalitarian socialist principles.

China's official sustainable development policy reveals in part the tensions pursuant to the recent economic reforms of the Open Door policy. As policy, it may not conform rigidly to the ways the concept has evolved in Western development literature. The attraction of the term is that it can incorporate both the current patterns of economic growth and the previous socialist rhetorical tradition, albeit with a tenuous balance demanding non-complacency from the Chinese authorities.

At a deeper level, the rhetoric of Chinese sustainability affirms that there is indeed a distinct cultural ethos to draw upon. Humanity, in the traditional Chinese scheme, was not conceived as transcendent over the rest of matter. Furthermore, the ethical system directing human moral behavior was tied to the practical exigencies of a constantly changing and transforming world. Following from these precepts, there exists a body of belief tied to placing individual desires within a wider framework of ritual customs and other-regarding behavior. From this body of belief, another construction of sustainable development is possible, one more compatible with the construction of the term outside its official context of *China Agenda 21*. This new construction of sustainable development, informed by a set of traditions that has maintained certain cultural processes over several millennia, offers a means to constructively deal with the tensions accompanying China's experience with market socialism, the unstable middle ground between tradition and modernity.

Acknowledgement

The authors wish to thank those involved in our discussions of sustainability, Achva Stein, Clarence Eng, Paul Driscoll, Steven Flusty, and Malinda Seneviratre, and the city officials of Changzhou, particularly vice-mayor Chen Sanlin.

References

Allison, Robert E. 1989, "An Overview of the Chinese Mind", pp. 1-25 in Allison, Robert E., (ed.), *Understanding the Chinese Mind: The Philosophical Roots*, Oxford: Oxford University Press.
Bartlett, Albert A. 1994, "Reflections on Sustainability, Population, Growth, and the Environment", *Population and Environment*, 16:1:5-35.
Beckerman, Wilfred, 1993, "Sustainable Development: Is it a useful concept?", *Environmental Values*, 3:3:191-209.
Chen, Feng, 1995, *Ideology and Reform: Economic Transition and Political Legitimacy in Post-Mao China*, Albany, NY: State University of New York Press.
China Agenda 21, 1992, Beijing: State Council of Science and Technology.
Daly, Herman and John Cobb, 1989, *For the Common Good*, Boston: Beacon Press.
Daly, Herman and Kenneth N. Townsend, 1993, *Valuing the Earth: Economics, Ecology, Ethics*, Cambridge, MA: The MIT Press.
Doob, Leonard, 1991, "Sustainable People", *Journal of Social Psychology*, 131:5:602.
Hall, David L. and Roger T. Ames, 1987, *Thinking Through Confucius*, Albany, NY: State University of New York Press.

Heikkila, Eric J. and Mark Griffin, 1995, "Confucian Planning or Planning Confusion?", *Journal of Planning Education and Research*, 14:269-279.

Jonas, Hans, 1984, *The Imperative of Responsibility*, Chicago: University of Chicago Press.

Kidd, Charles, 1992, "The Evolution of Sustainability", *Journal of Agricultural and Environmental Ethics*, 5:1:1-26.

Lao, Sze-Kwang, 1989, "Understanding Chinese Philosophy: An Inquiry and a Proposal", pp. 265-293 in Alison, Robert E., (ed.), *Understanding the Chinese Mind*, Oxford: Oxford University Press.

Linge, G.J.P. and D.K. Forbes, 1990, "The Space Economy of China", pp. 10-34 in Linge and Forbes, (eds), *China's Spatial Economy: Recent Development and Reforms*, New York: Oxford University Press.

Madsen, Richard, 1992, in Weiming, Tu, Milan Hejtmanek, and Alan Wachman, (eds), *The Confucian World Observed: a Contemporary Discussion of Confucian Humanism in East Asia*, Honolulu: The East-West Center.

Naughton, Barry, 1995, *Growing out of the Plan: Chinese Economic Reform, 1878-1993*, Cambridge: Cambridge University Press.

Nolan, Peter, 1995, *China's Rise, Russia's Fall: Politics, Economics and Planning in the Transition from Stalinism*, New York: St. Martin's Press.

Page, John M. 1994, "The East Asian Miracle: An Introduction", *World Development*. 22:4:615-625.

Pomeranz, Kenneth, 1993, *The Making of a Hinterland: State, Society, and Economy in Inland North China, 1853-1937*, Berkeley: University of California Press.

Rondinelli, Dennis A. 1994, "Asian Urban Development in the 1990s: From Growth Control to Urban Diffusion", *World Development*, 19:7:791-803.

Rubin, Vitaly A. 1976, *Individual and State in Ancient China: Essays on Four Chinese Philosophers*, New York: Columbia University Press.

Smil, Vaclav, 1984, *The Bad Earth: Environmental Degradation in China*, London: Zed Press.

Smil, Vaclav, 1993, *China's Environmental Crisis: An Inquiry into the Limits of National Development*, New York: M. F. Sharpe.

Smith, Howard D. 1973, *Confucius*, New York: Charles Scribner's Sons.

Solow, Robert, 1993, "An Almost Practical Step Toward Sustainability", *Resource Policy*, 19:3:162-172.

Ulrich, Werner, 1993, "Some Difficulties of Ecological Thinking, Considered from a Critical Systems Perspective: A Plea for Critical Holism", *Systems Practice*, 6:6:583-611.

Wang, Rusong and Zhiyun Ouyang, n.d. "Ecological Engineering for Sustainable Development: A Review of its Theory and Application in China", in *Eco-Technology for Sustainable Development*, Beijing: Research Center for Eco-Environmental Sciences, Chinese Academy of Science.

Weiming, Tu, Milan Hejtmanke, and Alan Wachman, (eds), 1992, *The Confucian World Observed: A Contemporary Discussion of Confucian Humanism in East Asia*, Honolulu: The East-West Center.

"When the Boomster Slams the Doomster, Bet on a New Wager", 1995, *The Wall Street Journal*, (June 5), pp. A1, A9.

Wildavsky, Aaron, 1994, "Green Reporting—Accounting and the Challenge for the Nineties", *Accounting Organizations and Society*, 19:4-5:461-481.

11 Tourist Facility Development and Coastal Zone Management in Costa Rica

BRIAN COFFEY, BRONWYN IRWIN, THERESA L. URBAN

Since the emergence of mass tourism in the 1950s and 1960s several developing countries have embraced tourism development as an engine for economic growth. In promoting the industry, governments have often provided incentives to tourist-related businesses to encourage investment. Incentives of this sort can take many forms and may include such measures as guaranteed or subsidized loans, tax breaks, and the provision of needed infra-structure.

Questions have been raised about the economic value of such approaches (Bowden 1995, 33). Further, studies indicate that the impact of tourism on national economies has been variable. In some cases these efforts have met with considerable success, while in others governmental policies and programs have failed to generate the expected jobs and foreign exchange (Mathieson and Wall 1982, 50; Lea 1988, 37-50).

While the issue of economic growth is subject to debate, there seems to be near universal agreement about the social and cultural impacts of tourism. Scholars have generally argued that the results have been negative, linking tourism to increases in crime and prostitution, cultural imperialism, economic disparity, and social discontent (Witt, Brooke, and Buckley 1991, 159). Such problems are magnified by rapid growth of the industry and the tendency for tourism to be concentrated in select areas resulting in the disruption of day-to-day life (Pearce 1987, 205).

Despite the socio-cultural problems associated with the industry and questions about the degree to which tourism promotes economic development, many governments continue to view tourism as both desirable and necessary and continue to support it through various incentive programs. Unfortunately, such programs are often implemented without adequate planning. Careful planning can reduce the negative social impacts of tourism development while at the same time enhance the economic benefits. However, despite calls for planning of the industry and the readily available literature on the process (Inskeep

1991), in many instances planning procedures are implemented after the fact in an effort to rectify past mistakes (De Kadt 1979), or they are ignored entirely. The present essay examines the methods used to promote tourism and the land use planning which is associated with tourist facility development in Costa Rica, one of several lower income countries pursuing tourism as an economic development tool.

Costa Rica is a country with a fragile natural environment and a fairly young tourist industry. Further, it is a country which is currently attracting large numbers of tourists. Thus, it is a logical candidate for study of tourist industry promotion, environmental protection, and planning processes within the context of a developing nation. In this particular case incentives for tourism development and their economic impacts are examined. In addition, planning as related to Costa Rica's coastal zone is considered and issues related to comprehensive tourism planning are discussed.

Costa Rican Tourism

Since the early 1980s the number of foreign visitors to Costa Rica has increased sharply, rising from approximately 333,000 people in 1981 to nearly 800,000 in 1995 (Table 11-1). Noteworthy among these statistics is that while visitors from Central America, the Caribbean, Mexico, and South America accounted for about 70 percent of all visitors in 1981, in 1995 they made up less than 40 percent of all arrivals. This is the case despite a 30 percent increase in their numbers.

The reason for this is that the industrialized world "discovered" Costa Rica in the late 1980s and early 1990s and the country saw a dramatic rise in visits from North Americans and Europeans. One factor which may have had an impact on increases in tourism is the fact that in 1987 Oscar Arias, then president of Costa Rica, was awarded the Nobel Peace Prize for his efforts to end conflict in Central America. This focused international attention not only on him but on his country as well. During this period Canadian visitors increased approximately tenfold, climbing from about 4,500 in 1981 to nearly 45,000 by 1995. Visitors from the United States more than quadrupled, with their numbers going from approximately 60,000 in 1981 to almost 300,000 during this period. A similar increase is seen with Europeans whose numbers rose from a 1981 total of 28,415 to 132,057 by 1995. In 1981 tourists from Anglo-America and Europe accounted for just over one-fourth of Costa Rica's foreign arrivals. By 1995 they accounted for nearly 60 percent of the total.

Table 11-1 Origins of Costa Rica's Foreign Tourists, 1981-1995

Country/Region	1981	1985	1990	1995
North America	73,062	89,825	191,284	349,307
Canada	4,544	6,291	30,892	41,898
United States of America	60,136	77,346	150,224	287,434
Mexico	8,382	6,188	10,168	19,975
Central America	191,654	112,623	139,913	218,023
Caribbean	6,707	4,294	4,192	7,125
South America	26,190	20,915	32,575	58,600
Europe	28,415	28,179	57,177	132,057
Other	7,074	5,716	9,896	29,498
Total	333,102	261,552	435,037	784,610

Source: ICT.

The marked increase in European and North American visitors has important economic implications for Costa Rica given that these groups will spend, on the average, considerably more money than will visitors from other Central American countries, the Caribbean nations, or Mexico. The impact of their numbers was clearly seen in 1994 when income from tourism surpassed that of the banana trade to become the country's leading earner of foreign exchange by generating more than US$600,000,000 (Instituto Costarricense de Turismo 1996, 62). In addition, in recent years arrivals from individual Asian countries have been tracked by the Costa Rican Tourism Institute (ICT 1996, 14). Japan, accounting for approximately 5,000 visitors in 1995, may well become an important source of tourist income in the future.

Much of Costa Rica's attraction lies in its natural environment. With more than 20 percent of the country's land area reserved for parks, wildlife refuges, and forest preserves, Costa Rica has been a global leader in environmental protection (Figure 11-1). The biological and physical diversity found in the nation has received a good deal of attention by the international media (Brombert 1996). Such press has raised awareness of the country's attributes in the minds of First World vacationers and the nation's natural environment is clearly a factor in the decision to make it a vacation destination.

The rapid rise in tourism is interesting given that in the early 1980s Costa Rica had little to offer tourists in the way of accommodation, particularly outside of the capital San José. Obviously such facilities are necessary if tourist numbers are to rise. Thus, one additional factor which served to increase tourism was the government's efforts to promote the industry's development within the country.

Incentives for Tourist Facility Development

As the number of visitors from higher income countries gradually began to increase in the early 1980s, the Costa Rican government realized that if tourism was to further expand, tourist facilities had to be improved. In 1984 the country's president declared that tourism must be considered in the development of the nation's economy (*La Nación* 1984, 4A). However, high tariffs on foreign goods hindered development of the industry. With increased costs due to these taxes businesses such as hotels, car rental agencies, and transport companies found it difficult to compete with other destinations in the Caribbean Basin.

In response to the president's call and the need to reduce development costs, the National Assembly passed a 1985 law providing incentives for the development of tourist-related enterprises (*Ley de Incentivos para el Desarrollo Turístico*). In an effort to capture a greater share of the international tourism market, the incentives are directed at six types of businesses: hotels, airline companies (both domestic and international), water transportation companies, car rental agencies, travel agencies which cater to international tourists, and restaurants.

Incentives differ from sector to sector. Hotels, for example, are exempt from all taxes and surcharges related to the purchase of imported or locally manufactured goods or materials necessary for the construction, improvement, expansion, or daily functioning of the business. Those hotels which are outside of metropolitan areas are exempt from property taxes for a period up to six years. Hotels are also permitted to purchase foreign exchange from tourists on behalf of the National Bank of Costa Rica and they receive accelerated depreciation on certain goods.

In other cases, airline companies receive accelerated depreciation of aircraft and can purchase fuel at the average price found on the international market. Water transportation companies are exempt from taxes and surcharges on any materials necessary for the construction or expansion of wharfs for tourist use. They also are exempt from taxes and surcharges (except tariffs) on the importation or local purchase of boats and they receive accelerated depreciation of their equipment. Car rental agencies are exempt from 50 percent of all taxes on cars purchased for tourists' use and travel agencies are

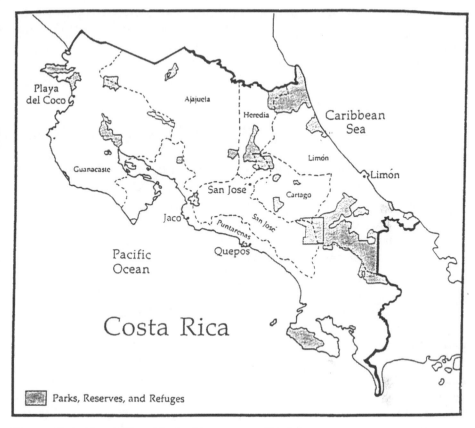

Figure 11-1 Costa Rica: Parks, Reserves and Refuges

exempt from all taxes and surcharges, except from tariffs on the importing of cars (ICT n.d., 25-30).

The incentives are granted through a tourism contract signed between the business and Costa Rican Tourism Institute (ICT) and approved by a regulatory commission. The commission is composed of one representative from the ICT; one representative from the Ministry of Finance; one representative from the Ministry of Industry, Energy, and Mines; and two representatives from private industry. When a proposal is submitted for approval the commission members consider such factors as the applying company's predicted contribution to the balance of payments, the extent to which national resources will be used, the direct and indirect creation of employment, the effect on regional development, the extent to which the business will modernize or diversify national tourist offerings, and the benefits accruing to other sectors of the economy.

To obtain a tourism contract the business must present completed construction plans for approval by the regulatory commission. In addition, the business must prepare a feasibility study which includes:
   • a detailed description of the project;
   • a market analysis justifying the creation of a new business in its economic sector and geographic region;
   • a marketing plan;
   • an organizational chart showing the number of local and foreign employees, their positions and functions, educational levels, and salaries;
   • a complete schedule of construction dates and costs along with costs of land and infrastructure;
   • a buying plan showing estimated costs and dates of purchases;
   • projected earnings, expenditures, and loan payments for a five year period;
   • justification of the project on economic and social grounds;
   • identification of those incentives being requested.

The response to the investment incentives law has been favorable. Since its implementation more than 700 businesses, including nearly 400 hotels, have enrolled in the program (Table 11-2) and levels of investment by participating companies have been high. For example, approximately $600,000,000 was invested by the hotel industry from 1986 to 1995 (Table 11-3). Of this, just over $100,000,000 was spent on tax-exempt imported goods. Other activities had much lower levels of investment with the total in all other sectors coming to about $100,000,000. Overall, approximately 25 percent of the nearly $700,000,000 invested by participating companies from 1986 to 1995 was spent on imports suggesting that a significant amount of investment affected the domestic economy (Table 11-3). During this same period sales and consumer tax exemptions amounted to more than 400,000,000 colones (approximately $2.2 million).

While businesses began to take advantage of incentives soon after the law was passed, much of the activity has occurred in recent years and accelerating investment is apparent as tourism numbers increase. For example, between 1986 and 1995 $594,835,690 was invested in hotels associated with the incentives program. Of this amount, nearly 80 percent was spent during the five year period of 1991 and 1995 (Table 11-4). The data indicate that 1991 ushered in an era of sharply increased investment. In that year nearly $90,000,000 in hotel investments were approved under the incentive program, or significantly more than the $38,000,000 spent in 1990 (Table 11-4). Since 1991 investment has remained significantly higher than was the case during the program's first five years. This surge in construction coincides with the

Table 11-2 Businesses Participating in the Tourism Development Incentive Program 1985-1995

	1985	86	87	88	89	90	91	92	93	94	95	Total
Hotels	4	57	20	21	26	49	41	54	44	46	29	388
Air transport.	0	5	0	1	2	1	1	4	2	1	0	17
Automobile	12	3	0	1	10	7	3	15	8	9	4	72
Travel agencies	0	13	4	8	8	16	19	25	31	22	15	158
Water transport companies	0	2	6	4	7	11	12	14	8	9	9	82
Restaurants	2	11	0	0	0	0	17	4	0	0	0	34
Total	18	91	30	32	53	81	93	116	93	87	57	751

Source: ICT.

Table 11-3 Investments and Imports, Tourism Development Incentive Program, 1986-1995 ($US)

	Imports	*Investment*	*Imports as % of Investment*
Hotels	$107,816,239	$594,835,690	18
Travel Agencies	8,502,307	23,519,162	36
Car Rental Agencies	31,222,874	37,315,094	84
Water Transport	19,569,646	26,847,631	73
Restaurants	3,587,800	4,221,907	85
Airlines	235,601	6,760,371	3
Total	$170,934,471	$693,499,857	25

Source: ICT.

increase in the numbers of tourists from North America after 1987 (Table 11-1).

One issue associated with this development which merits consideration is the matter of foreign control. While data are not readily available on the percentage of establishments which are foreign owned, ICT officials place the

Table 11-4 Hotel Investment Under the Incentives Program, 1986-
         1995

Year	Investment	% of Total	Cumulative %
1986	$43,635,089.29	7.0	7.0
1987	3,692,277.07	0.5	7.5
1988	15,340,290.00	2.5	10.0
1989	27,431,300.61	4.5	14.5
1990	38,442,270.00	6.5	21.0
1991	89,958,918.92	15.0	36.0
1992	110,674,318.40	18.5	54.5
1993	129,662,998.00	22.5	76.5
1994	64,580,501.93	11.0	87.5
1995	71,417,725.86	12.0	99.5*

Total $594,835,690.08

*figures do not total 100% due to rounding

Source: ICT.

figure at more than 95 percent. Carlos Muñoz, president of the Central Bank, warns that this is a problem (Muñoz 1989, 17). He argues that Costa Rican nationals must regain control of the country's tourism industry in order to maintain sustainable development with minimal adverse effects. He also fears that foreign control will prompt disdain for the tourist industry (and tourists) on the part of Costa Ricans and he calls for a policy to promote greater domestic ownership within the industry. Foreign ownership is an issue in many parts of the developing world and Muñoz's call merits serious consideration by the government.

Finally, it should be noted that not all hotels participate in the incentives program, often because of the quality of their accommodations. In fact, of the more than 1,500 hotels in the country fewer than 400 can take advantage of the incentives offered by the government. However, non-participating establishments tend to be smaller, averaging 10 rooms each, while participating hotels average about 35 rooms each (ICT 1996, 68). Thus, of the 25,000 rooms available in the country, 47 percent are in hotels which are associated with the incentives program.

In recent years much of the increase in lodging has occurred in the Pacific coast provinces of Guanacaste and Puntarenas (Table 11-5, Figure 11-1). Of all rooms added from 1990 to 1995, 49 percent were built in these two areas. In 1989 Puntarenas and Guanacaste accounted for 39 percent of all the rooms in the country. By 1995 this figure had increased to 44 percent, representing increased demand for coastal properties by the tourist industry and, hence, increased development pressure in this fragile environment.

The law most pertinent to the regulation of coastal zone development is Law 6043 (*Ley sobre la Zona Maritimo Terrestre*). In effect since 1977, the law regulates development of the country's coastal areas and provides the framework for Costa Rica's coastal zone management policy. While the law is structured to provide for a good measure of control over the coastal environment, problems associated with enforcement tend to dilute its effectiveness.

Table 11-5 Rooms in Establishments Participating in the Incentives Program, 1989 and 1995, by Province

Province	No. of Rooms 1989	% of Total	No. of Rooms 1995	% of Total
San Jose	2330	43	4176	35
Alajuela	129	2	789	7
Cartago	29	1	78	1
Heredia	486	9	957	8
Guanacaste	992	18	1868	16
Puntarenas	1117	20	3350	28
Limón	373	7	644	5
Total	5456		11862	

Coastal Zone Management in Costa Rica

Costa Rica's coastal zone management law mandates that the state has responsibility over the country's coastal zone, particularly those areas most suited for tourism. The law calls for protection of a 200 meter zone extending inland from the mean high tide. Within this zone the first 50 meters is designated as a public area wherein access is guaranteed and construction is prohibited except for special cases in which direct access to the sea is required (e.g. port facilities, aquaculture works, and marinas). The next 150 meters is a restricted zone in which construction can take place but only with approval of the local government and the Costa Rican Tourism Institute or the National Institute of Housing and Urban Planning. The law exempts urban coastal areas. However, the National Institute of Housing and Urban Planning works with coastal towns to develop similar protection plans (Sorensen 1990, 41). In addition, the flora and fauna of the maritime zone are protected under the law so that such activities as tree cutting, the extraction of products, and blocking access with fences or other barriers are prohibited.

Of particular importance in the coastal zone are those areas which have value as tourist beaches. For such beaches the law requires that detailed plans be created before development can take place. The Costa Rican Tourism Institute has a program to prepare plans for each beach of tourist quality. However, if a developer wishes to build in a tourist area for which no plan exists, a plan can be prepared by a private firm which is then presented to ICT for approval. The plans are based on the carrying capacity of the particular beach in question and take into account a range of factors. These include such things as water availability, transportation routes, local population, and sewer capacity. An example of such a plan is shown in Figure 11-2 which depicts the ICT's plan for Playa Real and Playa Roble, two adjacent beaches in north-western Costa Rica. The plan is quite detailed with specific locations designated for residential uses, hotel use, and condominiums.

In addition, the coastal zone management law limits development density by establishing minimum lot sizes for various uses. For example, in tourist zones residential lots have a minimum of 300 square meters and commercial uses have minimum lot sizes of 150 square meters. Maximum sizes are also stipulated.

Finally, in both tourist zones and non-tourist zones certain uses have priority with respect to development. In zones declared to be of tourist quality priority uses include tourist activities (hotels, restaurants, and related uses), recreational/sports activities, residential uses, commercial and handicraft activities, agriculture, cattle raising, fishing, and industry related to tourists' interests (e.g. ceramics, handicraft workshops, and toy manufacturing).

The law also stipulates penalties for violators. These punishments can be severe. An individual can receive six months to four years in prison for damaging the natural environment of the maritime zone. For unapproved construction the penalty is one month to three years in prison.

Although the law appears to afford considerable protection to the coastal environment, there have been some problems in the area of enforcement. While planning and development approvals are handled at the national level, local

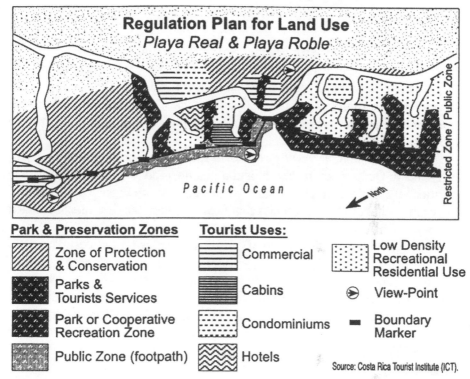

Figure 11-2 Regulation Plan for Land Use, Playa Real and Playa Roble

governments have responsibility for enforcing the law within their boundaries and it is often at the municipal level that problems arise. For example, local inspectors are sometimes lax in informing developers about restrictions and developers do not always conform to regulations when building. Additional problems relate to inadequate staff to inspect areas, overburdened prosecutors who assign such cases a low priority, lenient or ill-informed judges, and poor enforcement on the part of local government (Sorensen 1990, 43-47).

The impact of these problems becomes evident when visiting beaches. Observations at three beaches in 1993 revealed the need for more stringent application of the law. The sites visited, Playa Hermosa, Playa Flamingo, and Playa Espadilla, are located on the Pacific coast and all are experiencing development pressure.

The southernmost beach, Playa Espadilla, is a relatively long beach in the province of Puntarenas and the canton of Aguirre. The area is a popular destination due to the presence of Manuel Antonio National Park, a site which attracts thousands of visitors annually. Tourism development for this beach is found along a seven kilometer stretch from the city of Quepos to Manuel Antonio Park.

Inspectors who work for Manuel Antonio National Park estimate that 70 percent of the structures violate Law 6043 in some way. For example, an illegal campsite has been established in the restricted zone and numerous structures including cabins, restaurants, vendor stands, and tourist shops can be found in the public zone. Often, projects without permits are raised prior to ICT approval.

Blake and Becher's *New Key to Costa Rica* (1993) lists 53 lodging establishments in Quepos, on the beach, and along the road to Manuel Antonio. In addition, there are several bars and restaurants which cater to tourists and there are a number of tour agencies that offer kayaking, biking, and horseback riding in the area.

The fact that this area is one of the most popular tourist destinations in Costa Rica has taken a toll on the environment. Camping is no longer permitted on the beach because of environmental degradation and "the tourist explosion has outstripped the infra-structure of the area so that waste disposal, both liquid and solid, has become a problem" (Blake and Becher 1993, 234).

Investment in the region has grown steadily in recent years. Of 470 projects approved by the ICT between 1989 and 1993, 37 were for the Quepos/Manuel Antonio area. In the field survey more than 30 tourist-related establishments were counted in Quepos and more than 40 lined the road from Quepos to Manuel Antonio. Further, several construction projects were underway suggesting that development pressures remain high.

Planning for Playa Espadilla is somewhat difficult. Large land area, natural and environmental diversity, and proximity to Manuel Antonio National Park serve to complicate matters. Playa Espadilla's plan calls for 43 percent of the total area to be devoted to tourist-related uses. Another 56 percent is designated as public in the form of protected areas, the public beach zone, and green space. The remaining one percent is allocated for institutional uses. The large amount of reserved land is an indication of the fragility of the environment. However, limited development opportunities may also explain the large number of illegal construction sites.

Although approved development projects ideally comply with regulatory plans, in reality the plans are often compromised for the sake of economic gain. Local officials are under considerable pressure to develop the area making strict regulation difficult. Opportunities for economic growth and the increasing availability of jobs sometimes result in lenient interpretation of standards.

While such development is clearly a boon for the local economy, too much development will ultimately tax the infra-structure, pollute the environment, and diminish the area's aesthetic value. Development at Manuel Antonio and Playa Espadilla should be more carefully monitored to ensure sustainable development.

The second area observed, Playa Hermosa, is a small beach in Carrillo Canton in Guanacaste. The atmosphere of Playa Hermosa is tranquil with many Costa Rican visitors as well as international tourists. Although several tourist guides describe Playa Hermosa as clean, garbage polluted the beach during the field inspection. Only one garbage can could be found on the entire length of the beach.

The survey also revealed six lodging places as well as six restaurants. A seventh structure which appeared to be a hotel was under construction. Two of the restaurants viewed were modest establishments with limited menus. The remaining four, two of which were associated with hotels, were higher priced and oriented toward the growing tourist trade.

None of the tourism projects listed by the ICT for the years 1989-1993 were for developments at Playa Hermosa. This was the case despite the fact that at least three of the structures mentioned above were built during this period. One possibility is that the establishments did not apply for incentives because they are in violation of the coastal zone management law. A second is that they applied for incentives but were denied a contract because they did not meet certain standards. In either case, the ICT has no data on the improvements.

The regulatory plan for Playa Hermosa indicates that 45 percent of the controlled land area is reserved for hotel and commercial tourist services, while 24 percent consists of the public zone where improvements are forbidden. Private development which includes residences and commercial activities accounts for another 12 percent.

Several structures in Playa Hermosa have been built in the public zone in violation of the coastal zone law. Some are private houses which were likely built before the law went into effect. Others, however, are tourist related, including two hotels which violate the site rule. In other cases tourist related structures are in the restricted zone, but have fences and lounge areas which extend into the public zone, again a violation of the code.

Despite these problems, the public does have access to the beach for its entire extent. Also, none of the buildings are taller than ten meters (the maximum allowed) and none detract significantly from the natural surroundings. Garbage and clutter are associated with the two modest restaurants on the beach and the presence of chickens and pigs at these sites detracts from the overall beauty of the area.

Playa Flamingo (also called Playa Blanca) is also located in Guanacaste. Development here is fairly recent with much of it taking place in the late 1980s and early 1990s. Developments are largely foreign-owned, fairly exclusive, and cater primarily to a North American clientele.

Playa Flamingo has been a regulatory nightmare for the ICT. A good deal of construction occurred prior to the drafting of a regulatory plan. Thus, most of the facilities at Flamingo, including one of the largest hotels in the country, are not in conformity with the law.

The field investigation revealed plans for several development projects including hotels, casinos, condominiums, and private residences. Construction equipment occupied a section of the public zone and signs in the public zone declared property "private" and prohibited trespassing. To further complicate matters, oil and gas leakage from a marina at Playa Flamingo has polluted the bay resulting in the destruction of marine life.

The development problems at Playa Flamingo can be linked to local officials who approved projects in the absence of a regulatory plan. As previously indicated, local governments sometimes grant permits for projects which violate the coastal zone law. The creation of employment opportunities and the multiplier effect of increased development are often strong incentives to disregard legal requirements.

Ensuring Proper Development

Both the incentives program and the coastal zone management law are well designed to achieve their respective goals. However, issues surround each program which merit consideration.

With regard to the incentive program, the primary issue relates to its remarkable success. Given the pattern of facility development and the growth in international tourist numbers since the program became law, it appears that the incentives offered stimulated the early development of the industry. The growth of tourism facilities and services between 1985 and 1990 can be directly attributed to the incentive program. The result was an initial increase in tourism which swelled during the 1990s as the media began to report on the area and as word-of-mouth referrals generated increased interest in the country. One result of this has been a surge in tourist facility development since 1990. This later investment has been such that one suspects that much of it would have taken place without the incentives program.

Costa Rica now needs to be concerned about over-development in some areas and the socio-cultural impacts of large numbers of relatively well-to-do visitors on small coastal communities. Further, such development is a threat to the environment which initially drew the tourists of the 1980s.

While complete elimination of the program may not be advisable, the government may wish to consider a number of options. One is to eliminate the program in certain regions which, because of their attributes, would probably continue to receive investment without incentives. The Manuel Antonio region and coastal Guanacaste are two areas which appear to fit this category.

A second possibility is to drop certain businesses from the program (e.g. hotels), while retaining benefits for other facilities such as car rental agencies or restaurants. Yet another option is to provide benefits only for enterprises wholly owned by Costa Ricans. This would begin to address the issue of foreign versus domestic ownership.

Any of these steps clearly requires study and analysis before being implemented. However, growth is progressing so rapidly that something should be done to immediately slow development. Thus, it is recommended that a moratorium be placed on the incentive program until the current level of development can be assessed. It is far easier to resume the program (or portions of it) at a later date than it is to undo damage which might be brought about by over-development.

With respect to the coastal zone management program several matters should be considered. First, the program itself is well formulated. However, greater resources for inspection and enforcement are needed. Further, both the judiciary and local governments need to be better educated about the importance of the law. Without proper enforcement, regular inspection, and a realization on the part of permit-granting local officials that poor development will ultimately hurt the local economy the law will have limited impact.

Finally, related to both the incentive program and the coastal zone law is the need for a national tourism plan which must, by necessity, be linked to economic development planning, community planning, and rural planning. In short, Costa Rica needs a national comprehensive plan of which a tourism plan is an integral part.

A national plan would benefit Costa Rica in that a comprehensive analysis of national goals and a clear understanding of the various avenues for economic development would allow planners to strike a balance between tourism and other economic sectors. Social, cultural, and economic concerns could be addressed in the plan and specific guidelines could be established to achieve the goals necessary to create a diverse economic base.

Conclusions

To encourage most efficiently the development of a sound tourism industry, governments must regulate development, especially in environmentally sensitive areas. In Costa Rica the coastal zone management law attempts to do this. However, problems with the delegation of authority, enforcement, and the lack of broader planning objectives cripple its potential. Further, the rapid growth of the tourist industry is a matter of concern and the government should reassess the incentives program.

Within these frameworks sustainable tourism development is an important consideration. Too much dependence on the industry and too much development will have negative economic, environmental, and social effects. In this case the industry needs to be better controlled, well planned, and viewed as only one part of a developing economy. Clearly a strong economic base is important to a country like Costa Rica and tourism can play an important role in establishing that base. However, other factors such as the preservation of a national and cultural identity and environmental conservation are equally important. Thus, Costa Rica's tourism development program, while primarily rooted in economic development, must also be based on the nation's social and cultural needs.

References

Blake, B. and A. Becher, 1993, *The New Key to Costa Rica*, Berkeley, CA: Ulysses Press.

Bowden, D. 1995, "Angkor: Planning for Sustainable Tourism", *Expedition*, 37:30-41.

Bombert, B. 1996, "Costa Rica: The Rough, The Smooth", *The New York Times*, Sect. 5, (Feb. 18):1, 14-16.

Coffey, Brian, 1993, "Investment Incentives as a Means of Encouraging Tourism Development: The Case of Costa Rica", *Bulletin of Latin American Research*, 12:83-90.

de Kadt, E. 1979, *Tourism: Passport to Development?* Oxford: Oxford University Press.

ICT. n.d. *Normas que regulan las empresas y actividades turísticas*, San Jose, C.R.: Instituto Costarricense de Turismo.

ICT. 1996, *Anuario Estadistico de Turismo 1995*, San Jose, C.R.: Instituto Costarricense de Turismo.

Inskeep, E. 1991, *Tourism Planning: An Integrated and Sustainable Development Approach*, New York: Van Nostrand Reinhold.

La Nacion. 1984, "Presidente Menge reclama por fortalecimiento del turismo", San Jose, C.R.: *La Nación* (November 29):7A.

Lea, J. 1988, *Tourism and Development in the Third World*, London: Routledge.

Mathieson, A. and G. Wall, 1982, *Tourism: Economic, Physical and Social Impacts*, Essex: Longman Scientific & Technical.

Muñoz, C. 1989, "Los Responsables del Exito en Turismo", San Jose, C.R.: *La República* (March 2):17.

Pearce, D. 1987, *Tourism Today: A Geographical Analysis*, Essex: Longman Scientific & Technical.

Sorensen, J. 1990, "An assessment of Costa Rica's Coastal Zone Management Program", *Coastal Management*, 18:37-63.

Witt, S., M. Brooke, and P. Buckely, 1991, *The Management of International Tourism*, London: Unwin Hyman.

12 Reflections on Cuban Socialism and Planning in the "Special Period"

JOSEPH L. SCARPACI

Like it or not, the Cold War provided stability. Whether it meant justifying defense expenditure build-ups among "hawks" in the U.S. Congress, or identifying voting blocks at the United Nations, the world seemed a more predictable place because the global scene had more order than it does today. Political and ideological camps were well identified. For the Third World, the socialist camp and the Council for Mutual Economic Assistance (CMEA) provided a steady line of foreign aid.

Few countries have experienced the external shocks of the demise of these sources more than Cuba. The dissolution of the USSR and the CMEA sent profound shock waves that severely undermined the Cuban economy (Pérez-López 1995a; 1995b). Daily fuel shortages prevail, as do food rationing, power outages, and even "cracks" in the once renowned system of education and health care (Feinsilver 1993). This economic dilemma is called the 'Special Period' (*El Período Especial en el Tiempo de Paz*).

Measuring the economic malaise in Cuba is difficult. A telephone directory was not even published in Havana between 1983-1985. The statistical yearbook (*Anuario estadístico de Cuba*) was last published in 1989. Economic data, therefore, result from interviews and proxy measures. Conventional interpretation is made difficult by this dearth of information. What we do know at the macro-level is that since the late 1980s, the economy has stagnated. In 1990, estimates in the decline of Cuba's GSP (Gross Social Product) ranged between 4 and 7 percent. As the crisis of the Special Period accelerated, so did the GSP decline; between 15-27 percent, though it narrowed to between 7 and 15 percent for 1993 (Mesa-Lago 1994, 8).

The economic stagnation derives from at least three factors. One is the de-emphasis of material incentives in the workplace and the abolition of informal produce markets. Although those policies were reversed in 1994-95, it has been attributed to a decade-long drop in economic productivity (del Aguila

1992). Second, as noted above, the collapse of the former USSR interrupted considerable price subsidies, foreign aid, and about 300-odd industrial/public-works projects that were in production. Third, the continuation of the U.S. blockade has increased the cost of consumer and industrial goods. Many products could no doubt be secured more cheaply if imported from the U.S. versus third countries (Bahamas, Canada, Jamaica, Mexico and Venezuela). The blockade also deprives the Cuban economy of a lucrative tourist trade.

This chapter traces the evolution of Cuba's recent launching into a global market and identifies the ways in which the island's economy has changed over the past ten years or so. Special attention is given to the political economy in which these changes occur. Central planning is assuming new roles during the Special Period. It now contends with different kinds of international commerce, real-estate markets on the island, food consumption, and joint ventures.

It appears that the Castro government is taking his island into economic *perestroika* while tightly keeping a lid on any form of Cuban *glasnost*. The results are mixed and underscore a widening contradiction between socialist goals and the daily experience of Cubans. Cynics might argue that stripping away the rhetoric, very little of a socialist model prevails. Indeed, new planning problems have arisen. Residential segregation increases, purchasing power widens quantitatively and qualitatively in ways not seen since the 1950s. A new entrepreneurial class—both legal and illegal—is sprouting up in the major towns of the island. Income taxes surfaced in 1995 for the first time during the Revolution (Lee 1996). The Revolution is, therefore, over for all practical and theoretical purposes. The chapter concludes with some observations about the Eastern European experience and the challenges that await the Cuban Revolution in the next few years.

Background

For decades CMEA had provided preferential trade agreements among signatory partners, of which Cuba was a member. CMEA was a rather small, isolated market whose "backbone" was the USSR. Member nations, especially Cuba and North Korea, received special treatment because of their "developing nation" status. That meant that the USSR would buy Cuban sugar at prices three-times the going rate, and Cuba would receive discounted oil from USSR petroleum reserves. These favorable terms of trade allowed Cuba to satisfy about 90 percent of its energy needs (Mesa Lago 1994, 1). Cuba had the highest dependency rate of trade among CMEA members (84 percent), and 70 percent of the island's trade was with the USSR alone.

Even before 1989, the Soviet Bloc trading arena had provided Cuba with 85 percent of its foreign trade, and it did so with favorable prices subsidized by the Bloc. The collapse of that market, coupled with a lack of capital, credit, and energy, sent shock waves throughout the Cuban economy and society. Moreover, it was aggravated by the persistent U.S. blockade. By 1991, imports to Cuba had been almost cut in half of the 1989 level (Figure 12-1), including a drastic reduction in petroleum which was the only source of energy in Cuba. From 13 million tons imported in 1989, the figure slid to just 6 million in 1990. Fuel imports plummeted from $2.6 billion in 1989 to $720 million in 1994 (Figure 12-2). In the last trimester of 1990 Fidel Castro declared that Cuba was in the first phase of the "Special Period". This label refers to the hardships placed by the US economic blockade as well as the withdrawal of Soviet subsidies.

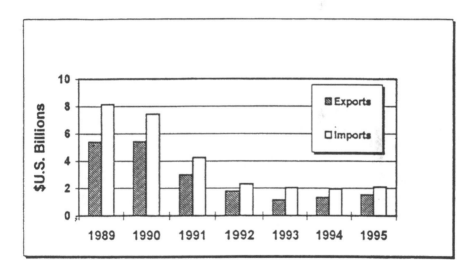

Figure 12-1 Cuba: Foreign Trade, 1989-95
Source: *Cuba: Handbook of Trade Statistics*, 1995. Washington, DC: CIA. p. iii.

The immediate effects of Soviet dis-investment were disastrous. In agriculture and construction, activities were paralyzed because of the lack of imported production inputs: petroleum, fertilizers, pesticides, fodder, machine repair parts, steel, building components, and semi-assembled products. Lacking these materials exposed the vulnerability of a model that was very dependent on external resources. Under the unquestionable scrutiny of the market

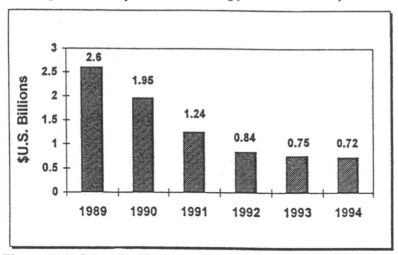

Figure 12-2 Cuba: Fuel Imports. 1989-1994
Source: *Cuba: Handbook of Trade Statistics,* 1995. Washington, DC: CIA. p. 11.

economy, these vulnerabilities exposed ecological, social, aesthetic, and cultural problems.

As a planned economy, ostensibly at least, the government reveals major investment decisions and priorities through its Five-year Plans. Neither the 1986-90 nor the 1991-1995 plans achieved their stated goals because of the Special Period. The 1986-1990 plan identified the following areas upon which the development base of Havana would concentrate: industrial employment, construction, transportation, tourism, direct services to the resident population, and science and technology (GDIC 1991).

Surviving in Cuba During the Special Period

Cuba will never be the same because of the Special Period. Life is hard. The Special Period has exposed weaknesses in a development model to which it has fiercely adhered, and which prevailed for three decades. Fuel shortages and power outages force industry to slow production, as does the lack of hard currency for the processing of imported raw materials. Another unfolding pattern is that in 1995 the construction and transportation industries suffered from fuel shortages. In 1986 in Havana, for example, buses accounted for 86 percent of total trips by motorized transport, and automobiles—never a significant mode of transportation—accounted for just 6 percent. Roughly 50 percent of all bus routes have been eliminated, consolidated or cut back, making lengthy waits at crowded bus stops a multi-hour endeavor. Workers in construction and transportation have been displaced elsewhere, remain under- or un-employed, or form part of the new self-employed labor pool.

A huge increase in bicycles responded to this transportation crisis. In 1990, *Habañeros* used their roughly 70,000 bicycles for recreation and sport. By 1993, there were 700,000 bicycles, mostly bought from China, and used for commuting. The figure climbed to 1 million in 1997. The Chinese models *Phoenix* and *Flying Pigeon* sell on installment plans. They range from $60 pesos for students to $120 pesos for workers (Alepuz 1993; Chavez 1992). In deflated real dollars in 1996 this ranged from about $2US to $5US. Increases in bicycle imports and use mean less air pollution in Havana, greater commuting time for workers, and a proliferation of private-sector bicycle repair and parking services (Scarpaci and Hall 1995).

The proportion of Havana's population working in agriculture, estimated at 15 percent in 1990, has also probably increased. Neighborhood gardens proliferate everywhere, but it is difficult to determine whether workers who tend to those gardens would be counted as agriculturists (Roca 1994). Intensification of truck farming has also increased in the southern portions of the Havana municipalities of Boyeros, Arroyo Naranjo and Cotorro, departing from a Third World pattern of decreasing peri-urban agriculture (Browder, Bohland and Scarpaci 1995). Urban "husbandry", once outlawed and subject to stiff fines and incarceration, now flourishes. Balconies throughout Havana hold chicken coops and make-shift sties where piglets are kept. For decades pork was unavailable in the City of Havana because of "pork fever" infection and because it was destined for tourist consumption in such typical dishes as roasted pork "Cuban style" (*lechón asado*).

Cuban economic development analysts and officials herald the rise in domestic livestock as a sign of local creativity. However, public health officials fret over the unsanitary conditions (Personal communication, Ministerio de Salud Pública, June 15, 1994). In the main, though, these examples underscore the likely decline in the proportion of workers in the productive sphere. Along with that shift has been a rise in black market (informal) activities and a new urban "subsistence" economy that is difficult to measure.

The tenacity of the *Habañero* in the face of the city's economic free-fall reveals itself at several levels. Their ability to overcome food shortages, power outages, public transportation problems and other obstacles is evident on every city block. Everyone is trying to survive: Communists, school children, retirees, college professors, manual laborers, and bus drivers. In the daily parlance of the *Habañero*, this means *resolviendo* or *solucionando* (problem solving), even if it implies breaking the law. Before 26 July 1993, possession of

U.S. dollars was illegal for all Cubans except special workers and diplomats. Violators were subject to incarceration. Despite the law, millions of Cubans had stashed away dollars for use on the black market. So desperate is the government for dollars or other convertible hard currencies, that Castro announced in his now famous 26 July 1993 speech the possibility of opening checking accounts in dollars. As well, more than 140 job categories now operate outside state control, and most can charge in dollars after securing the proper license and remitting the appropriate taxes.

What had been illegal and clandestine until 1993 was suddenly permissible and overt. Attendance at work and schools is off, though no reliable data exist to assess its magnitude. Street hustlers (*jineteras*) who work the tourists have increased notably. Before the Special Period, Cubans approaching tourists for a pen, chewing gum, or a coin would be scolded by a passerby or, worse yet, detained by police if they persisted. Some of the civil pride and discipline, once badges of honor for Revolutionary Cuba, are gone.

New "economic liberty" carries externalities. Examples include delinquency and street crime (Klak 1994), though it pales to the kinds of problems found in U.S. cities, or the rise of organized crime derived from the thawing of communism in Moscow. Havana was heralded during the first decades of the Revolution as being one of the safest cities in the Americas, even by harsh critiques of the regime (Mesa-Lago 1994, 12). Petty theft and assaults are no longer isolated, random events in Havana. Today, the state-run shops for Cubans are barren and the streets of the city, with their poverty, prostitution, shabby buildings, and begging, increasingly resemble other Latin American cities.

Black market activities prevail because central distribution systems cannot perform adequately. It is difficult to obtain soap, toothpaste, shoes, spare electrical or automotive parts, extra coffee, or fresh meat and poultry. *Habañeros* must invest time by going from market to market, house to house, or contact to contact. Almost every household has someone who devotes time checking out black market activity and the new state-sanctioned markets springing up around the city.

Both public and private services in Cuba are disorganized. Public services in Havana, for example, often come only when one has a "buddy" (*socio*) in government. The gap in purchasing power between skilled and unskilled laborers, and between those who have access to a dollar economy and those who do not, is wide. Taxi drivers earning dollar wages spend little of their monthly wages on basic foods, lard or soap. A physician and manual laborer, for example, would spend 16 and 100 percent, respectively, of their monthly wages on a one-liter bottle of lard, while a taxi driver would spend about 1 percent (Segre, Coyula and Scarpaci 1997).

Habañeros are employing new ways of subsistence that fall outside the state's 30-odd year practice of central planning. Although the state system of food allocation is heavily subsidized and affordable even for households earn-

Table 12-1 Monthly Food Allocation by Household or Person, According to the Cuban *libreta*, Selected Items, 1995

Item	*Monthly Allocation*	*Per Unit*
rice	5 lbs.	person
black beans	30 oz.	person
sugar (white)	3 lbs.	household
sugar (brown)	3 lbs.	household
chicken	3 oz.	person
fish (various)	one (6-9 oz.)	person
soy meal (*picadillo de soya*)	varies	person
salt	8 oz.	person

Source: Author's field notes.

ing between 120-200 pesos monthly (Table 12-1), it lacks variety and regularity. This pumps up demand for alternative food sources.

Tourism provides considerable promise as the influx of tourists has increased three-fold in a decade (Figure 12-3). Although Fidel Castro has fought hard to avoid turning Cubans into an island of waiters and waitresses, he has moved aggressively on this front. One nongovernmental organization in Havana, *El Grupo para el Desarrollo Integral de la Capital*, brings tourists to Afro-Cuban neighborhoods in the back-bay townships of Atarés. Residents perform *santería* ceremonies and mount back-alley *rumba* dances with foreigners. While the setting is a contrived inter-cultural experience, there is no doubt that a segment of the tourist market seeks those experiences. Carefully managed and locally based tourist services could prove to be a source of revenue for Havana's municipal governments. This strategy is also a major policy and planning goal elsewhere in Latin America (Scarpaci and Irarrázaval 1994).

Decentralizing economic decision-making in Havana does, at least for Communist Party hard liners, run the risk of the liberalization of political life and creating a new space for civil society (Pastor and Zimbalist 1995, 715). That possibility notwithstanding, Coyula (1992, 49) documents the advantages

Tourist Arrivals to Cuba, 1985 - 1995

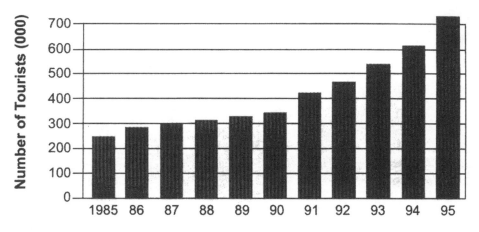

Figure 12-3 Tourist Arrivals to Cuba, 1985-1995
Source: *Business Tips on Cuba*, May 1995, p. 25.; *CubaNews*, Volume 4, No. 3, March
               1996, p. 1.

of bringing production and consumption issues back to the *barrio*. He argues
that the "neighborhood [is] an identifiable, manageable territory whose social,
cultural, and productive power has not been exploited".

New Real-Estate Market

While the market unfolds cautiously in socialist Cuba, the former bourgeoisie
residential neighborhood of Miramar is again changing. In the early years of
the Revolution, the state transformed mansions into boarding houses and shel-
ters for students with scholarships who came to the capital to study and work.
These old mansions are now being refurbished to accommodate joint-venture
businesses. In contrast, the traditional non-tourist commercial center of Havana (*Centro Habana*) displays empty stores and precariously adapted housing in substandard buildings.

These new functions in Miramar include the presence of embassies and
housing for foreigners. Along with these groups we also find high level government officials and research centers that co-exist with the populations which
in the 1960s and 1970s had received empty apartments. One still finds in

Miramar many of the workers and home attendants who worked with the students in the boarding houses in the 1960s and 1970s. These include the well known *tías* (literally "aunts" but in reality housekeepers) and caretakers who initially occupied the back garages and service rooms of their former employers. Eventually they took over the rest of the mansions. Relatives from the countryside would join them and live in the many rooms of these spacious homes. In this regard, Miramar has become the Havana neighborhood with the most change occurring in it since the triumph of the Revolution. Its relative location boasts access to the city center along the Malecón promenade and access to the beach. The beautiful image of 5th Avenue (*Quinta Avenida*) is practically the same. There is even less automobile traffic through there today. Although truck and car traffic have declined, bicyclists tenaciously peddle through the city on broad, central, and often tree lined roads that form the best bikeways of the city.

Other real-estate investments concentrate in the old city, *Habana Vieja*. Originally the walled city of colonial Havana, UNESCO declared *Habana Vieja* a World Heritage Site in 1982. The district encloses the largest collection of colonial architecture. The housing stock dates from the 16th-19th century and is the largest urban historic district of any city in the Western Hemisphere (Segre 1989). Two major renovations mark unprecedented investment in Havana real estate. One is the Italian-financed remodelling of a modern pier for docking Caribbean cruise ships. Another is the estimated $20 million leasing of the former Cuban Stock Market (*Lonja de Comercio*) by a Spanish bank. The revivals of the former "Caribbean Wall Street" and a new luxury passenger terminal call to mind the classic images of capitalist perils in the 1950s. Meanwhile, Habana Vieja is home to about 85,000 residents who crowd into quarters that planners contend should hold about 40,000 (Personal communication, 1995).

Joint Ventures

A growing number of outside analysts dismiss the economic policies implemented in Eastern Europe and Cuba because they are admissions that the creation of a socialist state is impossible (Skidelsky 1996). The reality of a post-Cold War world is more complex than that. In Cuba a key strategy to cope with the economic problems of the Special Period has been to allow a gradual opening of foreign investors to participate in joint ventures with the Cuban government. A second strategy is to stimulate the flow of funds from family members of Cubans who have emigrated. Until 1993, foreign remittances by Cuban exiles (*gusanos*, or worms) were illegal and punishable by imprisonment. State planning now emphasizes sectors such as tourism, biotechnology, and the pharmaceutical industry to generate their own foreign exchange. All of this takes place in a work setting in which moral incentives must prevail. Lacking capital and without access to credit from international organisms because of U.S. pressure to discourage such lending, the situation has become increasingly difficult in Cuba and has broadened problems of scarcity.

Throughout Cuba, but especially in Havana, efforts to generate hard currency accentuate differences among Cubans who have access to dollars and those who do not. Scarcity of basic foods, a large portion of which had come from the Soviet Bloc, began to affect the health of Cuban society. Ironically, this was a realm that had received international acclaim as an accomplishment of the Revolution even by the most ardent opponents. In 1993, the government provided massive doses of multi-vitamins to 11 million residents to compensate for a lack of protein in the diet (Stix 1995, 32). For the first time since the Revolution, food has displaced housing as the number one social problem (Segre, Coyula and Scarpaci 1997).

Cubans jest about the "second conquest" of the island when they refer to the surge in Spanish investment (Table 12-2). Havana is the target for much of this investment. A Spanish conglomerate refurbished the former Havana Hilton, one of the city's landmarks built in 1958 by Conrad Hilton. Although $100 million of Spanish investment has generated employment in the construction industry, most building materials, furniture, wiring, glass, and plaster come from other countries, depriving the city of the usual kinds of economic multipliers and integration among the building and home-furnishing trades. Within the first 90 days of operation in 1993, hotel management of the Habana Libre renamed the hotel the Havana Libre-Guitart (reflecting the name of the Spanish investment group) and reduced the hotel's work force from 1,200 to 400. Only a few main social spaces in Havana's tourist district remain isolated from this rising investment (Curtis 1993).

Cuba, however, is still not a sanctuary for "safe investment". *The Economist Intelligence Unit* ranked Cuba 116th out of 129 countries regarding investment safety. Tourism, though, remains a reliable commodity the Cubans can market. Mesa-Lago (1994, 19) reports that 460,000 tourists reached the island in 1992 and generated $400 million in gross revenues. After subtracting expenses, net revenues stood at just $240 million or 1 percent of the GSP. Risky investment notwithstanding, companies recently investing in Cuba are Western Mining (Australia), Sherrit (Canada), ING Bank (Netherlands), Grupo Domos (Mexico), Unilever (UK/Netherlands), Labatt (Canada), and Pernod Ricard (France).

Table 12-2 Selected Joint-Venture Hotel Chains in Cuba, c. 1995

Hotel Chain	*Country of Origin*	*Tourist Zone*
Guitart S.A.	Spain	Cayo Coco, Varadero, Havana, Santiago de Cuba
Super-Club	Jamaica	Varadero
LTI	Germany	Varadero, Santiago de Cuba
Iberostar S.A.	Spain	Havana
Sol-Meliá	Spain	Havana, Varadero
RUI Hotels, S.A.	Spain	Varadero
Raytur	Spain	Santa Lucia
Delta	Canada	Santiago de Cuba
Commonwealth	Canada	Marea del Portillo
Hospitality, LTD	Canada	Granma
Kawama Caribbean	Spain	Varadero
Hotels (KCH)	Spain	Havana

Source: *Business Tips on Cuba*, July 1994, page 21.

The Transition to Socialism

The traditional measures that define the role of the state show that Cuba is far from the socialist state it was ten years ago. Its means of production are no longer state-owned and the government expropriates surplus value for the simple result of profit. To be sure, the transition to socialism faces structural barriers. Interpreting Marx's creation of a socialist state is daunting, especially in the post-Cold War era. That socialist transitions unfold unevenly is not surprising. The Revolution took hold differently in Sandinista Nicaragua in the *sierra* than on the coast (Vilas 1989; Wall 1990). Vietnam—with a population six times larger than Cuba's—is firmly placed in a global market and it retains its socialist principles. Perhaps most troubling to Cubans and those who sympathize with the Revolution are changes in daily living (*la vida cotidiana*) including begging, prostitution, and petty theft. Before the Special Period, such behavior was anti-Revolutionary and unnecessary in a socialist state that provided everything. These explanations are no longer used so sweepingly because such behaviors are widespread. Cuba's transition to a post-socialist economy not only impedes market mechanisms, but it has led to a great demoralization of the workers. Moral incentives no longer seem relevant. The halcyon days of the 1960s and 1970s are over. This was when altruism, abnegation, sacrifice and discipline produced a "revolutionary conscientiousness", and when the state discouraged "critical thinking" (Medin 1990; del Aguila 1992). New economic liberties in Cuba brought changes in civil behavior. When a riot threatened on Havana's seaside promenade (the Malecón) in August 1994, Fidel Castro entered a crowd of several thousands to defuse the setting. The riot was in response to the scarcity of material goods, political repression, human rights abuses, and a loss of civil liberties.

It is unclear the ramifications this protest could have had if the rapid response brigades (*brigadas de respuesta rápida*) had not snuffed it out. Both this audacious protest and rise in the black market underscore the power of human agency. Many Cubans now question the dogma about how socialism and capitalism should co-exist in Cuba. This first massive public demonstration in 35 years of the Revolution shares a broad parallel with the IMF riots reported by Walton (1987) in other Third World countries; it marked the beginning of a movement of vocal opposition against the status quo despite the threat of state retaliation.

Autarky cannot be a development option for peripheral Third World economies (Fitzgerald 1986; Stallings 1986). Combining economic growth and equity while engendering viable participation (politics) remain perennial challenges (Fagen, Dear and Corragio 1986). Vietnam and China show that a capitalist sector can co-exist with socialist principles, but those principles seem secondary in the daily lives of most Cubans. The state controls surplus and decisions about accumulation. Orthodoxy in the socialist sphere has given way to a new hybrid model. Several think-tanks in Cuba had pushed for policy changes that create community participation (CEA 1995; Coyula, González, and Vincentelli 1993; GDIC 1991). In March 1996, however, the government closed two think-tanks (the Center for European Studies and the Center for the Study of the Americas). Raul Castro, head of the Armed Forces, claimed that they had strayed too far from the state's view of how a transition to socialism should result. Scholars in Havana remark that these centers were either pushing too quickly and publicly for market reforms, or that their researchers had become too visible in the international arenas as spokespersons for the socialist government.

Small internal and regional cooperation markets for most socialist nations (Sandinista Nicaragua, 1979-1989; Cuba; Grenada under Maurice Bishop's New Jewel Movement, 1979-83; and Chile under the administration of Dr. Salvador Allende, 1970-73) may be insurmountable obstacles in the transition to socialism (Wall 1990). Cuba aims to replace its former CMEA partners with Canadian and Western European (EU) trade. The Cuban army has re-tooled, given that Cuba is no longer the boot camp it once was for revolutionary struggles in Africa.

The Cuban experience suggests that state-ownership alone cannot define a socialist production model. Centralized planning politicizes the economy and it constrains market signals in the form of prices and seeks to keep producers from being motivated only by self-interest (Zimbalist and Brundenius 1989, 141). Deep changes must also take place in the labor process—bringing in worker participation—and in mass participation in forging state policy (López Segrera 1972). The Cuban state labor force, however, is downsizing. The implications these layoffs portend for Cuba's social relations of production are unclear.

Cuba's pattern of market reform has been erratic over the past two decades. Between 1973 and 1986, the state sold surplus products in government stores at fixed prices. Free peasant markets prevailed briefly in 1986 but were eliminated because of alleged high profiteering among middle brokers. Castro claimed that some small farmers where turning over only 10 percent of their crops to the state. The 1993 reform relating to self-employment, the circulation of the dollar, and the loosening restrictions on Cuban exiles visiting the island are the most recent policy changes on this front.

The Eastern European Example

Cuba's role in the international socialist alliance was unique. Unlike Afghanistan, it was not a buffer state between the USSR and economically and politically threatening nations in South Asia and China (Kakar 1995, Cordovez and Harrison 1995; Rubin 1995). Instead, it provided a powerful symbol of anti-U.S. hegemony in the "American Lake", and was therefore a bulwark of socialism for the poor countries of the Third World. In a post-Cold War era, this symbolism is anachronistic.

The experience of the former Soviet Union and Eastern bloc nations should be telling in the case of Cuba. Havana's micro-enterprises must cope with pricing, advertising, return business, buying in bulk, and access to credit. A new business class plods along, building its small businesses, and incrementally improving their entrepreneurial learning curve. *Parador* owners welcome their new-found occupational independence, and desire less of what the government wishes to impose over all self-employed: "greater discipline and control" (*Granma*, 1995; my translation). Perhaps the only obstacle in their horizon will be a battery of new government guidelines should the new entrepreneurs prove to be too profitable in the eyes of Cuban authorities.

Recent structural changes in Havana's political economy raise more questions about the nature of the socialist state than our theoretical tools can answer. Theorists have long been disenchanted with the nature of the state in Eastern Europe, and the former USSR. These doubts arose well before the socialist camp collapsed. Far from withering away in those societies as envisioned by Marx, the state took on an almost Orwellian role in its management of everyday life. Today, though, new free-market regulations in Cuba go into detail about who can become self-employed (Pérez-López 1995a; Mesa-Lago 1994, 32). *Perestroika* and *glasnost* in the former USSR ushered in rapid change as have market liberalization reforms in post-Mao China. An unanswered question is whether the introduction of the dollar has derailed the Revolution and, if so, whether it can get back on track.

The sudden shift towards limited capital accumulation in Havana, historically a process confined to the capitalist mode of production, is changing Cuba's transition to socialism. No longer is socialism the anti-systemic force theorists once considered it to be (Polanyi 1977). While the Eastern Europe experience proves that it is not alarming to find inequalities under State socialism (Szelenyi 1982), we can only hope that the Cuban experience can avoid the pitfalls that have plagued the former Socialist alliance (Tismaneanu 1992; Clawson 1992; Ickes and Rytergman 1992). Because of 39 years of socialism, Havana is ill-prepared to support a market economy easily. Budapest, Prague, Warsaw, and governments in other Eastern European nations permitted limited private ownership of businesses, self-employment in certain services, and farmers' markets alongside their centrally planned economies (Pérez-López 1995a, 249). To students of development studies around the globe, Cuba was going to provide a missing ingredient in Marx's incomplete theory of the state. That prospect has vanished as Cuba teeters at the edge of political and economic unrest. Austerity, in the name of salvaging the socialist Revolution, may not be tolerated much longer. If sustainable development is to be achieved, *Habañeros*, informal markets, clandestine trade, and public disdain for the standard bearers of the Revolution may have to raise their voices in timbres that prevail on the leaders of the Cuban Communist Party.

Eastern European governments have openly criticized Cuba since the mid-1980s. Although progress in fields of education and health care received high praise, at least three criticisms characterize this Eastern European view. First, Cuban firms have for too long operated at staggering losses. Approximately one-third of Cuban firms were losing money in the early 1990s (Mesa-Lago 1993, 7). Second, worker productivity has been slipping. As workers perceive the possibility of material rewards, or as materials shortages keep public workers from performing optimally, productivity continues to slip (Carranza, Guitérez and Monreal 1995). Third, sugar production has fallen back to under 4 million tons, a low-point not reached since the 1960s. Failure to pay attention to the nation's principal cash-earning crops has had dire consequences. A related criticism is Cuba's failure to diversify away from sugar monoculture. Lastly, Cuba had rigidly adapted socialist development models—including *perestroika* —producing unanticipated outcomes on the island. While the island touted its "real socialism" in Latin America, Eastern Europeans remembered Cuba's support of the Soviet invasion of Czechoslovakia. Cuba benefited from its position as an ideological radical because the island was "eating others' bread and building socialism at the expense of another country" (Varkonyi 1989, 2).

Conclusions

Fidel Castro will ride out the current economic storm. He sees these market reforms as a transition towards real Cuban socialism (Castro 1995). His advisors have argued that the Cuban leadership needs to return to a closer reading of Marx and Engels (Hart-Dávalos 1990a, 1990b). At a time when a year-long thawing in U.S.-Cuban relations seemed promising, two civilian aircraft piloted by Cuban Americans were shot down by MIG planes on 24 February 1996 in international waters. In the weeks that followed, Congress passed the Helms-Burton bill and the U.S. Treasury issued tighter restrictions on March 4, 1996 for all countries trading with Cuba. Castro used the event for saber-rattling; he documented the numerous incursions by the anti-Cuban organizations Brothers to the Rescue (Petkofsky 1996). On the day of the attack, he also arrested members of the moderate opposition group, *Concilio Cubano*. Eastern Europeans watch as Cuba charts its transition to socialism. Some are skeptical. As Vaclav Havel observed:

> In my view, Cuba, too, sooner or later will be affected in one form or another by what is historically inevitable...that the totalitarian systems of the communist type are at least being transformed into something else or disintegrating, that more democratic conditions are being formed (cited in Linden 1993, 17).

Just a decade ago, Cuba suffered from a kind of apartheid that separated foreigners, with their non-Revolutionary ideas, from Cubans. State planning and policies have changed since then. Since 1993, though, a host of private-sector self-employed ventures have proliferated in Havana and elsewhere, albeit with erratic "stops" and "gos", as the Cuban government defines and re-defines what can only be described as new type of *criollo* Keynesian economics. To some Cuban policy analysts and economic advisors to the Castro government, there is a way out of Cuba's quagmire, and its formula sounds much like those drawn from neo-classical economics. The prescription is as follows:

> a new economic structure, more dynamic, more diversified, less dependent on a single product, with very active international tourism; hundreds of joint-venture firms and associations and dozens of manufactured goods and agricultural products being exported; this could be Cuba's outlook at the end of the century (Figueras 1994, 181; my translation).

Recommendations like this were advanced by Cubans who analyzed the economy before and after the Revolution, but the latter analysis *always* came from the exile community (Marrero 1956, 1981a, 1981b). While few observers inside and outside Cuba would dispute the value of Figueras' prospect in the quotation above, the macro-policy constraints placed on the simplest of enterprises begs the question about whether such a prospect for the year 2000 is realistic. An increase in government regulation to equalize income distribution may dampen all self-employed enterprises (Alonso 1995). As Pastor and Zimbalist (1995, 705) note, "the recent spate of halfhearted measures...has likely worsened distributional inequities, distorted incentives, and failed to improve the macroeconomy".

Those observations by "sympathetic" U.S. economists, notwithstanding, Castro is determined to use the market to guide Cuba through this crisis in which global regulation imposes new forces and governance in Cuba. As the leading economic advisors to Castro recently concluded about the island's restructuring:

> The need for a conceptual paradigm that will serve as an organizing core in the workplace has made it necessary to adopt an unconventional perspective about socialism. To be more precise [it has meant a] new perspective on [socialism's] past, present and future. Building a market economy whose means of production are in the context of a socialist economy remains the main element of the program that defines the fundamental economic reform. In this model is the rebuilding of a viable socialist economy in Cuba, even though its viability will be hindered by external economic factors such as the political hostility of the government of the United States (Carranza, Gutiérez and Monreal 1995, 169, my translation).

The tenacity of the Cuban regime is noteworthy and the proximate causes of the demise of socialism elsewhere have not phased Castro. In Poland, and less so in other Eastern European nations, a domestic political alternative challenged the Communist parties. Cuban opposition is overseas, weak, and is effective only in blocking rapprochement with Castro. Eastern Europe could not withstand the crumbling of the USSR and Soviet acquiescence; Cuba moves onward nearly a decade later. Cuba has also remained insulated from serious military activity elsewhere. This includes the defeat of the Sandinistas, removing Noriega from Panama, the U.S. invasion of Grenada, the agreement with South Africa over Namibia, withdrawing Cuban troops from Angola and Ethiopia, and the disorder of the Left elsewhere in Latin America (Chile, Brazil, El Salvador). Perhaps the greatest menace to Cuba is disorder from within because of an inability to satisfy the needs of the people (Mesa-Lago and Fabian 1993, 365).

Critics of Havana's new economy argue that new laws stifle the kinds of multiplier effects we might envision in a less restrictive setting. Cuba lacks what the Chilean business community calls *transparencia*: clarity about norms, regulation and information that is unrestricted by cumbersome regulation. Like de Soto's (1989) description of pre-Fujimori Peru, state regulations hinder more than help Cuba's fledgling cottage industries. While it is perhaps not surprising that the normative features of a market economy have not yet taken hold in Cuba, increasing numbers of Cubans, as well as foreign scholars, keep searching for the socialist dream.

References

Alepuz, Manuel, 1993, "Bicycles overtake bus travel in Havana", *Urban Age*, 2:16-17.

Alonso, J.F. 1995, "The path of Cuba's economy today", *ASCE Newsletter*, Spring-June, 18-23.

Browder, John, James Bohland and Joseph Scarpaci, 1995, "Patterns of development on the metropolitan fringe: Peri-urban expansion in Jakarta, Bangkok and Santiago", *Journal of the American Planning Association*, 61:310-327.

Carranza, J., L. Gutiérrez, and P. Monreal, 1995, *Cuba: La restructuración de la economía*, Habana: Editorial de Ciencias Sociales.

Castro, Fidel, 1995, "Seguimos creyendo en los enormes beneficios del socialismo", *Granma*, December 30:2.

Chávez, Roberto, 1992, Non-motorized transportation in Cuba, ITDP Report AF6IN.

CEA (Centro de Estudios de las Américas), 1995, *Los consejos populares, la gestión de desarrollo y la participación popular en Cuba, Conclusiones preliminares*, Habana: CEA.

Cordovez, D. and S. Harrison, 1995, *Out of Afghanistan: The Inside Story of the Soviet Withdrawal*, London: Oxford University Press.

Clawson, Patrick, 1992, "Understanding the post-Soviet wasteland", pp. 49-58 in Vladimir Tismaneanu and Patrick Clawson, (eds), *Uprooting Leninism, Cultivating Liberty*, Lanham, MD: University Press of America and Foreign Policy Research Institute.

Coyula, Mario, 1992, "El veril entre los siglos. Tradición e inovación para un desarollo sustentable", *Casa de la Américas* (Habana), 189:94-101.

Coyula, Mario., G. González, and A.T. Vincentelli, 1993, *Participación popular y desarrollo en los municipios*, Habana: CEA.

Curtis, James, 1993, "Havana's Parque Coppelia: Public space traditions in socialist Cuba", *Places*, 8:3:62-69.

del Aguila, J. 1992, "Cuba", *The Software Toolworks Multimedia Encyclopedia*, Novato, CA: The Software Toolworks.

de Soto, H. 1989, *The other path: The invisible revolution in the Third World*, New York: Harper and Row.

Fagen, R., C.D. Deere, and J.L. Coraggio, (eds), 1986, *Transition and Development*, New York: Monthly Review Press.

Feinsilver, J. 1993, *Healing the Masses*, Berkeley: University of California Press.

Figueras, M.A. 1994, *Aspectos estructurales de la economía cubana*, Habana: Editorial de Ciencias Sociales.

Fitzgerald, E.V. 1986, "Notes on the analysis of the small underdeveloped economy", pp. 28-53 in R. Fagen, C.D. Deere, and J.L. Coraggio, (eds), *Transition and Development*, New York: Monthly Review Press.

GDIC (Grupo para el Desarrollo Integral de la Capital), 1991, *Estratégia*, Habana: GDIC.

Granma, 1995, Si de cuenta propia se trata, 15 June, 2.

Hart-Dávalos, A. 1990a, "Leamos de nuevo a Lenin", *Cuba Socialista*, 43 (January - March), 2-3.

_____. 1990b, "Volvamos a leer a Engels", *Cuba Socialista* 44 (April - June), 1.

Ickes, B. and R. Rytergman, 1992, "Credit should flow to entrepreneurs", pp. 69-84 in Vladimir Tismaneanu and Patrick Clawson, (eds), *Uprooting Leninism, Cultivating Liberty*, Lanham, MD: University Press of America and Foreign Policy Research Institute.

Kakar, M.H. 1995, *Afghanistan: The Soviet Invasion and the Afghan Response*, Berkeley: University of California Press.

Klak, Thomas, 1994, "Havanna and Kingston: Mass media and emperical observations of two Caribbean cities in crisis", *Urban Geography*, 15:318-344

Lee, S. 1996, "A comienzos del 96: Hablemos de presupuesto y de impuesto", *Granma*, January 5.

López Segrera, F. 1972, *Cuba: Capitalismo dependiente y subdesarollo (1510-1959)*, Habana: Casa de las Américas.

Linden, R.H. 1993, "Analogies and the loss of community: Cuba and East Europe in the 1990s", pp. 17-58 in C. Mesa-Lago, (ed.), *Cuba After the Cold War*, Pittsburgh: University of Pittsburgh Press.

Marrero, L. 1956; *Historia económica de Cuba: Guía de estudio y documentación*, Habana: Universidad de Habana: Instituto Superior de Estudios e Investigaciones Económicas.

_____. 1981a, *Geografía de Cuba*, San Juan, Puerto Rico: Editorial San Juan

_____. 1981b, *Cuba: Economía y Sociedad*, San Juan, Puerto Rico: Editorial San Juan.

Medin, T. 1990, *Cuba: The Shaping of Revolutionary Consciousness*, Boulder: Lynne Reinner.

Mesa-Lago, C. 1993, *Cuba After the Cold War*, Pittsburgh: University of Pittsburgh Press.

_____. 1994, *Are Economic Reforms Propelling Cuba to the Market?* Miami: University of Florida, North -South Center.

Mesa-Lago, C. and H. Fabían. 1993, "Analogies between East European socialist regimes and Cuba: Scenarios for the future", pp. 353-380 in C. Mesa-Lago, (ed.), *Cuba After the Cold War*, Pittsburgh: University of Pittsburgh Press.

Pastor, M. and A. Zimbalist, 1995, "Waiting for change: Adjustment and reform in Cuba", *World Development*, 23:705-720.

Pérez López, J. 1995a, *Cuba's Second Economy*, New Brunswick: Transaction Books.

_____. 1995b, "The Cuban economy in 1995", *ASCE Newsletter*, (Spring – June)24-29.

Petkovsky, A. 1996, "Attack may lift Castro", *Richmond-Times Dispatch*, February, p. A-6.

Polanyi, N. 1977, "The problem of the capitalist state", *Review*, 1:9-20.

Roca, S.G. 1994, "Reflections on economic policy: Cuba's food program", pp. 94-117 in. J.F. Pérez-López, (ed.), *Cuba at a crossroads: politics and economics after the Fourth Party Congress*, Gainesville: University of Florida Presses.

Rubin, B.R. 1995, *The Search for Peace in Afghanistan: From Buffer State to Failed State*, New Haven: Yale University Press.

Scarpaci, Joseph and A.Z. Hall, 1995, "Havana peddles through hard times", *Sustainable Transport*, 4:4-5, 15.

Scarpaci, Joseph. and I. Irarrázaval, 1994, "Decentralizing a centralized state: Local government finance in Chile within the Latin American context", *Public Budgeting and Finance*, 14:4:120-136.

Segre, Roberto, 1989, *Arquitectura y urbanismo de la revolución Cubana*, Habana: Editorial Pueblo y Educación.

Segre, Roberto, Mario Coyula and Joseph Scarpaci, 1997, *Havana: Two Faces of the Antillean Metropolis*, New York: John Wiley.

Stallings, B. 1986, "External finance and the transition to socialism in small peripheral societies", pp. 54-78 in R. Fagen, C.D. Deere, and J.L. Coraggio, (eds), *Transition and Development*, New York: Monthly Review Press.

Skidelsky, R. 1996, *The Road from Serfdom: The Economic and Political Consequences of the End of Communism*, New York: Allen Lane, Penguin Press.

Stix, G. 1995, "Ban that Embargo", *Scientific American*, 272:3:32-34.

Szelenyi, I. 1982, *Urban Inequities Under State Socialism*, Oxford: Oxford University Press.

Tismaneanu, Vladimir, 1992, "Between liberation and freedom", pp. 1-48 in Vladimir Tismaneanu and Patrick Clawson, (eds), *Uprooting Leninism, Cultivating Liberty*, Lanham, MD: University Press of America and Foreign Policy Research Institute.

Varkonyi, T. 1989, "The Grandma Phenomenon", *Magyar Nezmet*, (July 18) 2.

Vilas, C.M. 1989, *State, Class and Ethnicity in Nicaragua*, Boulder, CO: Lynne Reinner.

Walton, J. 1987, "Urban protest and the global political economy, The IMF Riots", pp. 364-386 in M.P. Smith and J.R. Feagin, (eds), *The Capitalist City*, London: Basil Blackwell.

Wall, D. 1990, The Political Economy of Sandinista Industrialization, 1979-1989, Department of Geography, Doctoral Dissertation, University of Iowa.

Zimbalist, A. and M. Burundeis, 1989, *The Cuban Economy: Measurement and Analysis of Socialist Performance*, Baltimore: Johns Hopkins University Press.

13 The Promotion of Neoliberal Industrialization in Third World Countries

THOMAS KLAK, GARTH MYERS

Development planning in Third World countries is becoming homogeneous. A heavy emphasis on exports, an outward orientation to the economy, a rollback of government intervention in economic matters, a prescribed devaluation of national currencies, a welcoming stance toward foreign capital and a devotion to free markets - all unite under a ubiquitous development model termed neoliberalism. Neoliberalism instructs Third World governments to deregulate their economies and labor markets in order to attract foreign investors, boost exports, and generate growth (Williamson 1990; Krueger 1993; Colclough 1991; Balassa, 1993). The apparent congruity in Third World development planning policies results from pressures from the World Bank, the International Monetary Fund and western countries for economic reform, the narrow range of policy options available to governments which sign adjustment and reform agreements with lenders, and the present shortage of clearly articulated alternatives to the capitalist road to development (Peet and Watts 1993). Neoliberalism is shaping the prevailing ideas about development in most parts of the world, in rich and poor capitalist countries as well as in formerly socialist ones. In marginalized countries of the global economy, where development planning options have narrowed most dramatically, neoliberalism's impacts are quite striking.

The triumph of neoliberal thought is especially notable in industrial development planning on the margins of the global economy. The strategy is heavily reliant on foreign direct investment in export processing. Foreign-led export-oriented industrialization increasingly takes place in specially designated and demarcated areas commonly called export processing zones (EPZs). In this chapter, we describe the orientation in countries of the global periphery--the minor players in the world economically and industrially, who nonetheless represent the majority of countries. Peripheral countries contain the greatest human needs in the world today, and therefore sound development planning is essential across the spectrum of social concerns.

EPZs take several forms across countries of the global periphery. These include publicly- and privately-run zones; full scale EPZs and industrial estates, in which only a portion of the new industrial export incentives apply; government-designated sites and dispersed sites at which foreign investors lobby for EPZ status. In this chapter we use the terms export processing zones (EPZs) and free zones as shorthand representing all empirical variations within neoliberal industrial export policy.

Transition to Industrial Export Processing

The thinking behind industrial export processing is hardly new. The brand of industrial development inviting foreign capital to produce for export has been debated in Third World settings for nearly half of the twentieth century. The crucial question for the purpose of understanding the forces behind industrial export processing, is why the policy has gained so many new advocates recently. In 1960 export processing zones did not exist in the Third World, and by 1970 only ten countries offered them. As of 1996 some 90 countries representing all Third World regions had designated over 200 such "free zone" sites (Chandra 1992, 100; Ramsaran 1992, 150; Islam 1996). Why has export-oriented industrialization through foreign investment spread worldwide over the last two decades? Answers are linked to forces primarily exogenous to the countries demarcating these EPZs. For its part, foreign capital has vertically disintegrated production processes and begun to search globally for profitable but low-risk manufacturing sites. For Third World governments, heavy foreign indebtedness and deteriorating terms of trade with respect to the developed world has left little leeway for endogenous policy creation, since increasing exports is perhaps their main vehicle for repaying that debt and redressing trade imbalances. Hence their global position leaves them with few options other than to seek to attract foreign industrial capital by any means necessary.

The current transition towards export industrialization has been more of an abrupt response to fiscal crisis, as Third World governments fell farther and farther into debt, than a theory-based, planned transition from import substitution industrialization (ISI) to industrial export processing. Falling output from and demand for traditional primary product exports and foreign debt led governments into structural adjustment agreements and consequently into their component industrial export policies. Modernization theorists contend that too much protectionism and too much government generally caused the debt crisis. Dependency theorists and critics of the modernization approach contend that the crisis emerged from a confluence of externally-controlled factors (Brown 1995; Klak and Rulli 1993). Leaving these academic debates aside, it is our intent here to examine the post-facto planning dimensions of this transition out of ISI development and into industrial export processing. Plan-driven development, or indeed theory-driven development, would appear to have taken a backseat to a bandwagon-like industrial strategy driven by the desperate search and appearance of success and replicability as much as by any evidence of it in reality.

Under the currently transcendent neoliberal development concepts, whichever countries appear to be "winning" become the models for aspiring countries to follow. By the 1980s the alluring model for small Third World governments had been relocated primarily to East Asia, but the issues of what exactly happened to foster industrialization in these contexts and its replicability elsewhere are woefully neglected (Lindauer and Roemer 1994). For most smaller Third World countries such as those in the Caribbean basin and littoral Africa, economic desperation, in the context of the disintegration of most state socialist countries, have led political leaders to cry "there's no other way" (Klak 1995b). Yet what exactly the "way" is, and the evidence that it will lead peripheral countries to development, is seldom scrutinized. In short, current government development planning activity in the periphery now seems to have little to do with the actual experience in the model countries.

Most East Asian NICs have found success in export-oriented industrialization with a strong government role in economic planning (especially for industrial infra-structure), high degrees of protectionism, intense government-corporate links at the root of economic growth, and limited political freedoms (Lindauer and Roemer 1994; Dunning and Narula 1996; Naya and McCleery 1994). The neoliberalism presently espoused by the World Bank and other advocates, while promoting export-oriented industrialization, relies instead on a severely downsized role for government and stiff international exposure at the initial industrialization stage. Further, neoliberalism's espousal coincides with a time of political opening in many countries. A stark discrepancy exists between the means (development planning in practice) and ends (the countries supposedly being emulated).

These points lead to the impression that peripheral countries may be setting themselves up for failure by not doing their homework. Further, it would appear that the industrial export processing pursued by peripheral countries as "development planning" is planning out of the planners' control.

Foreign Investment Promotion as Third World Development Planning

Now that most policy makers and planners in the Third World have embraced the neoliberal approach to development, there follows the important step of convincing the foreign investors upon which neoliberalism so heavily relies that countries long wary of their intentions are now firmly pro-capitalist. Given that there has emerged a chorus of governments of the global periphery seeking to lure in the same investors, it becomes necessary for countries to craft quite an elaborate story-line in an effort to appear distinctive and worthy of receiving the fastidious international capital. In essence, countries need to aggressively advertise, promote and sell themselves.

Eight small countries in the Caribbean Basin and outlying islands of Africa: Barbados, Cabo Verde, Costa Rica, El Salvador, Haiti, Jamaica, Mauritius, and Zanzibar, are representative of small and economically marginal countries of their respective regions. The eight are also fairly representative of the Third World more broadly: they have limited-to-moderate strategic importance in global geo-politics, and have economies which at the outset of their industrial experiences could be categorized as underdeveloped.

Investment promotional materials are only one part of a multi-faceted venture by investment-hungry governments to sell their countries as neoliberal sites. Besides the guidebooks, governments devote much money and effort towards advertising themselves at trade shows, opening and operating investment promotion offices in the countries of desired investors, and placing materials in single-page ads or even entire supplementary advertisement sections in business news magazines. Several Third World countries operate World Wide Web sites advertising their free zones.

Neoliberal development policy advertisements attempt to lure export manufacturers as well as investors emphasizing other economic sectors. Investment in tourism, non-traditional agriculture, and off-shore bank accounts are often sought as well. Still, despite these inclusions, the investment promotion materials do not tell the whole story about Third World development. They communicate little explicitly about such crucial elements of local political economy as the built environment (outside the EPZs themselves), social welfare programs, locally-destined production, and informal or non-commodified activities. The guidebooks are nonetheless crucial statements of how a Third World government wishes to present the country to the outside world. They are also representative of the new pro-investment and outward-oriented messages that governments are conveying domestically. The guidebooks can therefore be read as guides to the hopeful positioning on the neoliberal bandwagon by governments in the global periphery.

As a rule, what one finds in the investment guidebooks ignores the actual context of the country. It could have been written from a textbook outline of features known to be attractive to foreign export-oriented investors (Chandra 1992:75-8): political stability, pro-investor policies, infra-structure, cheap labor, incentives, and preferential market access (Table 13-1). Context is generally highlighted only to the extent that it adds charm or additional opportunities for profit to the generic features that are pro-foreign industrialist. One needs to read carefully and critically to discover scattered glimpses of the often turbulent and divided societies beneath the media imagery.

The "Investment Incentives" and "Ideal Locations" Themes

Fiscal incentives like tax holidays, tariff exemptions, and low wages form the core of the promotional message of the eight cases selected. After industrial infra-structure, the availability of investment incentives is the most frequently mentioned theme in the guidebooks (Table 13-1). These lures are reinforced by paeans to economic liberalization, free trade, deregulation and commitments to private sector development. These are categorized under the heading "open economy", the fifth-most frequent theme in Table 13-1. The investment guidebooks for all countries except Haiti (whose recent history disallowed it) make grandiose claims about being the ideal, competitive location for profitable investments (the eighth theme in Table 13-1, "profitability"). Again as though following a standard list from economic geography texts (Chandra 1992, 100-2; Dicken 1992), the typical set of EPZ investment incentives provided by these Third World governments include:
- provision of factory space and industrial infra-structure
- personal assistance with establishing operations and residency, and in identifying markets
- a "one-stop" streamlined permit acquisition process
- no or low tariffs and taxes for a multi-year period or even in perpetuity
- duty free and unrestricted import of production inputs into the zones
- unrestricted foreign ownership and exchange
- no export duties
- unlimited capital and profit repatriation
- nationality blind treatment of investors for loans and services
- worker training programs
- various explicit or effective operational subsidies
- access to the markets of emerging regional free trade blocks
- guarantees against nationalization

Table 13-1 Frequency of Themes in Investment Guidebooks, by State

Theme	Barbados	Cabo Verde	Costa Rica	El Salvador	Haiti	Jamaica	Mauritius	Zanzibar
Infra-structure	4	4	4	4	1	4	4	4
Incentives	4	3	2	3	4	4	4	4
Labor Quality	3	4	4	3	2	3	4	2
Ideal Location	2	4	4	2	1	4	4	3
Open Economy	3	2	2	3	2	2	4	3
Peace and Stability	3	3	2	1	2	2	4	1
Tropical Paradise	1	2	3	1	2	2	2	1
Profitability	4	1	2	1	0	1	3	2
Environmentalism	0	0	1	0	1	0	1	2
Social Welfare	0	0	0	0	1	0	2	1

All guidebooks offer these thirteen incentives in one form or another, although the precise levels and duration of subsidies in each category show some variation. Besides the explicit benefits listed in the guidebooks, in practice informal ones extend the incentives to include commitments that those at the highest levels of government will respond promptly and decisively to their concerns (Klak 1994; Mwakanjuki 1995). Peripheral governments provide these incentives because foreign investment is construed by neoliberalism as, in the words of Cabo Verde's materials, "the engine of the country's economic growth". Given the freedoms from taxation, options for repatriation of profits, and limited development of backward and forward linkages to the local economy evidenced for most Third World EPZs, though, it would appear to be a problematical engine indeed.

Exporting from less developed countries also qualifies investors to take advantage of various trade agreements, such as the Lome Convention, the Caribbean Basin Initiative, the Multi-Fibre Arrangement, and the General System of Preferences, and thereby gain preferential access to the world's largest markets. Foreign investors are invited to fill the special market-access quotas allotted to Third World countries, but which they themselves cannot meet. Note the imbalance in the current arrangement, whereby peripheral countries are offered special access to markets in wealthy countries, but these quotas are largely filled by firms from the wealthy countries.

Each government highlights its relative position proximal to potential investors and markets; this strategy yields the fourth-most frequent theme in Table 13-1, "ideal location". In so doing the guidebooks demonstrate that with creative cartography and embellishment virtually any place on earth—even Mauritius in the middle of the Indian Ocean—can be deemed centralized and accessible. Several of the guidebooks go farther and employ a tactic whereby the country is deliberately grouped for analytical purposes with the East Asian NICs. For instance, Zanzibar's material says several times that it is "like Singapore", while Jamaica's mentions "its Far East counterparts". For countries seeking investment, the references to East Asia are salesmanship-by-inference for geographical position, potential market access, political stability, openness to investment, and the cutting edge of the twenty-first century global economy. Of the eight countries, however, only Mauritius can claim figures which even approach the NICs' for economic growth (World Bank 1995). The unreality of the comparisons for the other countries weakens this aspect of the promotional package.

At first glance one might find these embellishments unremarkable. Given how television commercials and other advertisements inundate us daily with false and hyperbolic claims, the distinction between rhetoric and reality is blurred, and media manipulation has become accepted as normal. But these are governments selling their countries like private sector marketing departments sell consumer commodities. Until a decade ago it was not this way. Across the Third World there has been a dramatic reversal in the direction of development policy and the attitude towards foreign investors.

All guidebooks promise investors an efficient welcome mat, security from nationalization threats, and a future free of social upheaval. This notion of "Peace and Stability" is the sixth most frequent theme in the guidebooks (Table 13-1), and combines aspects of neoliberalism with the strategic omission of contextual information: the past matters only inasmuch as it is alluring, and is generally ignored or recast in a peaceful light in the promotional materials. The investment promoters are onto something, though, in that perceived political stability drives investment decisions more than the actual political circumstances. A Mexican reporter recently observed that "an adverse editorial in the *New York Times* or *Washington Post* can do more damage to a Latin American government than a thousand large-scale demonstrations" (Rubio 1995:4). Development planning promoters run up against the problem of how to construct the perceived stability foreign investors insist on, when they are working with societies which are by-and-large highly unequal, still smarting over recent repression, and now served by electoral processes through which deep discontent can be aired.

The "Profits in Paradise" Theme

The warm climate that these eight tropical governments advertise refers to both their natural environments and their new-found attitudes towards foreign investors. The guidebooks describe the attractiveness and benefits of a tropical locale (the seventh most-frequent theme in Table 13-1). One cannot address industrialization in eight insular or coastal tropical countries without recognizing that tourism is itself a vital, if not the principal, source of foreign exchange. As tourism-alone-as-development began to decline in stature (after being proclaimed a kind of "manna from heaven" for Third World development in the 1960s), it came to be linked more coherently to the free-zone industrialization model (Crick 1989, 314; de Kadt 1979; Bowman 1991, 132-7). Now, most guidebooks imply a policy of seeking wealthy tourists. According to the model, many of these tourists will be business persons, and will note the opportunities for investments during their vacations. Thus tourism will feed the EPZ plan. This theme is abundantly clear in the choice of photographs for the guidebooks. Beaches empty except for perhaps a European couple, beautiful unspoiled mountain vistas, and various other recreational settings, are all depicted as generally devoid of local people who might tarnish the promotional image of serenity and solitude in a vacation get-away. Such touristic scenes comprise the second-most prominent category of visual imagery in the guidebooks (Table 13-2).

While tropical bliss is emphasized, the heart of the profitability pitch resides in portrayals of the work force (the third most frequent theme in Table 13-1). Most investment guides actually boast of the low wages paid to their EPZ workers. At the same time that tropical labor is advertised as an investor's bargain, guidebook-writers generally feel it necessary to stress that low wages buy an energetic and productive workforce, perhaps to counteract the stereotypical idea of tropical "slothfulness" (Richardson 1992; Crick 1989).

The "Science and Technology" Theme

Illusions of scientific grandeur punctuate much of the photographic and textual information in the guidebooks. Many photos display either free zone buildings and grounds or transport facilities. But the largest share of photographs illustrate "high technology" industries (Table 13-2). Computers and computer-training facilities, as well as other high- technology equipment related to industries such as medical equipment engineering or pharmaceutical production, figure prominently in nearly every guidebook. All guidebooks with photos display satellite dishes, presumably with the intent of assuring investors that the peripheral country is networked into the global telecommunications systems. The message to investors is that they can relocate labor-intensive portions of production processes to bargain-wage locations that may be remote, but are highly accessible thanks to advanced telecommunications.

Although presented as factual, the guidebooks convey less of the existing reality of EPZs than of a fictional (and improbable) technological future to which they aspire, often based on patterns laid down by the East Asian NICs. As Dicken (1992, 181) recounts, "[i]n 1986, following the unexpected recession of 1985, the Singapore government announced ... measures to move away from being merely a labor-intensive assembly site to a high-technology, knowledge-based economy". By the mid-1990s, the investment guidebooks of Mauritius and Barbados in particular appear to mimic Singapore's earlier reorientation. Unfortunately for the many Third World governments aspiring to the East Asian model, rhetoric alone will not attract the cutting-edge of global trade. Few aspirants have been able to approximate the NICs' rapid policy adjustments in response to changes in the world economy.

Other visual images largely omitted from the guidebooks are also quite telling. Their absence directs potential investors' attention away from what we consider crucial contextual features of the eight countries. The guidebooks suggest that part of the promotion of stability and a pro-investment cli-

Table 13-2 Types of Photographs in the Investment Guidebooks of Seven Small States

Type of Image and Setting	Percent	Number
High Technology Industrial Laboratory or Telecommunications Facility	25.2	41
Natural Beauty or Tourism-Related	20.9	34
Free Zone Building and Grounds	14.1	23
Transportation or Industrial Infra-structure	13.6	22
Garment Production, Electronics Assembly or Agro-Processing	13.6	22
Agriculture	4.9	8
Other	7.7	13
Total for Seven States	100	163

mate involves avoiding politically-related photographs. The reality of regime turnover need not dampen the image of long-term political stability. Only one national leader, President Salmin Amour of Zanzibar, appears in any guidebook, and it is no coincidence that Zanzibar's pro-investor presentation is one of the least convincing in our sample. Further, in spite of the fact that each of our eight cases represents a country where fishing and/or monocrop plantation agriculture form the historic basis of the national economy, agriculture-related photographs are nearly absent from the guidebooks (Table 13-2). In most cases, the agricultural processes shown are those associated with neoliberalism's non-traditional exports, such as hydroponic gardening and cut-flower "factories in the field."

Such strategic omissions from the photographs parallel others in the textual materials. Primary product industries receive substantial verbal attention only from Haiti and Cabo Verde (Table 13-3). Other than those two cases, the guidebooks generally underemphasize existing traditional and lower-technology economic activities, while they exaggerate the extent to which the country has already attracted high technology firms. While most guidebooks promote basic manufacturing (e.g. garments, electronics assembly), references to it, when measured as a percentage of themes mentioned, (Table 13-3) are far below its actual share of EPZ activity in all eight countries. Tourism is similarly under-represented relative to its major economic significance to these countries. In contrast, scientific, high-technology manufacturing, information processing and financial services, which are rare within the sample group and absent from the EPZs of several countries, are the most frequently emphasized sectors for four of the eight guides (Table 13-3). The countries that are economically farther ahead in the group (Barbados, Jamaica, Costa Rica, and Mauritius) realize that to continue playing and to benefit from the EPZ game, they must position themselves as ripe for investment associated with the information age, however unrealistic their aspirations may be.

General Absence of "Social Welfare" and "Environmental" Themes

Besides investigating the way in which global neoliberalism is transcribed into the investment guidebooks of peripheral countries, we are also interested in the current status of concerns about human welfare and the environment. We therefore searched the materials carefully for references to environmental protection, social justice, human welfare, and poverty reduction, and found they are themes rarely deemed appropriate to the model of EPZ promotion (see rows 9 and 10 in Table 13-1). The promoters know, and if they do not, international advisors will soon inform them, that investors are wary of environmentalist agendas (Mwakanjuki 1995; Vanderbush and Klak 1996). Investors are seldom sold on a country because locating there will help reduce poverty. Hence, with a few exceptions, the guidebooks generally minimize references to the country's environmental laws, and say little about investors' other social responsibilities. To the contrary, seven of the eight guidebooks explicitly attempt to sell their countries on the basis of remarkably low cost of labor and/ or lack of trade union power. The guidebooks' omission of social responsibility is ironic in that the governments are sold on the industrial export processing model—and in turn sell it to their constituents—as a means of poverty reduction and expanding opportunities. But that speaks to audiences other than the fastidious international investors. Rather than social concerns, what investors are expected to want to read in the guidebooks are incentive packages and other give-aways that will make their investments super-profitable, often at the expense of local people.

Table 13-3 Economic Sectors Emphasized in the Texts of the Investment Guidebooks (by percentage of country total)

Economic Sector	Barbados	Cabo Verde	Costa Rica	El Salvador	Haiti	Jamaica	Mauritius	Zanzibar
Primary Production	0	45.0	5.5	12.0	40.0	0	0	7.9
Basic Manufacturing	24.3	9.0	16.6	48.0	40.0	36.4	12.5	42.1
Scientific or High-Tech Manufacturing (including information processing)	44.6	9.0	44.4	28.0	0	27.3	62.5	18.5
Transport and Shipping	0	9.0	11.1	4.0	0	18.2	0	21.1
Financial Services	28.4	9.0	11.1	4.0	0	18.2	18.8	5.3
Tourism	2.7	18.0	11.1	4.0	20.0	0	6.3	5.3
Total	100.0	100.0	100.0	100.0	100.0	100.0	100.0	100.0

All numbers are rounded to the nearest 10th

Two Readings of the "Geography Matters" Theme

The neoliberal model establishes the basic requirements of a Third World export platform (e.g. incentives, docile labor, ability to repatriate profits), and these are standard items in the guidebooks. There is also much consistency to promotional embellishment, but important differences in tone, emphasis, and sophistication separate the promotional strategies of the eight cases. Each government's promoters blend in materials that are contextually distinctive, and these take two forms.

One form describes local distinctive features, some of which are cultural, that investors will find attractive and that may set them apart from the competition. Often such descriptions seem designed to exempt the country from conventional Eurocentric images of the Third World as chaotic, disordered, unfamiliar, unpredictable, revolutionary, and ruled by the law of the jungle (Said 1993). In some cases countries choose to counter negative images abroad that are specific to their own recently troubled histories.

While the first form of contextually distinctive material attempts to bolster the standard investment incentive package, a second contrasting form begins to reveal some of the non-correspondence between neoliberal image and local reality. Some aspects of the distinctiveness are prisoner to the internal dynamics of these countries, as in the poor grammar and sometimes incomprehensible English translations in the Haitian and Zanzibari guides, or the Ghandian "social justice" rhetoric of Mauritius. Although writers are trying to "talk the talk," in some cases it becomes apparent that neoliberalism is not really their native tongue. Some investment guidebooks, such as Zanzibar's, ultimately seem incapable of fully embracing the necessary "politics of forgetting" of their local historical geographies (Slater 1993). Zanzibar's guidebook is virtually alone in the sample group in inviting productive activity that will help in "reducing the scourge of poverty," the lofty goals of a regime whose television station logo still reads "Revolution Forever." There is no mistaking Zanzibar's reticence and reluctance to fully embrace what neoliberalism entails. In order to enter the EPZ game, the island's leaders are making the required neoliberal ante of incentives, but ultimately their position proves to be far more equivocal.

These differences in priorities and tone are emblematic of the importance of examining the regional and local contexts of promotion and investment under the global neoliberal banner. One form of difference springs from each government's drive to create competitive advantage. A second form emanates from their historical socio-political development patterns. Either way they crystallize the specifics of place despite the fact that so much of what Third World governments seek and promise is homogeneous. These differences are most closely revealed by development-related motivations, chief among which are those tied to regional dynamics within countries.

Regional Development Planning under Neoliberalism

Many governments in the global periphery, where spatial inequalities are often as stark as the social inequalities, seek to use industrial export processing to redress regional imbalances in development and industrial growth. Central governments commonly demarcate EPZs in economically weaker regions for the practical political reasons of expanding their legitimacy and support, and cultivating a sense of nationalism. Yet attempts to use EPZs to ameliorate regional inequalities generally fail.

Three cases, Costa Rica, Jamaica and Zanzibar, will serve to illustrate why. Costa Rica is relatively developed by Third World standards and relatively successful with EPZ promotion, Jamaica represents an intermediate level of economic and EPZ development, while Zanzibar lags considerably behind on both counts. Together the three examples represent a spectrum of experiences across the global periphery.

Six constraints work against successful deployment of export processing zones for the purpose of fostering regional development in the Third World. Most if not all of these constraints help to explain the poor results of any specific peripheral site chosen for EPZ development.

First, choosing marginal locations to create attractive and competitive EPZ sites requires far larger commitments of government resources for infra-structure, including both basic and more high-technology services, than in already more developed locations. Even if governments are able and willing to marshal such resources, heavy public investment of this sort tilts the balance of costs and benefits away from the host country and towards the foreign investors.

Second, it is relatively expensive to raise workers' quality standards and skill levels in marginal regions, since these are regions where educational investment and achievement have been weakest. Rollback of government expenditures makes extension and expansion of educational infra-structure unlikely. Such work force development would take years to accomplish at any rate.

Third, ulterior motives of short-term political gain often operate behind the scenes in the site selection process, at the expense of successful long-term development. The mere declaration of EPZs can itself create a public perception that development is occurring. Whether or not this conforms to reality now or soon is of less interest to political leadership than the claim of an EPZ.

Fourth, in many cases, export-oriented factories are located where the investors prefer them, regardless of the actually demarcated EPZ boundaries. EPZ-like incentives are often offered extra-territorially to foreign investors, undermining the integrity and attractiveness of specially designated zones in peripheral regions.

Fifth, overlapping or conflicting authorities governing EPZs, controlling industrial land use, and attracting foreign investments loom largest in the more marginal and less investor-friendly settings. The apparent intractability of the public agency labyrinth decreases the odds that an EPZ situated for regional development purposes will ever be fully deployed.

Sixth, and as the following three case studies make clear, regional planning and development in the neoliberal era is heavily weighted towards investors' interests. More than in previous decades, investors have power over planners to dictate not only the location of EPZ development, but also the distribution of its benefits.

The Costa Rican Example

In the early 1980s, the Costa Rican government built its first free zone near its major port of Limón, on the Caribbean coast. This coastal region certainly qualifies for regional development assistance. It contains the majority of Costa Rica's Black population, is weakly developed, and has high un- and under-employment (Augelli 1962; Barry 1991). In 1992, private investors opened a second peripherally-located EPZ, this time on the Pacific coast near the country's secondary port of Puntarenas. This region, too, lags far behind the national core region of the central valley both economically and socially, although not as far behind as the Limón region. To further encourage investment in such peripheral regions, the government has set a lower minimum wage than that which applies in the core region of the central valley. Actual pay rates are similarly lower in the peripheral regions.

Despite all these regional development incentives, investors have shown very little interest in the two coastal free zones. Since its founding, Limón's EPZ has been scarcely occupied, and through 1994 had but one tenant. The Puntarenas free zone has similarly had few tenants and many vacancies. Their locations require that they truck raw materials and finished products two to three hours to and from Limón, or secondarily Puntarenas, up over mountains to reach the San José metropolitan area. Alternatively, some investors are willing to pay even more to fly materials and products in and out of the central valley. In Costa Rica, industrialists overwhelmingly prefer the seven privately-owned free zones around San José.

This remarkable economic geography can be explained by negative investor perceptions of the coastal areas, which show no signs of being swayed by the government's regional development efforts. One survey of industrialists summarizes their views as follows. Investors assessed the business prospects of the two coastal zones as very poor, owing to "a poor work ethic, lack of an 'industrial culture' and work stability, union tendencies (Limón), and the hot weather" (Altenburg 1993, 24). In Costa Rica's central valley, industrialists can employ relatively educated workers of primarily Spanish descent, where labor unions are virtually non-existent. Employer-sponsored worker organizations, called "*Solidaristas*", provide a forum for worker-management dialogue, but they are prohibited by law from collective bargaining. *Solidaristas* long ago effectively substituted for trade unions in the central valley, and virtually all free zone workers there belong to one. Given the long history of government favoritism towards the central valley and the lack of organized threat from the local work force, industrialists cannot be swayed to participate in current regional development efforts.

The Jamaican Example

The geographical impacts of export industrialization on Jamaica's urban system demonstrate many similarities to Costa Rica, and one crucial difference. Like Costa Rica, Jamaica offers investors investment incentives and locations in both core and peripheral regions, ranging from the primate city of Kingston, to the secondary city of Montego Bay and many rural locations. In both countries, investors have shown greatest interest in factory sites located in or near the primate city, and very little interest in factory sites in peripheral and/or rural regions, despite government efforts at regional development.

The crucial difference between the two cases is in labor market conditions at the various levels in the national regional hierarchy. In Costa Rica, labor is considered most compliant in the core region and therefore investors generally avoid all peripheral sites. In Jamaica, labor is relatively uncooperative in the national core compared to the secondary city of Montego Bay. The result is a displacement of some investment away from the primate city and towards the secondary city. Recent government investments have made manufacturing-for-export in previously-unindustrialized Montego Bay possible, but the main cause of its growth is Kingston's unattractiveness rather than regional development planning.

Although only about one-quarter of Jamaica's public industrial space is located in the ten rural parishes, about two-fifths sat vacant during the early part of this decade. Most of this vacancy is associated with an idle designated free zone and an unused industrial park available for EPZ-type investment in the parish immediately west of the Kingston metropolitan area. Because factories exporting from these sites would use Kingston's port facilities, the lack of demand for these rural sites reflects negatively on the national capital. Most other industrial sites in rural Jamaica were developed during the last two decades as political "pork" projects by local members of Parliament seeking to demonstrate their commitments to developing their constituent regions. These rural sites have had high vacancy and tenant turn-over rates. Industrialists that have tried them have been mostly ISI-oriented Jamaican citizens rather than export-oriented foreigners. The rural sites are marred by investor concerns about labor availability and quality, and about infra-structural deficiencies and transportation costs, a pattern with parallels in other Third World countries (World Bank 1992).

The size of Montego Bay's factory space is dwarfed by that of Kingston, but it has the only fully occupied sites on the island. Montego Bay also offers investors the normal free zone incentives. However, industrialists' demand for the Montego Bay Free Zone is in part attributable to the decision on the part of political and economic leaders in the early 1980s to make it the site of a *digiport*, a facility offering satellite-based high-speed telecommunication worldwide. The digiport is partially funded and owned by Britain's Cable and Wireless and the U.S. AT&T. The digiport has drawn firms specializing in information and computer image processing, and using some innovative flexible production methods (*The Gleaner* 1990). The communications facility attracts firms seeking to reduce production cycle time and to move data back and forth from the U.S. Prior to the digiport and to a lesser extent since, information-processing firms congregated in Kingston, the logical location for such enterprises given its industrial infra-structure and large labor supply.

The digiport partly explains investors' interest in Montego Bay, while the remainder largely stems from Kingston's growing unattractiveness compared to the second city. Although the digiport's operating license restricts it from serving non-information based firms, this has not stopped garment firms from filling virtually all of Montego Bay's remaining industrial space. Plant managers and government officials suggest that Montego Bay is attractive because of relatively high worker productivity, low absenteeism and turnover, and an absence of work stoppages. In stark contrast with the vacant factory shells in all other parts of the island, including Kingston, in Montego Bay the government has not built additional industrial sites quickly enough to keep up with the growing demand (*The Gleaner* 1993).

Some political and economic leaders in Montego Bay wonder if the public agencies directing industrial export development from Kingston have intentionally slowed growth in the second city. While Montego Bay's factory shortage frustrates potential investors, the work force in this industrial boomtown has more immediate problems. Montego Bay, with its tourism-based land and real estate prices, does not provide affordable housing for a majority of its EPZ workers. They must make a commute on unreliable mini-buses of up to two hours one-way from surrounding towns, an arrangement beginning to alarm public and private managers worried about their labor supply (Dixon 1994; Kelly 1989).

EPZ development in Jamaica illustrates that disincentives for investment in the primate city can steer investors towards sites down the urban hierarchy. However, this should not be understood as an example of successful regional development so much as a case of unsuccessful primate city development. There is a tremendous need for formal sector employment in Kingston. The overall impact of disincentives to investment in the primate city is to reduce total country-wide investment. Secondary city development here amounts to a siphoning off of some of the investment from the main city without net national benefits.

The Zanzibari Example

Zanzibar is a semi-autonomous island group within the United Republic of Tanzania. Zanzibar is a far more recent entrant onto the industrial export processing bandwagon than Costa Rica or Jamaica. The Zanzibar government formally began its own EPZ policy in 1992 independent of the Union (or "mainland") government of Tanzania. Three export processing zones have been demarcated in the Zanzibar islands since 1992 (COLE 1994). The Zanzibar regime, heavily dependent on foreign assistance to operate even at a basic level, used its limited sphere of autonomy to attempt to bring development to Pemba, the poorer of Zanzibar's two main islands (Myers 1996). Other economically marginal areas on the main island, known locally as Unguja but internationally as Zanzibar island, were targeted in the investigation of possible EPZ sites.

Two of Zanzibar's EPZs were demarcated with a clear eye toward internal politics and development in the islands, rather than being sited for investor interests or planning logic. The government declared an EPZ on a barren coralline peninsula at the northeastern tip of Pemba island in a village called Maziwang'ombe (Cow's Milk). President Amour overrode the arguments of consultants and planners in an attempt to bring a development showcase to what had once been his ruling party's only region of support on Pemba. The ruling party's decisive loss there in October 1995 in the islands' first multiparty elections since 1963 suggests that the ploy made little difference. Besides its notoriety for political opposition to the government, Pemba's labor force has a lower educational base and skill level than that of Unguja, and its basic infrastructure lags behind (Myers 1996). Investors were not interested in going to this "periphery of the periphery".

Zanzibar's second EPZ is where the industrial export processing strategy has become a novel centerpiece of the territory's development dreams. It is also where a dualistic institutional structure of investment promotion and complex overlapping responsibilities have created the largest potential tensions. The Zanzibar Free Economic Zone Authority (ZAFREZA) has declared the entire Fumba peninsula to the south of the city an economic free zone and launched an industrial park and resort complex it calls Star City. The Authority claims that Star City's scheduled investments total over a billion U.S. dollars, and that it will include 160 different factories, provide 10,000 jobs and still have room for two giant resort hotels and a golf course (Lorch 1995). The government plan is to turn Fumba Bay into Unguja island's second free port, replete with container terminals and warehouses.

Yet Star City is already a falling star. Its exorbitant financial demands on the government for infra-structural development are wildly unrealistic given the severe cutbacks on government expenditures required by Tanzania's and Zanzibar's long list of creditors (Aslam 1996). The Free Zone Authority has run roughshod over the National Land Use Plan created by other government agencies. Two building materials industry investments underway as a part of the Fumba peninsula EPZ actually lay outside of its declared boundaries, and one is in the reserved expansion zone for the Zanzibar international airport. No environmental impact assessments have gone forward in Fumba. The Free Zone Authority has also infringed on the land rights and cultural values of the current residents of the peninsula, as well. Other government officials could barely contain their outrage at ZAFREZA for doing these things in the election year, and ZAFREZA now routinely withholds important data from its rival agencies.

In contrast to the vacancy that characterizes the two EPZs just described, Zanzibar's third one is nearly full of tenants and operational in most respects. It is a minor outfit with just 18 small factories or warehouse facilities. It was a simple step to convert what had been the government's Small Scale Industries center into an EPZ. The center's warehouse and workshop space, a vestige of Zanzibar's 25-year experiment with an ISI approach to industrialization, was by the late 1980s little used, well serviced, and efficiently located just a mile from the international airport in a pleasant suburb. Zanzibar city, with a 7:1 primacy ratio, is overwhelmingly the island's primate city. Yet even as the investor-friendly spatial logic of the EPZ has worked, the largest foreign industrial employer in the zone has been plagued by work stoppages since opening day (*Nuru* 1994). This work stoppage problem in Zanzibar appears to be parallel to Kingston's bad investor climate in terms of labor relations, with the added dimension of East Africa's long-simmering tensions between Asian and Arab holders of capital and African laborers. Indeed, the Zanzibar case exhibits a poorer, smaller-scale and newer-model version of the regional development failures in both Costa Rica and Jamaica.

Public industrial site vacancies in rural and small town Jamaica, Costa Rica, and Zanzibar suggest that investors are unenthusiastic about governments' attempt to use manufacturing promotion to develop their peripheries (World Bank 1992). Industrial export investment is more urban-oriented and centralized than its state sponsors intend, and is more strongly determined by investor than state interests. The failure of regional development planning under industrial export processing is a microcosm of the weakening of government as a shaping force for Third World development during the neoliberal era. No longer can planning alter the internal geography to the extent it did in regional development's heyday (Friedmann 1966; Bromley and Bromley 1988; Sidaway and Simon 1990). It is therefore unsurprising when World Bank analysts resign themselves to advising African governments that perhaps excessive urban primacy is not such a bad thing (Becker, Hamer and Morrison 1994); after all, African governments, like many Third World governments, seem to have little choice but to accept the regional development imbalances that the neoliberal approach reinforces.

Conclusion

Thanks to a collapse in traditional export markets and an overexposure to foreign indebtedness, Africa, Latin America and the Caribbean experienced economic decline and swelling poverty in the 1980s. Governments responded, with some prodding from the IMF, the World Bank, and Western governments, with a rapid economic opening and the provision of incentives to attract foreign capital to produce for export, in a manner claimed as representative of East Asia's industrialization trajectory. Indeed, there would at first glance appear to have been a remarkably rapid unification of policy around a neoliberal development model across the Third World.

As our analysis of investment guidebooks and our regional development case studies suggest, there is actually an expanding disjuncture between traditional development theories, current industrial models and the actual practice of industrial planning today. It is perhaps misleading to term the industrialization program we label industrial export processing as "planning", because that suggests more strategy, coordination and long-term goals than currently characterize Third World industrial policies. The myriad of subsidies and enticements that governments hold out to fastidious investors have become virtually ubiquitous. Countries therefore cancel out each other's attractiveness. Perhaps Third World planners consider the generous incentives as short-term public investments for longer-term national development gains. In practice they function more like mandatory antes for entry into a high stakes poker game against a sea of international competitors, most of whom are ill equipped and can scarcely afford to play. Pressures are great to pile the incentives ever higher, still with no guarantee of payoffs. In many cases, poor countries have given away the bank before the investors even show up.

Investigation also suggests that there is little capacity for real development planning today from within the Third World countries, because external interests hold most of the cards. East Asia's NICs developed their successful export-oriented industrial bases with a strong role for government. Most contemporary Third World governments are weak from debt and traditional export collapse, yet they offer incentives and subsidies which they can ill afford. Virtually no alternatives to the neoliberal approach are being articulated, and investors have many options to choose from both at the international scale and within the national territory. Both national and regional development are thus foiled by the relative weaknesses of governments of the global periphery vis-a-vis their international counterparts (multilateral agencies, core governments, and international investors).

EPZs and their promotional policies are quintessential representations of the current experiments with neoliberal development planning, but evaluation of them suggests that these are flawed experiments. Their obvious contradictions and instability make EPZ policies dangerous. They contribute to creating and entrenching a new and qualitatively different set of economic conditions in their host countries. These countries are exposed to greater trade dependency and international vulnerability after a long and troubled history of such arrangements. EPZ policies thereby contribute to propelling countries down a flawed development policy path, while constricting options for alternative and more stable and sustainable approaches. The promotion of export manufacturing demands a high price from Third World governments and workers, in exchange for fleeting and inadequate levels of foreign investment.

References

Altenberg, M. 1993, Unpublished Report on *Investors and Free Zones in Costa Rica*, Obtained from La Corporación de la Zona Franca de Exportación, San José, Costa Rica.

Aslam, A. 1996, "Tanzania-Finance: Mkapa in U.S. to Woo Investors", *PeaceNet World News Africa Digest*, (electronic), (October 8).

Augelli, John, 1962, "The Rimland-Mainland Concepts of Cultural Areas in Middle America", *Annals of the Association of American Geographers*, 52:119-29.

Balassa, B. 1993, *Policy Choices for the 1990s*, New York: New York University Press.

Barry, T. 1991, *Costa Rica: A Country Guide* (3rd edition), Albuquerque, New Mexico: The Resource Center.

Becker, C., A. Hamer and A. Morrison, 1994, *Beyond Urban Bias in Africá: Urbanization in an Era of Structural Adjustment*, Portsmouth, New Hampshire: Heinemann.

Bowman, L. 1991, *Mauritius: Democracy and Development in the Indian Ocean*, Boulder: Westview Press.

Bromley, R. and R. Bromley, 1988, *South American Development: A Geographical Introduction*, Cambridge: Cambridge University Press.

Brown, M. 1995, *Africa's Choices after Thirty Years of the World Bank*, New York: Penguin.

Chandra, R. 1992, *Industrialization and Development in the Third World*, New York: Routledge.

Colclough, C. 1991, "Structuralism versus Neo-liberalism: An Introduction", pp. 1-25 in C. Colclough and J. Manor, (eds), *States or Markets? Neo-liberalism and the Development Policy Debate*, Oxford: Clarendon Press.

COLE (Commission for Lands and Environment - Zanzibar), 1994, *The Zanzibar National Land Use Plan*, Zanzibar: Revolutionary Government of Zanzibar.

Crick, M. 1989, "Representations of International Tourism in the Social Sciences: Sun, Sex, Sights, Savings and Servility", *Annual Review of Anthropology*, 18:307-344.

de Kadt, E. 1979, *Tourism: Passport to Development?*, Washington: World Bank.

Dicken, P. 1992, *Global Shift*, New York: Guilford Press.

Dixon, A. 1994, *Greater Montego Bay Strategic and Development Plan, 1994-2014*, February 14 Draft.

Dunning, J. and R. Narula, 1996, *Foreign Direct Investment and Governments: Catalysts for Economic Restructuring*, New York: Routledge.

Friedmann, J. 1966, *Regional Development Policy: a Case Study of Venezuela*, Cambridge: MIT Press.

Gleaner, The. 1993, "No room for factories", Kingston, Jamaica: (May 26), p.1.

Gleaner, The. 1990, "Data entry boost for Montego Bay", Kingston, Jamaica: (Feb. 5), p.2.

Islam, S. 1996, "Labor: EPZ Abuses Denounced by Union Body", *PeaceNet World News Service, Africa News Digest*, (electronic), May 1.

Kelly, D. 1989, *Report of Board of Inquiry into the Garment Industry*, Unpublished report solicited by & prepared for Jamaica's Ministry of Labor, by Dan Kelly, Attorney at Law, February.

Klak, T. 1994, "Maidenforming the Caribbean: Concerns about Jamaica's Industrial Export Promotion Policy", *Caribbean Geography*, 5:102-116.

Klak, T. 1995a, "A Framework for Studying Caribbean Industrial Policy", *Economic Geography*, 71:297-316.

Klak, T. 1995b, "Squandering Development Possibilities from Within or Without? Problems with Neoliberal Industrial Export Policy in Jamaica", Paper presented at the Association of American Geographers annual meeting, Chicago, March 14-18.

Klak, T. and Rulli, J. 1993, "Regimes of Accumulation, the Caribbean Basin Initiative, and Export-Processing Zones: Scales of Influence on Caribbean Development", pp. 117-150 in E. Goetz and S. Clarke, (eds), *The New Localism: Comparative Urban Politics in a Global Era*, Beverly Hills: Sage Publications.

Krueger, A. 1993, *Political Economy of Policy Reform in Developing Countries*, Boston: MIT Press.

Lindauer, D. and M. Roemer, 1994, *Africa and Asia: Legacies and Opportunities in Development*, San Francisco: ICS Press.

Lorch, D. 1995. "Zanzibar Journal: Where Life Has No Spice, a $1 Billion Pick-Me-Up", *The New York Times*, (November 6), A3.

Mwakanjuki, D. 1995, Director of ZIPA (Zanzibar Investment Promotion Agency), personal interview with Garth Myers and Makame A. Muhajir, Zanzibar, July 8.

Myers, G. 1996, "Democracy and Development in Zanzibar?: Contradictions in Land and Environment Planning", *Journal of Contemporary African Studies*, 14:221-245.

Naya, S. and R. McCleery, 1994, *Relevance of Asian Development Experiences to African Problems*, San Francisco: ICS Press.

Nuru. 1994, Mgogoro wa Zanzibar Trends: Kamisheni ya Kazi yaipinga Maeneo Huru (The Strike at Zanzibar Trends: Labor Commission Obstructs EPZ), *Nuru (The Light, Zanzibar)*, (December 16), 1.

Peet, R. and M. Watts, 1993, "Introduction: Development Theory and Environment in an Age of Market Triumphalism", *Economic Geography*, 69:227-253.

Ramsaran, R. 1992, *The Challenge of Structural Adjustment in the Commonwealth Caribbean*, New York: Praeger.

Richardson, B. 1992, *The Caribbean in the Wider World, 1492-1992. A Regional Geography*, New York: Cambridge University Press.

Rubio, L. 1995, "Gobierno y corresponsales extranjeros", *La Jornada Semanal*, (June 25) 4.

Said, E. 1993, *Culture and Imperialism*, New York: Knopf.

Sidaway, J. and D. Simon, 1990, "Spatial Policies and Uneven Development in the 'Marxist-Leninist' States of the Third World", pp. 24-38 in D. Simon, (ed.), *Third World Regional Development: a Reappraisal*, London: Paul Chapman.

Slater, D. 1993, "The Geopolitical Imagination and the Enframing of Development Theory", *Transactions, Institute of British Geographers*, 18:419-437.

Vanderbush, W. and T. Klak, 1996, "'Covering' Latin America: The Exclusive Discourse of the Summit of the Americas as Viewed through The New York Times", *Third World Quarterly*, 17:3:537-556.

Williamson, John, (ed.), 1990, *Latin American Adjustment: How Much Has Happened?* Washington: Institute for International Economics.

World Bank, 1992, *Export Processing Zones*, Policy and Research Series #20, Washington: The World Bank.

World Bank, 1995, *World Development Report 1995*, Washington: The World Bank.

14 Urban Planning and Development in Calcutta: Local and Global Issues

CHRISTOPHER CUSACK

Growing concerns exist over whether Western planning paradigms have global relevance and, hence, applicability to non-Western settings. Universal application of Western practices has been criticized for overlooking the potential contributions of non-Western planners. This chapter provides an overview of Western planning paradigms as applied in Calcutta.

One of the largest metropolitan regions in the world, Calcutta serves as a global example of the problems and potential facing mega-cities in developing nations. The evolution of Calcutta has resulted from a number of influences, leading to many current development issues facing it today. From 1962, Calcutta has utilized American planning techniques to guide its physical development which have largely proven unsuccessful.

The objectives of this case study of Calcutta are to: 1) describe the formidable problems confronting Calcutta, so as to reveal the context in which efforts at planning and development must take place; 2) review the promotion and application of American and Western planning paradigms on this decidedly non-Western city; and 3) provide a feasible planning alternative which integrates foreign and indigenous planning while viewing the city in a global context.

Western Planning: Historical Dominance

The colonial era initiated a dominance of Western planning on a global scale that Western planners have yet to relinquish. Early planning programs developed in the West, particularly Great Britain, were implemented throughout the colonial world. In these colonies, "the planning systems adopted are almost always identical to those of their former colonial motherland" (Choguill, 1994, 942). Examples of this phenomenon include: legislation in Nigeria that has noticeable similarities to the British 1932 Town and Country Planning Act and, the Malaysian 1976 Town and Country Planning Act, which is virtually identical to the British Act of 1971. Furthermore, the Calcutta Improvement Act of 1911 was adopted directly from the Glasgow Improvement Trust of 1866 (Choguill 1994, 942-943).

The modern history from the colonial era to the present of "exporting planning", has been divided into three phases:

1. A period up to the early twentieth century when settlements, camps, towns and cities were consciously laid out according to various military, technical, political and cultural principles, the most important of which was the principle of securing military-political dominance.

2. A second period, beginning in the early twentieth century, which coincides with the development of formally-stated 'town planning' theory consisting of survey and analysis techniques and professional skills and legislation.

3. A third period of post- or neo-colonial developments when cultural, political, and economic links have, within a larger network of global communications and a situation of economic dependence, provided the means for the continued export of values, ideologies and planning models (King 1980, 203).

Western planning practices may be considered the dominant planning paradigm throughout modern history. Initially, British planning had the most prominent role, as it dominated the colonial era and based its development plans on military-political pre-eminence. After a time, this approach was succeeded by the town-planning paradigm, which also had British origins. Most recently, however, the dominant paradigm has been that of the American rational planning model. Although approaches may have shifted or been replaced over time, one factor has remained constant: each dominant paradigm has been of Western origin, and has been implemented on a universal basis.

The rational method defines planning as "the art of making social decisions rationally" and views planning as a "form of technical reason" which "attempts to link scientific and technical knowledge to actions in the public domain" (Friedmann 1987, 36-38). Although rational planning has come under criticism over the past several decades for its claims of technicality, the model continues to be heavily drawn upon by planners in the United States.

Rational planning was adopted for logical decision making and successful plan implementation in developing countries. As a result, the planning process in developing countries noticeably emulated the rational planning model developed through Western theory and practice. This imitation of, and belief in, the rational planning paradigm contributed to the universal acceptance of the idea that "there was only one way—the Western way—to plan and develop" (Sanyal 1990, 23).

Western Planning: Universal Application?

The dominance of the Western planning paradigm has been such that it has been considered to have universal, or at least near universal, application. However, Western concepts of planning are imbued with Western perspectives. Thus the transfer of such planning techniques to non-Western countries also necessarily requires a transfer of these Western beliefs as well. Indeed, in Western democracies,

> The very notion of planning presupposes the existence of: a political market; interest group representation; a culture of 'rationality'; and a bureaucracy committed to Weberian reason, or a rational application of means to receive ends (Dyckman, et al. 1984, 214).

Third World countries are faced with different planning needs under different planning settings. Thus application of Western planning techniques to non-Western settings may often be inappropriate.

> Developing countries may lack any well-developed system of local government. They may have political institutions which do not recognize the democratic claims we take for granted. Their view of rationality may be unrecognizable to Western planners. Environment may be seen as beyond man's capacity to dominate. 'Justice' may have traditional rather than procedural connotations. Almost certainly, such countries will lack a professional bureaucracy. Even as they crave 'modernization' they may resist many of its implications (Dyckman, et al. 1984, 216).

Dominant planning concerns in the West have at times been applied to developing countries with an assumption of relevance and applicability. For example, when slum clearance was the current trend in America, squatter clearance and relocation were advocated for the Third World. Then, when neighborhood improvement programs were promoted in America, similar slum upgrading programs were implemented in the developing world (Qadeer 1990, 201).

While the Western planning paradigm has been implemented across the globe, its success in non-Western situations has often been limited. Though there are a number of reasons for this lack of success, the most prominent is a dearth of adequate capital and resources in the developing world necessary to implement Western-style programs. The problem is then further exacerbated by the requirement of the "acceptable standards" held in Western models.

The irrelevance of such standards, is evident. For example, "at one time urban houses in Nigeria were required to have roofs capable of supporting the weight of snowfall that would be expected in an average British winter" (Choguill 1994, 938). Similarly, concepts such as lighting standards, ventilation, and over-crowding, are based on foreign notions unsuited to many Third World conditions "where cultural and climatic conditions are different, and where the interior of the house plays a role with a different emphasis from that of the house in the West" (Kironde 1992, 1278).

Other areas in which Western concepts and theories may have created rather than solved problems in the developing world include; slum clearance programs of the 1960s and 1970s, direct housing programs, and sites and services programs (Okpala 1987, 144).

As a result, Western planning theories have been challenged for their relevance to conditions in developing nations. At the same time that the universal application of Western planning has been called into question, the need for contextual understanding of local situations by local planning has been championed.

> There is desperate need for indigenous thinking about urban planning, housing designs, and construction material and about urban administrative structures and social organization (Nagpaul 1996, 143).

While Western planners may offer valuable contributions to non-Western settings, a need exists for indigenous input and a true interdependency between Western and non-Western planning. Through such an approach, new concepts and methodologies may be developed from a combination of local experience and foreign expertise.

Case Study: Calcutta

Initial efforts at planning in Calcutta, India following Independence represent an attempt to implement Western, specifically American, planning methods in a non-Western situation. The city of Calcutta is unlike any Western city in its population trends, political setting, social and cultural issues, environment, and service needs. Indeed, Calcutta differs from Western cities in nearly every facet of urban life. A review of the problems and prospects facing the city will illustrate the point, followed by an analysis of the application of Western planning to this decidedly non-Western city.

Physical Geography Considerations

Calcutta's position within the Ganges delta plays a significant role in the city's development. The primary city in its region, Calcutta has grown in a linear fashion along the Hoogly River which divides the city roughly in half. Walled in by the river on the west and salt marshes on the east, Calcutta has had no alternative but to develop in a north-south direction. As a result, the city is approximately 64 kilometers long, but only 6 to 10 kilometers wide (Singh 1968,343). The city has also had to develop around the obstacles provided by numerous lakes, ponds, marshes, and other bodies of water which are found in the delta. The aqueous surroundings of Calcutta have had a profound and primarily detrimental impact on the development of the city.

Most of the Calcutta area receives its fresh water from the Hooghly river, whose supply from the Ganges-Padma system had long been diminishing. In 1975, with construction of the Farakka barrage across that system, the supply to the Hooghly has increased significantly (Dutt 1978).

Ground water also supplements the surface supply. Currently, uncertainty exists concerning the future availability of ground water supplies in the central city, while water in the east and southeast is rapidly becoming polluted and unhealthy for consumption. The monsoon significantly contributes to the city's average annual rainfall total of 64 inches per year (Earthwatch 1992,91).

The geography of the region serves as the determinant of the supply and drainage of water in Calcutta, and is the source of both beneficial and detrimental impacts on the city. While the future prospects of water supply are not unfavorable, the natural problems posed by the situation of Calcutta have been inflated by the city's population size, an outworn water distribution system, and increasing sources of pollution (CMDA 1986, 5). Flatness of the land "leads to insufficient drainage throughout the area particularly during July-September which has 80 percent of annual rainfall total" and, thus, waterlogging is common, which "causes difficulties for transportation and general day-to-day activities" (Dutt 1978, 266-267).

Population Trends

Prior to the late 1800s, estimates of Calcutta's population growth remain indeterminate. By the end of the 19th century, Calcutta had grown into the most important colonial trading post and second biggest city of the British empire (Munsi 1975,78). Since Independence, Calcutta has continued to experience a rapid population increase, primarily as the result of significant immigration from the surrounding countryside (Figure 14-1).

Calcutta
Population Growth

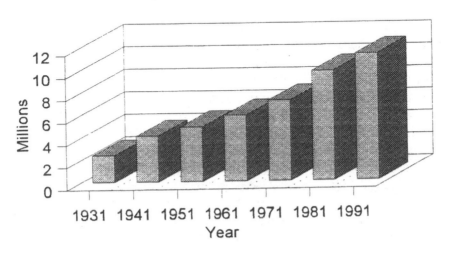

Figure 14-1   Calcutta: Population, 1901-1991
Source: Nagpaul, 1996.

Calcutta's population, (10,916,272 according to the 1991 census) which continues to grow by some 200,000 to 400,000 people every year, is expected to reach 12.9 million by the year 2001. So sizeable is Calcutta's expected population that by the year 2000 the city is anticipated to rank as the 5th largest city in the world (Darnay 1992, 616).

This population growth, coupled with severe geographical limitations, has resulted in a deficiency of open space in the city. Indeed, the open space per person in Calcutta at present is approximately 21 square feet, while the international standard is 290 square feet (Munsi 1990, 10).

With more than one-third of its approximately 11.5 million people packed within the core 40 square miles, creating a density of 56,927 per square mile, the alleviation of such extreme congestion is a primary planning concern. By comparison, densities of other mega-cities are significantly lower than that of Calcutta. For example, Mexico City has a population per square mile of 40,037; the New York City population density is 11,480; and London's 10,429 (Droste

1994, 463). Failure to accommodate this almost unmanageable population has resulted in the explosion of slums, which currently number over 4,000 and are "home" to over one-third of the population of the city (Banerjee 1990, 29).

Despite a continuing overall increase in Calcutta's population, the growth rate of the Calcutta Municipal Corporation area (CMC) has decreased over recent decades. This decline has been offset, however, by the continuing high population growth rates of the larger Calcutta Urban Agglomeration (CUA), which comprises the collection of towns surrounding Calcutta along both banks of the Hooghly river. The city of Calcutta has been evolving as the center of a significant and ever-growing peripheral area.

The Urban Structure

The number of inhabitants of the city places a severe strain on the city's ability to function. With over 11 million people existing in a city built originally for 1 million, Calcutta often cannot provide the most basic of services. For example, the most visible element of the infrastructure, the street network, is extremely inadequate. Only 6.5 percent of the area of a city faced with seemingly insurmountable congestion is occupied by roads. In comparison, cities with more than 250,000 population in the United States have, on average, 28.9 percent of city surface devoted to streets (Munsi 1990, 11).

Additionally, competing for transit on Calcutta's streets are excessive numbers of vehicles.

In Calcutta, on such a limited space ply 750-800 private buses, about 800 mini buses, 600 state buses, 9 thousand taxis, over 80 thousand private cars, 26 thousand lorries and trucks and 33 thousand two wheelers (Munsi 1990, 11).

The problem is further exacerbated by slow-moving traffic, including pedestrians, handcarts, rickshaws, and animal-drawn vehicles, as well as a severe shortage of available parking space.

Rather than showing signs of abating, conditions will worsen before they improve. Total motor vehicle registrations more than doubled between 1981 and 1989, and the trend is likely to continue to the year 2000, with the number of motor vehicles doubling every six years or so (Earthwatch 1992, 98).

These trends not only serve to increase traffic congestion, they also produce significant air pollution. Calcutta has the dubious distinction of being one of the most polluted cities in India, with an estimated 60 percent of the population suffering from some kind of respiratory disease due to air pollution (Nijkamp 1994, 263).

The primary source of air pollution in Calcutta is suspended particulate matter (SPM), which may be attributed to the high level of coal used by industry in the city. Additionally, Calcutta must contend with potentially hazardous levels of carbon monoxide (CO) and nitrogen dioxide (NO_2) (Earthwatch 1992, 20). Calcutta is a mega-city with mega-problems. Indiscriminate growth leading to a myriad of economic, social, and environmental problems burdens the city with seemingly insurmountable obstacles. Yet still the city survives, and in some sectors, thrives. The city must not be judged by Western standards, as Calcutta's capacity for survival makes it different from any Western city (Biswas 1992, 485).

Politics in Calcutta

The political scene in Calcutta has been one of conflict and change since India's Independence. Politics has played a profound and primarily detrimental role in planning efforts in Calcutta. Following Independence, Calcutta politics was dominated by the Congress party. Comprised of members of the city's elite, Congress controlled the municipal corporation and nearly all of the legislative constituencies in the state capital (Chatterjee 1990, 31).

However, mass in-migration and burgeoning social and economic problems throughout the 1950s enabled leftist political parties to gain influence in local politics. Their build-up culminated in 1967, with the defeat of the Congress party.

The next several years witnessed one political party supplant another in volatile exchanges of power. This instability fostered a new political force in the form of the leftist Naxalite movement that aimed at capturing the state through violence. As a result, the national government intervened and suppressed the Naxalite party. The elections of 1977 vindicated the left, as the Left Front government was elected and has since dominated politics in Calcutta (Chatterjee 1990, 32).

More than just a source of violence and confusion, the political instability of the 1960s and 1970s proved a hindrance to planning efforts in the city. Initially, the 1967 defeat of the Congress party, which had provided the principal support for planning efforts, served to detract from the attention and consideration being given to planning.

Subsequent involvement of the national government led members of the leftist parties, as well as local planners, to suspect the work of the planning department as a plot to obstruct the growth of Marxism in Calcutta (Banerjee and Chakravorty 1994, 75). These developments, illustrating nicely the relationship between planning and politics, have served to hinder successful planning in Calcutta.

Economic Trends

Though its own economy has virtually stagnated since the 1960s, Calcutta continues to serve an important, although diminished, economic role in the national economy. Several explanations have been given for the decline of Calcutta as the foremost economic force in India, as there are a number of fundamental and interrelated agents of the city's economic stagnation.

The first can be attributed to persistent drought during the first half of this century in farming areas of the city's hinterland. As a result, Calcutta became the migration destination for vast numbers of unskilled, or only semi-skilled, migrants from the surrounding countryside. This led to the rapid proliferation of slums, or *bustees*, which had a negative impact on the real property values of the city. (Editors' note-See Chapter 7 for a discussion of these slums.)

A second detrimental element is the city's political turmoil, which has not only hindered planning efforts, but also played a substantial role in Calcutta's economic decline. The impact of political instability was particularly noteworthy during the late 1950s and 1960s, when conflict between political parties created an extremely unstable situation. The results of such a situation were massive capital outflows from Calcutta and an escalation of urban under-employment (Goswami 1990, 94).

In addition to political instability and overwhelming rural immigration, other factors have contributed to economic decline in Calcutta. Primary among these is the manufacturing base of the city which did not diversify and subsequently stagnated. Particularly harmful for the city's manufacturing sector is the general stagnation of the jute industry, which historically had served as the major modern industry in Calcutta. Currently, six major jute mills in and around Calcutta are being operated at heavy loss by the government-owned National Jute Manufactures Corporation (Datta 1990, 101). The leftist government from 1977 to 1991 discouraged foreign capital and made it difficult for the native entrepreneurs as well. As a result few new industries were established and almost total industrial stagnation set in. The change of policy since 1991 has started to attract new investments.

Calcutta's economy remains tied to industrial output and has not yet climbed aboard the post-industrial engine which is driving today's economic forces. Yet with a cluster of jute mills among its manufacturing industries, Calcutta is in position to move from output to research and development in this historically productive industry. Calcutta must diversify its economic base and work to develop its own niche in the global economy in order to secure a more vibrant economic future.

Planning in Calcutta

Initial planning efforts in the city of Calcutta date back to 1961 and the creation of the Calcutta Metropolitan Planning Organization (CMPO). The CMPO, which was given responsibility for the development plans of the city, was established as a result of governmental concern for the city's many notable environmental and health problems. Specifically, a devastating cholera epidemic which plagued the city in 1958 prompted then chief minister of West Bengal, B.C. Roy, who was a physician, to invite the World Health Organization to inspect and evaluate the horrific living conditions in Calcutta (Bagchi 1987, 597). The international agency recommended several measures to deal with these concerns, and the CMPO was set up to implement these proposals.

In 1966, the first significant product of the planning efforts for Calcutta, the *Basic Development Plan* (BDP), was published. During the preparation of this plan, the CMPO was substantially assisted, even directed, by the Ford Foundation, a United States advisory group. The *Basic Development Plan*, concerned with development between 1966 and 1986, was an effort at providing a planned solution to a series of urban crises.

This plan, "outlined a broad perspective for the future growth of the region and a five-year development program" (Angotti 1993, 179). The primary strategy recommended in the Plan was for the transformation of Calcutta into a bi-polar metropolitan structure. This objective was to be achieved by promoting Kalyani-Bansberia, located north of the core, as a counter-magnet to the Calcutta-Howrah center. Though significant infrastructural improvements were made to both the urban center of Calcutta, as well as to Kalyani-Bansberia, inadequate funds were provided, and the region failed to develop as planned (Bagchi 1987, 598).

Not only did the BDP not succeed in its primary objective, but it also encountered a number of additional criticisms. Among these was the charge that the Plan lacked potential for increasing revenue, as well as giving scant attention to the role of the informal market in economic development. Perhaps most meaningful were the criticisms that the Plan overlooked potential contributions from local planning agencies and the local government (Banerjee and Chakravorty 1994, 74).

In such a complex situation, the Ford Foundation consultants alone cannot be held responsible for the inadequacy of the BDP. A dearth of necessary funds was a critical contributor to the failure of the plan. Additionally, the political scene at the time was one of uncertainty and conflict. A critical element missing from the mix was a sustained effort on the part of the Ford Foundation. American expertise brought to Calcutta through the Foundation was a one-time event and therefore neglected to consider the changing dynamics of Calcutta over time. This combination of factors was largely responsible in ensuring that the Plan would not succeed.

Western Influence and Planning in Calcutta

The consequences of the basic development plan warrant review. The unique approach of the BDP may be considered more meaningful than the plan's actual results. This approach, an attempt to implement American planning methods in a Third World city, may be viewed with a wider lens as an attempt to promote a universal paradigm. Successful application of the dominant Western approach to a non-Western city would give strong support to the universality of the Western paradigm.

Though the BDP was a plan tailored to meet the unique needs of the city of Calcutta, the prospect of receiving credit for the development of a worldwide planning approach may have distracted the Ford Foundation. Calcutta's planning experience also serves as the basis for an essential theoretical review. The involvement and authority granted to the Ford Foundation demands retrospective analysis. The focus of such an analysis must be placed on the potential transfer of Western planning concepts to non-Western countries.

The Ford Foundation was comprised primarily of American planners. Thus the assumptions and influences brought to Calcutta by the Foundation were more specific than simply Western, rather they were distinctly American. This perspective was influenced by the American experience of large-scale metropolitan growth and the involvement of the social sciences in planning and model development.

> Physical land-use planning was considered deterministic, and the 'master plan' too rigid. There was growing advocacy for a flexible, policy-planning approach. The social sciences introduced new planning methods, such as a multidisciplinary approach and mathematical modeling to solve urban problems. These were major American planning innovations which the Ford Foundation brought to Calcutta in their briefcases (Banerjee and Chakravorty 1994, 76).

Furthermore, the three guiding central tenets of the Ford Foundation: regional planning; a strategic policy orientation; and a strong reliance on data analysis and modeling were accepted without reserve by the local planning organizations (Banerjee and Chakravorty 1994, 76).

The Ford Foundation, comprised primarily of American experts, attempted to implement a Basic Development Plan for Calcutta, an Indian mega-city. Members of the Ford Foundation were adherents of American culture, a democratic government, environmental regulation, and a dependence on the rational method of planning. The social and political environment of Calcutta was unlike any city they had encountered before.

The Ford Foundation, which may have underestimated the differences and difficulties with which they would be faced in Calcutta, sought to provide assistance without thoroughly valuing local planning knowledge.

> This export of planning expertise to the Third World can be seen as a form of technology transfer, in which the technology can be a brand of planning currently practised in the Western world. The implications of uncritical technology transfer are many and are often painful to the receiving country (Dyckman, et al. 1984, 216).

What then is necessary is a more unified approach in which Western planners bring their technology to non-Western countries with the intent of working with, rather than without local members of the planning profession.

The transfer of Western technology, especially in computer "know-how", may serve as a vital tool to non-Western countries. However, fundamental theories based on the Western city experience simply may not apply to cities in underdeveloped states. Western planners must not enter non-Western countries with the sole intent to teach, but rather they also must be ready to learn.

The transfer of Western beliefs, strategies, and technologies to a city such as Calcutta has provided a framework for an examination of the potential for a universal urban planning paradigm. Questions remain as to whether the Ford Foundation truly understood and accepted the context in which their efforts were to take place. It is unfortunately all too obvious, however, that in this particular instance, planning strategies are not universal. Whether the efforts of the Western planners have had any long-term negative effects is also open to debate, though there seems little doubt that the Ford Foundation left no sustained positive accomplishments.

These results indicate that there is no universal antidote for urban problems across the globe. Each city is a dynamic entity with unique prospects and problems. Plans should custom-fit each city under consideration by enlisting the assistance of local members of the planning profession. It is a combination of the technological capability of foreign experts and the contextual understanding of local planners which together may offer the greatest potential to the city.

Strategies for the Future

Previous attempts to alleviate the problems facing Calcutta have proven largely unsuccessful. Yet there is, as always, still hope. What is necessary is the development of new and realistic strategies. While many plans may appear potentially beneficial, without proper implementation and financial backing, they too will result in failure.

Currently, a number of plans have been adopted to provide balance between the metropolis and hinterland of Calcutta. Among these programs are: the creation of new towns and development of growth centers; the promotion of growth of the rural economy; the diversion of water from the Ganges River to the Bhagirathi-Hooghly system; an augmentation of power generation; and the development of urban renewal programs (Banerjee 1990, 32).

Though these programs have enjoyed at least some degree of success, further assistance is imperative in order to stem the continuing deterioration of Calcutta. Additional decentralization strategies have been advocated, and include: the development of urban corridors (Dutt, et al. 1994, 155); and the promotion of district towns and rural municipalities (Dasgupta 1987, 343).

It is the relatively recent trend towards a global economy and the corresponding unrestricted movement of transnational corporations (TNCs) which offer Calcutta new potential for planned development as opposed to indiscriminate growth. Indeed, Calcutta may develop the potential of its existing socio-economic foundation and thereby enter the post-industrial society and compete globally by attracting TNCs to the province of West Bengal.

In today's post-industrial society, locational decisions of businesses differ from the criteria used by earlier production-based industry. No longer does proximity to market or raw materials overwhelmingly dominate locational decision-making of business and industry. As such traditional advantages are declining in importance, human and cultural resources are becoming more critical to successful urban and regional development. Although cities have traditionally been defined from a geo-political perspective, the new challenge is to consider the city from a socio-cultural perspective (Knight and Leitner 1993, 29).

Calcutta must strive to enter fully the global economy by strengthening and promoting its knowledge base, by which is meant "the particular skills, the expertise and knowledge of an entire region and special, innovative achievements in areas of culture and research" (Knight and Leitner 1993, 7). Further-more, Calcutta must make known that it possesses the amenities sought by TNCs. Included among the factors considered by TNCs is the "quality of educational, cultural, recreational, and medical services" provided by the city (Knight 1989, 238). Fortunately, Calcutta already has in place a potentially amenity-rich base, which should enhance its ability to compete globally.

Indeed, the city holds a wealth of educational, research, and innovation potential. Calcutta, which remains one of the academic centers of India, is set apart by "the unusual interaction between academic life and the wider intellec-tual and artistic life of the city" (Chaudhuri 1990, 196). Such synergy, which may be utilized to promote the city, is enhanced by the fact that in Calcutta is,

> the largest number of scientific and technological institutions in the country, and tradition continues to produce individual scientists and technologists of out-standing caliber (Ghose 1990, 214).

An abundance of libraries and the convincing claim to being the music and literary capital of India further establishes Calcutta's potential for knowledge-based development in the post-industrial society. Yet this potentially rich base remains largely untapped. Calcutta has not fully recognized that its future lies not with production, but with its own technological and innovative niche in the global economy. Perhaps even more troubling than the city's ebbing industrial output, is the corresponding decline in entrepreneurial interest in its mills. It is notable that today's post-industrial economy has not reached the jute mills, where "attempts are rarely made to develop new products, and the new re-search laboratories have not been very effective" (Datta 1990, 101).

Recent economic indicators suggest that some elements of Calcutta's economy may be venturing into the post-industrial age. Specifically, govern-ment sponsorship and collaboration has resulted in an "electronics complex" in the Bidhan Nagar area (Datta 1990, 108). It is this type of activity that may provide the Calcutta region with an entrance to the global economy and a boost to the region's overall economic vitality.

It is imperative that local involvement provides the springboard to the post-industrial society. Input from all sectors of society, including the private, public, and civic, is required for the Calcutta region to achieve some measure of socio-economic success. As noted,

> no foreign aid and technical know-how can bring permanent solutions unless indigenous foundations are laid for the provision of basic amenities within the broader framework of socioeconomic planning (Nagpaul 1996 144).

Only by following a regional approach, which recognizes and integrates the importance and potential contribution of the city's existing knowledge-base, including the vast informal sector of the economy, may Calcutta progress.

Individual initiatives within the informal economy could provide successful product development and innovative techniques. One example of such a contribution, established by the predominantly illiterate villagers downstream of Calcutta, is the development of a solar-driven alternative technology to electrically-powered sewage treatment plants (Meier and Abdul Quium 1991, 136). The success of such decentralized efforts may influence local planners and government officials to benefit Calcutta on a regional scale.

The potential of Calcutta to overcome its seemingly insurmountable obstacles seems dim. However, a fatalistic perspective will only ensure that the city continues in a downward spiral. The economic potential of the West Bengal region remains potentially impressive. Thus what is necessary for the regeneration of Calcutta is a regional planning approach which actively and realistically invests in the region's knowledge-base.

References

Angotti, Thomas, 1993, *Metropolis 2000: Planning, Poverty and Politics*, New York: Routledge.

Bagchi, Amaresh, 1987, "Planning for Metropolitan Development, Calcutta's Basic Development Plan, 1966-1986: A Post-Mortem", *Economic and Political Weekly*, XXII:14:597-601.

Banerjee, Bireswar, 1990, "Problems of Calcutta Metropolis and Strategy for the Future", *Geographical Review of India*, 52:2:26-33.

Banerjee, Tridib, and Sanjoy Chakravorty, 1994, "Transfer of Planning Technology and Local Political Economy: A Retrospective Analysis of Calcutta's Planning", *Journal of the American Planning Association*, 60:1:71-82.

Biswas, Oneil, 1992, *Calcutta and Calcuttans: From Dihi to Megalopolis*, Calcutta: Firma KLM Private Limited.

Calcutta Metropolitan Development Authority, 1986, *CMDA*, Calcutta: Calcutta Metropolitan Development Authority.

Chatterjee, Monidip, 1990, "Town Planning In Calcutta: Past, Present, and Future", pp. 132-147 in Sukanta Chaudhari, (ed.), *Calcutta: The Living City*, Calcutta: Oxford University Press.

Chaudhuri, Sukanta, 1990, "Education in Modern Calcutta", pp. 196-206 in Sukanta Chaudhuri, (ed.), *Calcutta: The Living City*, Calcutta: Oxford University Press.

Choguill, C.L. 1994, "Crisis, Chaos, Crunch? Planning for Urban Growth in the Developing World", *Urban Studies*, 31:6:935-945.

Darnay, Arsen J., (ed.), 1992, *Statistical Record of the Environment*, Detroit: Gale Research.

Dasgupta, Biplab, 1987, "Urbanisation and Rural Change in West Bengal", *Economic and Political Weekly*, XXII:8:337-344.

Datta, Bhabatosh, 1990, "The Economy of Calcutta: Today and Tomorrow", pp. 97-108 in Sukanta Chaudhari, (ed.), *Calcutta: The Living City*, Calcutta: Oxford University Press.

Droste, Kathleen, (ed.), 1994, *Gale Book of Averages*, Detroit: Gale Research.

Dutt, Ashok K. 1978, "Planning Constraints for the Calcutta Metropolis", pp. 258-280 in Allen G. Noble and Ashok K. Dutt, (eds), *Indian Urbanization and Planning*, Delhi: Tata McGraw Hill.

Dutt, Ashok K., et al. 1994, *The Asian City*, Dordrecht: Kluwer Academic Publishers.

Dyckman, John, Alan Kreditor, and Tridib Bannerjee, 1984, "Planning in an Unprepared Environment", *Town Planning Review*, 55:2:214-227.

Earthwatch, 1992, *Urban Air Pollution in Megacities of the World*, Oxford, England: Blackwell Reference.

Friedmann, John, 1987, *Planning in the Public Domain: From Knowledge to Action*, Princeton, NJ: Princeton University Press.

Ghose, Partha, 1990, "Scientific Research in Twentieth-Century Calcutta", pp. 209-214 in Sukanta Chaudhari, (ed.), *Calcutta: The Living City*, Calcutta: Oxford University Press.

Goswami, Omkar, 1990, "Calcutta's Economy, 1918-1970: The Fall From Grace", pp. 88-96 in Sukanta Chaudhari, (ed.), *Calcutta: The Living City*, Calcutta: Oxford University Press.

King, A.D. 1980, "Exporting Planning: the Colonial and Neo-colonial Experience", pp. 203-221 in Gordon E. Cherry, (ed.), *Shaping an Urban World*, New York: St. Martin's Press.

Kironde, J.M. Lusugga, 1992, "Received Concepts and Theories in African Urbanisation and Management Strategies: The Struggle Continues", *Urban Studies*, 29:8:1277-1291.

Knight, Richard V. 1989, "City Development and Urbanization: Building the Knowledge-Based City", pp. 223-242 in Knight, Richard V. and Gary Gappert, (eds), *Cities in a Global Society*, Newbury Park, CA: Sage Publications.

Knight, Richard V. and Kurt Leitner, 1993, *Werstattberichte: The Potentials of Vienna's Knowledge Base for City Development*, Vienna: Institute for Higher Studies.

Meier, Richard L. and A.S.M. Abdul Quium, 1991, "A Sustainable State for Urban Life in Poor Societies: Bangladesh", *Futures*, 23:2:128-145.

Munsi, Sunil K. 1975, *Calcutta Metropolitan Explosion: Its Nature and Roots*, New Delhi: People's Publishing House.

Munsi, Sunil K. 1990, "Land Development & Environmental Issues in Calcutta", *Geographical Review of India*, 52:2:1-11.

Nagpaul, Hans, 1996, *Modernization and Urbanization in India: Problems and Issues*, Jaipur: Rawat Publications.

Nijkamp, Peter, 1994, "Improving Urban Environmental Quality: Socioeconomic Possibilities and Limits", pp. 241-292 in Ernesto M. Pernia, (ed.), *Urban Poverty in Asia: A Survey of Critical Issues*, Hong Kong: Oxford University Press.

Okpala, Don C.I. 1987, "Received Concepts and Theories in African Urbanisation Studies and Urban Management Strategies: A Critique", *Urban Studies*, 24:2:137-150.

Qadeer, Mohammed A. 1990, "External Precepts and Internal Views: The Dialectic of Reciprocal Learning in Third World Urban Planning", pp. 193-210 in Bishwapriya Sanyal, (ed.), *Breaking the Boundaries: A One-World Approach to Planning Education*, New York: Plenum Press.

Sanyal, Bishwapriya, 1990, *Breaking the Boundaries: A One-World Approach to Planning Education*, New York: Plenum Press.

Singh, R.L. 1968, *India: Regional Studies*, Calcutta: Indian National Committee for Geography.

15 Planning for the City Efficient: The Hong Kong and Macau Experience

BRUCE TAYLOR

Urban planning is no longer considered a purely technical exercise. Urban planners are part of a broader socio-political system and their labors go only part way to determine a city's built-form. Urban change is the product of a complex set of factors, varying widely from technical innovation, to economic influences, to social and political matters.

Planners approach the task of regulating and governing urban change as but one of the agents with the capability of influencing it. If the planners are employees of the state, as is true almost universally in the cities of Hong Kong and Macau, then in their work they are encumbered with a state ideology that may work either in harmony with, or in opposition to, forces emanating from other sources. (Editors' Note: this chapter was originally prepared in 1996.)

When, for instance, we see a thoroughgoing transformation of a city's physical form in which architecturally pleasant, though dated, low-rise arcaded colonial buildings are replaced by sleek, glossy high-rise office towers and multi-story residential blocks, we can be quite sure that this outcome has not occurred by purely laissez-faire happenstance. Rather, an array of forces acting in concert has created an irresistible dynamic leading to the sweeping away of the existing cityscape and its replacement by an altogether different urban form. Public planners, in their capacity as representatives of the state, may have been central to this process, or to the contrary may have been marginalized by the strength of forces working in opposition to their concerns. Of course, they may have been disinterested in the outcome, mere spectators in what sometimes appears in Hong Kong at least to be a surreal, life-size version of a children's game of blocks, where towers and other structures appear and disappear with startling rapidity (Abbas 1994).

This chapter examines aspects of the transformation of the two cities of Hong Kong and Macau into contemporary urban centers. Its focus is on the philosophical vantage point from which planners in these two cities view the urban areas they serve, the purposes served by state intervention into city development processes, and their own role in guiding the physical manifestation of economic transformations.

The metaphor of the "city efficient" proposes three motivating factors that taken together have led planners in both places to put efficiency to the fore as a cardinal virtue in urban development. An illustrative planning document is Hong Kong's "Metroplan", as an example of how such philosophical perspectives are operationalized in practice. Two areas of planning practice, relating to conservation of the "colonial" landscape and the reclamation of new land for building purposes, are case studies of how a preoccupation with creating an efficient venue for the operation of consumptive and productive processes is translated into a defined spatial outcome. The chapter ends with observations on how a focus on urban efficiency has marginalized more holistic concerns with the living environment and quality of life of individuals who must experience the city day in and day out.

Context: Planners and the "City Efficient"

Historians point out that Macau in the 16th century and Hong Kong in the 19th, came into being as places to carry out trade. This, of course, has not been their only function. The Portuguese in Macau (though not as much the Chinese) are fond of citing their community as an exemplar of East-West cultural interchange. Undoubtedly, in the eyes of the most influential of their citizens the two European-administered territories have served economic purposes above all others. Planning for the ongoing development of Hong Kong and Macau, then, occurs in a context where the city is viewed through an economic lens.

The physical city, its buildings, highways, utility systems, telecommunications networks, and so forth, represent both the agency through which economic processes are carried out and the vessel which contains, and sometimes constrains, them. Clues to this abound in statements used in Hong Kong and Macau which purport to characterize the public interest served by the activities of urban planners working for the state. Such statements attempt to justify the government's involvement in land regulation, development control, and other such matters. Consider, for instance, the implications of the following statement of objectives for the Hong Kong Government's involvement in matters related to land, public works, and utilities: The main objectives of the government's lands and works policies are to ensure an adequate supply of land to meet the needs of the public and private sectors, to optimize the use of land within the framework of land-use zoning and development strategies, and to ensure the coordinated development of infrastructure and buildings (Hong Kong Government 1994, 200).

This statement, although obviously very general, contains important clues as to the motives for state intervention in land development and redevelopment processes. For instance, anticipating the demands for land supply from various users, including the state itself, is crucial. In times past such demands might have been for harborside quays or for industrial sites on which multi-story "flatted factories" were built to house the workshops that brought Hong Kong to prominence as an export manufacturing center. Today's demands are more likely to be for container storage facilities, or attractive centrally-located commercial building sites, or medium-priced (by local standards!) flats for the growing "sandwich class" market, so-called because the households comprising it earn incomes that are too high to qualify for low-rent public housing, but cannot afford apartments available on the private market.

The public sector has its own space requirements, for everything from subsidized rental housing, to convention centers, to Hong Kong's recently built first public golf course. Another key concern is to optimize the use of available land. The notion of what constitutes "optimal" use is, of course, subjective. In the past, one definition of the term applied frequently in Hong Kong is that land is used optimally when it is developed up to the maximum limit provided for in the relevant area plan. Not surprisingly, this interpretation leaves little room for the preservation of traditional buildings, which fail by a large margin to make use of the development potential allotted to their sites.

Another definition of "optimal" used in some circles relates to the revenues earned for the public purse from the leasing of building sites. All land in Hong Kong is owned by the Government and is made available to users only on a leasehold basis. For most private users the rights to develop land are acquired through competitive bidding at Government-administered land auctions. These are an important source of revenue for the Hong Kong Government. A similar system is now used in Macau. A land use pattern is optimal if it brings in the highest quantity of revenues to the government. Critics of government land policy point to its monopoly control over land supply and accuse it of acting in the same fashion as a profit-seeking private monopolist: maximizing revenue by restricting land availability in the face of high demand.

Finally, the statement above includes concerns over "coordination" in development. This can be taken to mean simply that new development, or redevelopment, must be in keeping with the available infrastructure resources; it may also mean that once infrastructure is in place development should be allowed to proceed to the full extent of its capacity. This latter notion is very much in keeping with the first sense of "optimality" referred to in the last paragraph. These objectives are fully in keeping with a philosophical standpoint that views a city such as Hong Kong as fundamentally a location for, and a facilitator of, the consumptive and productive processes engaged in by its citizens. For governments and their planners who adopt this point of view, planning for a city's development becomes fundamentally a process of supporting, through physical means, the efficient operation of the economy. The planners working in such a city bend their efforts toward creating a setting where site demands, especially of the investor community, are anticipated as much as conditions permit; movement of goods and people is convenient and, where possible, cheap; a minimum of red tape interferes with public or private investments in the city's infrastructure; the work force is housed in at least minimally decent conditions; activities are free to locate, within environmental or other limits, at sites best suited to their needs; and administrative activities are routinized as much as possible.

This point of view is termed by Kevin Lynch (1981, 81-88) as "the city as a practical machine". A more succinct phrase is "the city efficient", a paraphrase on the "City Beautiful" movement that flourished in turn-of-the-century America. In such a context, prominence is necessarily given to working alongside, and in ways that reinforce, land and property markets, rather than working in opposition to them. This is not to say that the market for property in cities like Hong Kong operates unregulated. Rather, it suggests that planners do not act aggressively to introduce controls or restrictions that might act to limit the potential for economic growth. Also, it calls into question the perceived desirability, from the planners' perspective, of acting to overturn market-derived solutions to the allocation of building sites, as opposed to acting to restrain market-derived solutions within reasonable limits, by such means as limiting the building density permitted on a given site.

This is one of the conclusions reached by Samuel R. Staley (1994) in defense of the current system of land-use regulation in Hong Kong, which he believes to be a cornerstone of the territory's economic success. For an analysis of Hong Kong's development control system that offers quite different suggestions for policy, see Cuthbert (1991).

The Motivation for Efficiency

The question may be asked, what has motivated planners and other officials to give pride of place to efficiency as a main object of their activities? Three motivating factors are of special significance. The first is physical; the second, philosophical; and the third, political. There is, first, a physical motivation, relating to the rather obvious fact that Hong Kong and Macau are small and crowded places, with tremendous pressure on developable land. Such pressure has its origins in continued population growth and is exacerbated by continued economic diversification. A broader range of land uses is trying to squeeze onto limited sites; per capita incomes are rising, coupled with rising expectations as to living conditions; and the profitability of land and property speculation in relation to other avenues of investment must be considered.

Both territories have a long history of reclaiming new land from the sea (Hong Kong Government 1995; Alves 1990, 59-77), and both have pushed the limits of the built-up area far into formerly-rural territory (Bristow 1989). Nonetheless these "safety valves" have not kept the value of centrally-located and accessible sites from skyrocketing to some of the highest levels in the world. In such a context, efficiency in the sense of being able to supply building land to facilitate continued growth is seen as a necessity. Put another way, demands by business, by public agencies, and by the community at large can only be satisfied, planners may easily think, if special prominence is given to meeting them and if policy towards the use of the two territories' land resources reflects this need.

The second, philosophical, motivation relates to the general perception in Hong Kong and Macau of the proper role of the state in the community. Put perhaps a little crudely, government is seen as a supporter and facilitator of economic activity which originates in the private sector and is driven by decisions taken in that sector. Hong Kong is not the laissez-faire economy it is sometimes made out to be; it cannot be when all land is nationalized and massive investments of public funds are made in subsidized housing. It is fair, however, to call Hong Kong a private-sector driven economy, with minimal state involvement in setting its direction. As Hong Kong has moved from trade entrepot, to center of light industry, to regional headquarters center, to nerve center of an emerging Greater China, the role of the state has been to facilitate and support these changes rather than to guide or direct them. Efficiency as a prime objective in planning is very much in keeping with this perspective of governance with a light touch. Planning for the city efficient does not threaten to overturn the status quo of a private-sector led economy. It helps to reconcile an inherent tension which has permeated all local planning: that of intervening in land and property markets on behalf of a state which itself accepts the primacy of private-sector decisions. If there is to be state intervention into the workings of development processes, so the thinking might go, let it be of a nature which promises to bring benefits to those who do most to fuel Hong Kong's or Macau's economic growth.

A third motivation, more evident to date in Hong Kong than Macau, re-lates to the future role and status of the two colonial cities with reintegration into China. Continued uncertainty exists as to how the two cities will fare politically and economically. Shanghai, for one, is being promoted by powerful interests in China as a competitor to Hong Kong for the location of service and information industries. Other urban centers in the region (e.g., Singapore and Taipei) are positioning themselves to "get a piece of the action" should Hong Kong falter in the transition. To maintain their status, public officials including planners in Hong Kong and Macau typically make the assumption that the two cities must demonstrate their "usefulness" to China. The term "useful", like the term "optimal" noted earlier, is interpreted subjectively. Most commonly utility is viewed in the sense of serving as a conduit for foreign capital into China, suggesting a role as regional headquarters for financial and business services; or as a center for trans-shipment of Chinese exports to foreign markets; or as a center for entrepreneurship which can support investment in China. "Usefulness", then, becomes viewed in economic terms: Hong Kong in particular is seen as the locomotive pulling South China in its wake.

A focus on efficiency as an objective in planning fits well with such a view, where Hong Kong becomes the fulcrum for integration of China into a global-ized spatial economy. It allows Hong Kong officials to point to the territory's container berths and office buildings, high-capacity telecommunications links and high-technology industrial parks, convention and exhibition centers, and so forth and imply "look at what we, a free-wheeling, business-focused, globally connected territory can offer to the modernizing, emerging economic power that is contemporary China if we can be left alone to get on with the job". In the local vernacular, the city efficient, together with the policies that make it so, become part and parcel of the "two systems" which, according to Deng Xiaoping's dictum of "one country, two systems", are to be preserved with Hong Kong and Macau becoming Special Administrative Regions of China.

Example: The Metroplan

The preceding analysis might seem to generalize too broadly from what is at best a vaguely-worded guideline as to the intentions of state policy in Hong Kong with respect to the land development process. For an example of how such ideas are translated into operational principles governing the activities of urban planners, it is instructive to look at one of the most extensive planning exercises carried out in Hong Kong in recent years. This is the so-called "Metroplan", a plan for restructuring the entire urban heart of Hong Kong in the years up to 2011 in order to improve living and working conditions for the area's 4 million or so residents.

According to the Hong Kong Government's publication outlining for a public audience the planning strategy adopted for the metropolitan area, the aims of the Metroplan are:

- to enhance Hong Kong's role as an international port and airport; as an international business, finance and tourist center; as a center for a diverse range of light manufacturing industry; and, as the center of Government for the territory;
- to provide opportunities to satisfy, as far as practicable, housing needs according to what people can afford and where they would like to live;
- to achieve a more balanced distribution of jobs relative to population concentrations, the locational preferences of new enterprises and the ease of travel;
- to relocate activities which cause severe environmental problems;
- to reduce population densities by such means as spreading development on to adjoining harbor reclamations and comprehensive urban renewal;
- to provide conveniently located community facilities which aim at new town standards;
- to create an urban form that will foster a sense of community identity;
- to conserve and enhance major landscape attributes and important heritage features;
- to provide a multi-choice, high capacity transport system that is financially viable, energy efficient and makes provision for the safe and convenient movement of pedestrians; and
- to provide a strategy that can be carried out by both the public and the private sectors under variable circumstances, particularly with respect to the availability of resources and significant changes of demand. (H.K. Government 1991, 2-10).

These are basically unexceptionable points, as is common in a statement of aims; it is unlikely that anyone in Hong Kong would voice opposition to them although some might quibble with their ordering. They imply a broad focus on matters relating to efficiency and to the provision of a range of urban services, although the fourth point refers to resolving environmental incompatibilities and the eighth point suggests an interest in conserving "heritage features" (i.e., traditional buildings) and elements of the natural landscape that contribute so greatly to the territory's scenic appeal.

It is the more detailed objectives that the Metroplan Steering Group put forward as instruments for guiding its choice among the many possible alternative development patterns for the metropolitan area, that reveal much more about the philosophical vantage point held by the planners who directed this exercise. A total of forty-eight objectives are specified, twenty-six of which are considered to be particularly important (Table 15-1). The majority focus on matters relating to economic growth, accessibility and "balance" in the pattern of development, cost minimization, and flexibility in plan implementation, all different aspects of the city efficient. A smaller number relate to introducing desired improvements to the living environment. Only one of the twenty-six high priority objectives relates directly to conservation of features in the environment and it is the natural landscape (e.g., the profile of the mountain ridge above Hong Kong harbor), rather than the human-created one, that merits priority. It is evident that at least in this planning exercise, pride of place is given to encouraging new development and redevelopment that contributes to

Table 15-1 Hong Kong: Priority Objectives for Assessing Land Development Patterns

Economic enhancement
- Maximize and promote the continued economic growth of Hong Kong
- Maximize the efficiency of the port and airport
- Provide adequate land in appropriate locations for the expansion of commercial office development

Housing provision
- Ensure adequate land is provided to meet the specific requirements of the Housing Authority's comprehensive redevelopment program

Balanced development
- Encourage the redistribution of employment opportunities from the Metro Area, particularly jobs in manufacturing industry
- Encourage the redistribution of jobs within the Metro Area to suitably accessible locations
- Encourage the development of export-oriented industries and appropriate container trade back up uses in close proximity to the container port

Environmental improvement
- Reduce existing and minimise future environmental problems
- Ensure Metroplan proposals would facilitate the implementation of a viable Sewage Strategy and provide sites for solid waste management facilities
- Minimize/reduce the juxtaposition of incompatible land uses

- Provide opportunities for the relocation of bad neighbour uses
- Maximize opportunities arising from comprehensive redevelopment schemes to restructure district land use patterns
- Reduce population and building densities in obsolete areas
- Plan existing and new development areas to new town standards, as far as is practical

Conservation

- Conserve/enhance the value of natural landscape features
- Maximise the amenity and recreational value of the harbour and its shoreline and of urban fringe foothills

Transport

- Minimize the need for travel
- Maximize the operational efficiency of the transport infrastructure
- Minimize land take from the community for transport infrastructure consistent with serving the social and economic demands of the community
- Promote off-street modes of travel and exclusive rights of way for public transport

Implementation

- Ensure strategies can be implemented in a logical sequence of development compatible with other projects
- Ensure programme for individual project components can be brought forward or delayed to keep pace with changes of demand or resources availability
- Ensure strategies can function effectively at different stages of development
- Ensure strategies can be implemented without overheating the economy and the construction industry
- Minimize capital and associated recurrent costs to the public sector
- Minimize need for complex institutional mechanisms to implement and operate strategies

Source: Adapted from Hong Kong Government, *Metroplan: The Selected Strategy-- Executive Summary* (Hong Kong: Government Printer, 1991), pp. 19-21.

the efficient operation and management of the city and at the same time, when feasible, alleviates environmental problems for city residents.

Where these objectives, and the priority given to them, come to matter is in cases where conflict arises between the attainment of two or more of them, and choices must be made between development proposals that show a tendency to favor one objective or the other. For instance, Metroplan gives priority to "reducing population and building densities in obsolete areas". Although presumably reduction of building density could be accomplished by selectively thinning out the crop of undistinguished buildings of the 1950s and 1960s, that are ubiquitous in the lower-income districts of Hong Kong's urban core, pre-war traditional buildings (admittedly for the most part in poor condition) also disappear in the course of such exercises in neighborhood density reduction. In the unequal competition between objectives, conservation of heritage invariably loses out as an impediment to the realization of the city efficient.

Likewise, the public clamor for more housing has been addressed in some instances by turning over the sites of established schools and children's playgrounds to the Housing Authority and other sponsors for building new high-rise tower blocks. Conserving local amenities takes a back seat in these cases to siting new housing at a location which addresses planners' concern with "balance" in development. The Metroplan, as a statement of public policy towards development in Hong Kong's metropolitan core, gives a distinct impression that given the prevailing emphasis on sustaining the value of Hong Kong as a locus of production and consumption, government planners in Hong Kong are most willing to direct their energies to other objectives 1) when achieving them facilitates the attainment of one or more broad efficiency aims; 2) when achieving them does not interfere with the perception of state support for economic growth; and 3) when the costs of meeting them, both direct costs, and opportunity costs in the form of foregone revenues from development, are seen to be manageable. It does not require a great leap of imagination to discern that, in actuality, the instances where all three of these conditions are met are quite few in number. Macau has no equivalent of the Metroplan, but as the next section suggests the area plans for Macau's new reclamations betray a very similar point of view.

Practice: Conservation of the Historic Landscape

Some human settlements are instantly recognizable by the nature of their overall visual environment, including structures large and small, the nature of their juxtaposition, and the way they are interspersed among other man-made elements (roadways, parks) as well as elements of the natural physical setting. Greek island villages, the boulevards and monuments of Paris, and the skyline of Manhattan all come to mind. Hong Kong, too, enjoys instant recognition owing to the presence of Victoria Harbor and the backdrop of mountains looking over Hong Kong Island as the central defining elements of its visual image.

In recent years, the highly dramatic physical setting has been punctuated by the appearance of landmark buildings, such as Norman Foster's Hong Kong and Shanghai Bank headquarters building and I.M. Pei's Bank of China. Yet the drama of Hong Kong remains largely natural, not human-created; the defining elements of the image left by Hong Kong are as much created by an intangible "vitality" and "dynamism" as they are by any aspect of the physical city; and even the landmark buildings, to say nothing of the much more numerous undistinguished, utilitarian structures of the last thirty years, are expected to be profitable above all else.

The architectural heritage of Hong Kong contributes very little to an observer's sense of the cityscape. Except in a few of the oldest districts of the city, where preservation of older structures owes much to benign neglect, and in a few locations in the New Territories where an indigenous heritage is the object of concerted efforts at preservation, Hong Kong's cityscape is a triumph of the city efficient. It is the odd traditional structure left in place for one reason or another that looks wholly out of context in present-day Hong Kong. The few exceptions to this pattern are usually either publicly-owned properties or buildings used by non-profit organizations with objectives other than, or in addition to, profit maximization. These are unrepresentative of the local development pattern in that their sites are not allocated through competitive bidding in Hong Kong's hotly-contested land auctions.

An "exception to the exceptions" might be the Peninsula Hotel, built in 1928 and preserved today. However, preservation of the old structure has come at a price: a thirty-story tower has been grafted immediately onto the rear of the original building. Two of the most prominent building anachronisms are important symbols of the state: Government House, the home of Hong Kong's colonial governors, and the Legislative Council building, originally the Supreme Court. The Legislative Council building will continue to function as the meeting place for the post-1997 legislature. However, the Chinese Chief Executive has refused to move into Government House, and its use is uncertain. Other structures are home to various government departments that for various reasons, including perhaps inertia, have never seen fit to redevelop their premises: included in this category are the headquarters of the Marine Police and the Royal Observatory in Tsim Sha Tsui and the Wanchai police station on Hong Kong Island. Among the non-profit organizations one might include churches such as St. John's [Anglican] Cathedral; temples such as Man Mo Temple; and the headquarters buildings of charitable bodies such as the Po Leung Kuk.

Only in isolated cases, most notably the Western Market, a shopping and entertainment center opened in 1991 in a restored public market building, has the architectural character of a building become part of the raison d'etre for its retention and renovation. Only in the case of the Tsim Sha Tsui Cultural Center's clock tower, a disembodied relic of the demolished railway terminus that once occupied the site, and set incongruously in front of the fortress-like concert hall building, has historic preservation in the Western sense come to the fore as a value in its own right, motivating the retention of a traditional structure purely for the purposes of retaining a link to past history. The controversies surrounding the demolition of the railway terminus are discussed by Cuthbert (1984).

Everywhere else, the story is the same. The traditional cityscape has been swept away by the economic and commercial pressures noted earlier that lead to irreversible changes in the character of the built environment, with planners and other representatives of the government basically standing to one side. The conditions noted in the last section for state support of preservation as an alternative to attaining broad efficiency aims, are virtually never realized in modern-day Hong Kong.

Why is this so? To answer this question it is most instructive to turn not to Hong Kong but to Macau. Here conservation of the city's architectural heritage enjoys, on paper at least, considerably greater state support than in Hong Kong, extending in some cases to the expenditure of considerable public resources, as in the pedestrianization, paving, and landscaping of the central square, the Largo do Senado. However, it is fair to say that, taken as a whole, the cityscape of Macau gives the same impression of being designed to facilitate the efficient operation of economic processes as does the cityscape of Hong Kong, albeit on a smaller scale. The appearance of Macau likewise illustrates, though perhaps not as completely as does Hong Kong, the triumph of market-oriented redevelopment processes at the expense of much of the city's cultural heritage.

To explain this outcome, one must look first at the nature of preservation activities carried out with public support in Macau. For the most part, these have concentrated on the preservation of landmarks and monuments rather than the maintenance of the cityscape as a whole. Two exceptions are the row of preserved residences (now government offices) along Avenida Conselheiro Ferreira de Almeida set off to maximum advantage by an adjacent sports field and the central Largo do Senado which is the focus of a recently-created pedestrian zone. Despite these, successful efforts at district-wide preservation in Macau are rare.

Some landmarks are important symbols of the (colonial) state. Examples are the Palacio do Governo on the Praia Grande (the office of both the Governor and the Legislative Assembly) and the elegant building of the Leal Senado, now Macau's municipal council. Others are significant because of their place in Macau's long and sometimes turbulent history, such as the fortresses that dominate Macau's heights. Still others are publicly-owned historic structures providing various amenities, such as the Dom Pedro V Theater and the Military Club. Perhaps the purest example of landmark preservation is the facade of St. Paul's Church, the most enduring symbol of Macau, which has been preserved despite the destruction by fire of the rest of the building.

The sheer number of these landmarks means that more vestiges of Macau's traditional form continue to exist than is true in Hong Kong. But in the traditional city, the number of landmark properties with important symbolic or historical associations were far outweighed by other structures, individually of no great significance, but collectively of great importance in establishing the harmonious, human-scaled cityscape associated with *"aiel d Macau"*. It is these buildings that have been irretrievably lost to redevelopment, virtually unhindered by Macau officials, through the operations of the same constellation of forces that has been responsible for changing the face of Hong Kong. Only the street pattern survives from the traditional city in many Macau neighborhoods. Even those street patterns sit uneasily in the modern cityscape where eight- or ten-story blocks are the rule, rather than the two- or three- story structures that characterized the historic city.

Adding to the perception that Macau's cityscape reflects modern-day emphasis on the city efficient are the new developments grafted onto the face of Macau, as a result of new reclamations. These are wholly modern in their concept, and are of such a scale as to visually overwhelm those landmarks of the historic city that remain. For instance, the seaward approaches from Hong Kong are dominated today by the squat rectangular towers of the "NAPE" (*Novo Aterro do Porto Exterior*) reclamation area erected in 1993-95, backed by the more eclectic buildings on the older "ZAPE" (*Zona Aterro do Porto Exterior*) built, for the most part, in the 1980s. The Guia Hill and lighthouse, long a landmark for Hong Kong visitors, receives barely a second glance. In the same way, reclamation of much of the Baia da Praia Grande will destroy forever the seafront vista that was such an importance part of the ambience of the historic Praia, despite designers' best efforts to make the relocated, reconstituted Praia a visually attractive setting. The inescapable conclusion is that, as in Hong Kong, preservation has made only a limited contribution to the character of the Macau cityscape as experienced today.

Apart from the focus on landmarks rather than streetscapes already noted, several other reasons can be cited:

• *Economic incentives.* As in Hong Kong, the government of Macau benefits when building sites are redeveloped to a higher density or, in the case of new reclamations, when permitted densities are set at a high level. It is unlikely that such economic incentives would be sufficient to overcome an inherent hesitancy to demolish an historic landmark such as the Monte Fort. For hundreds of Macau's other traditional buildings without the same individual significance, the opportunity cost arising from foregoing the chance to redevelop looms quite significant in officials' calculations.

- *Lack of overall planning*. There remains no legally enforceable general plan covering the whole of peninsular Macau, though there are such plans for the newest areas of reclaimed land, and for a few other districts. Development control like redevelopment itself, proceeds on very much a piecemeal basis, and there is little in the way of overall vision that would guide an official in determining whether to permit the demolition of a traditional structure.

- *Fragmented nature of redevelopment*. New development areas in Macau are designed, and their layouts established, with little reference to the character of areas that adjoin them. Thus, developments planned with economic efficiency uppermost in mind with tall tower blocks and wide, evenly-spaced streets can sit cheek-by-jowl with older districts with a much different development pattern. The most glaring example of this is on the island of Taipa, where Taipa Village and the new high-rise buildings of the Baixa do Taipa sit in uneasy proximity.

- *Limited vision*. Owing perhaps to a focus on individual landmarks, the beneficiaries of preservation are in the end limited: visitors to Macau, for one, and also a governing elite group that has the strongest interest in the maintenance of a Western (Portuguese) cultural legacy. A broader vision, embracing the interests of the individual and the community at large, might suggest a different approach to preservation in which a larger number of traditional buildings, and a bigger variety of them, would be targeted as suitable for retention in the city's urban fabric (Taylor 1993).

- *Speculative pressure*. The property development and construction industries are economically very significant in Macau, and the enclave has represented a destination for speculative capital originating in, among other places, mainland China and Taiwan. The property price spiral of recent years generated, in part, by the inflow of speculative capital, has led smaller property holders, of which Macau has an abundance, to take advantage of opportunities to maximize the value of their land grants. Macau's relative poverty in past decades had the side effect of preserving much of the cityscape in its historic form. Prosperity has permanently altered this state of affairs.

Taken together, these factors have marginalized the interests in Macau that support preservation and bolstered those that support continued redevelopment in modern style, perpetuating the transformation of Macau's spatial form in a way that supports the efficient operation of consumptive and productive processes. While the "new Macau" that results may indeed be efficient, and profitable to those who benefit from the process of transformation, one cannot help but feel that an opportunity has been lost to shape the urban form of a city from a more humanistic vantage point, with the focus not as much on efficiency or profitability, but on the enhancement of individuals' welfare and the overall quality of urban life as experienced by residents. Again non-economic objectives have lost out to the drive to build "the city efficient".

Practice: Reclamation of Building Sites

An earlier section noted that reclamation of land from the sea has been a constant throughout Hong Kong's and Macau's colonial history. In the past few years, however, reclamation has also become a politically-charged issue in Hong Kong, which pits planners as representatives of the state and main promotors of the city efficient against other self-proclaimed representatives of the community and guardians of the natural landscape. Even more than conservation of the landscape, the debate over reclamation epitomizes the conflicts that occur over planning objectives and the consequences of a single-minded focus on pursuit of the city efficient.

At the heart of the Metroplan's proposals for restructuring of the urban heart of Hong Kong, referred to earlier, are extensive land reclamations along the shores of Kowloon and Hong Kong Island. These reclamations serve several purposes. They provide locations for essential infrastructure, such as the subway link from Hong Kong's CBD to its new airport. They offer sites for new housing within easy commuting range of existing centers for service employment. They allow for open space and community facilities to be provided within reach of some of Hong Kong's most densely populated districts, meeting the Metroplan's aims of restructuring land use patterns in these areas and reducing their densities. Not coincidentally, they are also very attractive to developers, who have offered vast sums to the Hong Kong Government for the rights to develop especially attractive sites, such as above subway stations. All of these are evident in Figure 15-1, which shows the mixture of land uses proposed for the largest of the planned reclamations in West Kowloon. This is one of the less controversial proposals, and is at present well on its way to completion.

The public debate has centered on three of the proposed reclamations, labeled in Figure 15-2 as Kowloon Point (which adjoins West Kowloon), Green Island, and Southeast Kowloon. These share the trait of being located along the shores of Victoria Harbor, the city's traditional heart and the most impor-

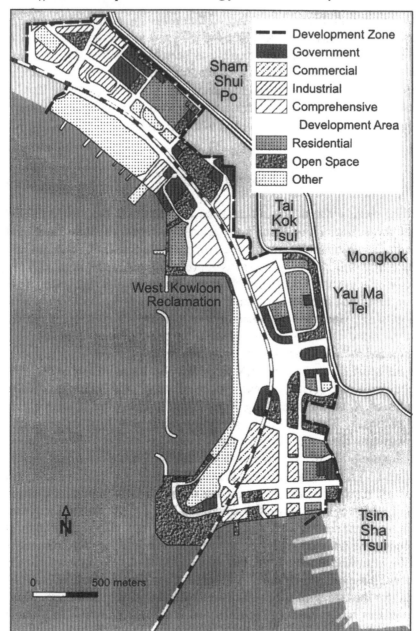

Figure 15-1 Planned Land Uses on the West Kowloon Reclamation

tant defining feature of Hong Kong's cityscape. Their effect would be to narrow the harbor to about 800 meters at its narrowest point (compared with

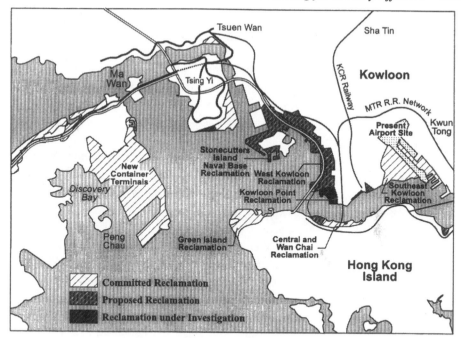

Figure 15-2 Planned Reclamations in the Hong Kong Harbor Area

1200 meters at present); equally importantly, they would narrow the harbor along basically its entire length. It is not surprising that disparaging references to the "Victoria River" have cropped up in the press, and pundits wonder tongue-in-cheek whether the ultimate aim is to dispense with the harbor altogether.

To defend these proposed reclamations, planners, led by the Principal Government Town Planner, the most vocal public spokesman on this issue, have offered a number of arguments:

- *Expediency.* Compared to other ways of acquiring land to meet both private and state demands, reclamation is quick. Urban renewal of older tenements is hampered by difficulties in site assembly. Building on rural land in the New Territories faces the same problem; in addition, some rural land has symbolic significance well beyond its utilitarian value and resistance to its expropriation is fierce among tradition-minded New Territories residents.
- *Locational preferences.* It is argued that housing built in the urban area is preferred to housing built in New Towns or elsewhere on the fringes. Such a preference definitely exists; ironically, however, it is in no small part based on

past failures by planners to create new communities where population and employment opportunities are in balance.

- *Community improvement.* The Metroplan aims to improve the living environment in densely-settled older districts. For example, a portion of Green Island or Southeast Kowloon reclamation could be devoted to facilities and services that would benefit nearby communities where conditions fall well below standard. These benefits to a reclamation site's neighbors are not available from other forms of development.
- *Infrastructure support.* The new reclamations, it is argued, allow for infrastructure to be sited at the points where it is most needed to alleviate some of Hong Kong's worst deficiencies. New road and rail crossings of Victoria Harbor are cited as a case in point. This also results in value for money: a single major reclamation project can support expansion of vital infrastructure at the same time as it offers sites for needed housing or attractive community facilities.
- *Attractiveness.* Although often spoken *sotto voce*, the argument is made that reclamation more than any other development option offers land where it is wanted by those who are willing to pay to develop it. Builders of office and commercial complexes want to be near their markets and to concentrations of workers; hotel builders prefer sites contiguous to established tourist districts (e.g., Tsim Sha Tsui in Kowloon); and developers of luxury apartments favor harborside sites or at least ones with sea views. Sites on reclamations contiguous to the main urban area offer these benefits; building sites in the New Territories, the main alternative, do not.

Opposition to the reclamation proposals has been spearheaded by a lay member of the Town Planning Board, a body which reviews all development proposals, and by a local legislator. Opponents make one economic argument of their own: the possible effect of restrictions on maritime traffic in a narrowed harbor on Hong Kong's role as a leading port. This argument does not withstand close scrutiny, as container vessels, which carry the bulk of cargo destined for Hong Kong or exported from there to other markets, normally do not traverse the stretch of Victoria Harbor in question. The main container berths are located well to the west, in the Kwai Chung and Tsing Yi areas (Figure 15-2).

Their other arguments, though, stem from quite different concerns:

- *Congestion.* By increasing the concentration of people and workplaces around the harbor, building on new reclamation will contribute to worsening congestion that already is threatening to strangle the built-up area.
- *Environmental concerns.* Narrowing of Victoria Harbor may interfere with tidal flushing of the harbor basin, which despite improvements to the sewage dis-

posal system still receives a fair proportion of Hong Kong's liquid waste. The implications for water quality, despite several attempts at modeling water flows in the harbor, are uncertain. The extreme environmental degradation of Macau's Baia da Praia Grande before its recent enclosure and partial reclamation -- the result of ill-conceived reclamations dating to the 1920s -- is pointed to as an example to avoid.

- *Visual image.* In narrowing the harbor, it is argued, reclamation destroys or at least diminishes the central defining element of Hong Kong's visual image: Victoria Harbor and its backdrop of hills. This is the best known and arguably most spectacular panorama in Hong Kong and is in itself a tourist attraction. Noting the significance of tourism to Hong Kong, opponents of the reclamation plan question the wisdom of tinkering with the cityscape in ways that diminish its tourist appeal.

The contrast between the sets of arguments put forward by proponents and opponents is instructive. Each of the points put forward by the Government's planners, with the possible exception of "locational preferences", is grounded in a concern for efficiency: the process of making building sites available to meet private and government demand would "work better" (i.e., be more rapid and profitable, and meet community needs more effectively) if the reclamation plans are supported. Opponents of the plans, even if they acknowledge elements of this, hold that efficiency needs to be balanced by other objectives such as, the protection of the environment and of the territory's visual image. The debate over reclamation plans is, as much as anything else, a debate over values.

This debate has spilled over into Hong Kong's political arena with the introduction of a bill into the Legislative Council that would require the Council's assent to all future reclamations. This is not required at present; nor is there a requirement for public participation in the sense understood by American or British planners. Under the Foreshore and Seabed (Reclamation) Ordinance, only persons or organizations directly impacted by a proposed reclamation have any legal standing to oppose it. Government spokesmen oppose the bill to require Council's assent. The concern, echoed by property developers, is over the possible delay introduced into the planning system through involvement of the legislature, at a time when public demand for housing in particular has reached new heights. But at heart this concern is grounded in worries that legislators, or other lobbying groups perceived as having special interests, will take it upon themselves to intervene in decisions regarding urbanization and planning, perhaps with the intent of scoring political points with constituent groups. In such a scenario the city efficient must compete for the public's allegiance with other notions of what makes a desirable city.

Looking Toward the Future

Regarding the conservation of the historic landscape, planners have been disinterested spectators, focused mainly on other "bread and butter" matters. When, as in Macau, some state support has been given to conservation, intervention has had only the limited effect of retaining a range of landmarks with important symbolic or historic associations, fragments, basically, of the historic cityscape -- rather than the sought-after qualities (visual harmony, human-scaled development, and the like) of the traditional city. This is a non-interventionist approach. The place of "optimality" of land usage as a cardinal virtue in urban development is defined mainly by others. Planners mainly set the parameters within which cityscape transformation takes place, provide a few ground rules and guiding principles, and administer the development system's technical aspects.

In the case of reclamation, the planners' role is different. Hong Kong's government planners are themselves the strongest supporters of plan-making for the purpose of supplying land quickly and profitably. They are set against opponents who espouse other objectives and who suggest that, in the end, values may conflict: efficiency runs into conflict with other desired ends held by those with different vantage points concerning the appropriate purposes of state intervention into urban development processes.

Certain lessons can be drawn from Hong Kong's and Macau's experiences in urban transformation. These relate to the values held by planners and other participants in the transformative process. If planners are part of, or indeed captive of, a socio-political system that imposes constraints on their influence on the course of city growth and development, the path of least resistance for them is to adopt wholeheartedly the priorities held by other actors within that system promoting the city as, in essence, a machine that facilitates the operation of consumptive and productive processes, for example. The evidence suggests that, by and large, this is what has happened in Hong Kong and Macau.

What is missing, and what in other parts of the world is provided by democratic political systems, participatory mechanisms, public hearings, and other means which serve to broaden the range of interests involved in plan-making and plan review, is a consideration of other sets of values, focused perhaps on the living environment of individuals who must experience the city day in and day out. When an urban area is viewed as a finely-tuned machine to facilitate economic processes, the individual users of the area are too often seen as an undifferentiated mass: cogs in the machine, perhaps. When the purpose of plan-making is construed as smoothing the operation of such processes, the view of the individual as an actor within the city has been ignored or at best marginalized.

Preservation of a city's heritage, for instance, is a means of retaining ties to a perhaps idealized past, to be sure. But it is also a means of retaining certain aspects of the cityscape which individuals value, all too often missing in the controlled, regulated, fundamentally sterile environments that characterize the city efficient. Protecting the natural environment is another area that can easily become an afterthought: the focus of remedial action, that is, ameliorating environmental degradation after the fact, rather than forward-looking planning. The opponents of Hong Kong's harbor reclamations claim to speak for those who experience the city, as resident or visitor, and argue that a single-minded focus on efficiency is short-sighted and neglects equally important contributors to the quality of urban life.

Plan-making in Hong Kong and Macau can potentially be much more than attempts to guide and regulate a headlong rush to reconstruct the built form of the two territories in support of the needs of today's technologically-advanced industrial and service economy. Planners in Hong Kong and Macau, and in East Asia more broadly, might benefit from pausing to reflect whether cities as presently developed serve to enhance other values, needs, and desires, held perhaps on a highly individualized basis by a city's residents. The values exemplified by heritage conservation, environmental protection, and landscape preservation, which each are addressed in this chapter, are among those having an important bearing on the holistic aim of improving the overall quality of urban life. It is time, perhaps, for planners to direct their activities more towards the support of these values, alongside their continued support of efficiency in a city's functioning.

References

Abbas, Ackbar, 1994, "Building on Disappearance: Hong Kong Architecture and the City", *Public Culture*, 6:441-459.

Alves, Manuel, 1990, "O Espano Territorial de Macau", in D.Y. Yuan et al., (eds), *Population and City Growth in Macau*, Macau: Centre of Macau Studies, University of East Asia.

Bristow, Roger, 1989, *Hong Kong's New Towns: A Selective Review*, Hong Kong: Oxford University Press.

Cuthbert, Alexander, 1984, "Conservation and Capital Accumulation in Hong Kong," *Third World Planning Review*, 6:95-115.

Cuthbert, Alexander R. 1991, "For a Few Dollars More: Urban Planning and the Legitimation Process in Hong Kong", *International Journal of Urban and Regional Research*, 15:575-593.

Hong Kong Government, 1991, *Metroplan: The Selected Strategy-Executive Summary*, Hong Kong: Government Printer.

Hong Kong Government, 1994, *Hong Kong 1994*, Hong Kong: Government Printer.

Hong Kong Government, 1995, *Hong Kong 1995*, Hong Kong: Government Printer.

Lynch, Kevin, 1981, *A Theory of Good City Form*, Cambridge, MA: MIT Press.

Staley, Samuel R. 1994, *Planning Rules and Urban Economic Performance*, Hong Kong: Chinese University Press.

Taylor, Bruce, 1993, "Assessing the Contribution of Historic Preservation in Macau to the Quality of Urban Life", pp. 241-255 in Bruce Taylor et al., (eds), *Socioeconomic Development and Quality of Life in Macau*, Macau: Centre of Macau Studies, University of Macau and Instituto Cultural de Macau.

16 Singapore, the Planned City State: Government Intervention in Nation Building

VICTOR R. SAVAGE

With its separation from Malaysia in 1965, Singapore became an independent country and the process of nation building began. Over the last 30 years, the Government has been involved relentlessly in the nation-building process that has been undertaken in three different arenas: physical infrastructure of the city; social engineering; and political and economic development.

All three nation building processes have been so intertwined that a discussion of each cannot be done separately. The purpose of this chapter is to assert that Singapore's nation building has been predicated on a consciously planned process that has entailed the planned development of a city state, social engineering of Singapore's society and a unique political structuring of one-party dominance within a democratic political system.

When the Peoples Action Party (PAP) took over power in 1965, it was confronted with major problems in nearly all areas. Singapore's viability as a country was seriously questioned and many political commentators wondered whether Singapore would be able to survive as a state. The city like other Third World cities was faced with major housing, health, educational and infrastructural problems. Slum residents and squatters accounted for nearly one-third of the city's population. Singapore was faced with a fragile political system undermined by a communist infested labor union on the one hand, and a fragile multi-ethnic urban population where racial tensions ran high on the other. Population was growing at an extremely rapid rate. Between 1947 and 1957 the average annual growth rate was 4.4 percent; between 1957 and 1962 it was 3.9 percent, and between 1963 and 1965 2.5 percent. Unemployment was high (in 1960 8.9 percent of the total work force or 44,000 persons), and the population was generally unskilled and lacked formal education.

All these negative issues and problems are often forgotten today when visitors, tourists and younger academics come to Singapore. Without this historical perspective, it is difficult to comprehend the transformation that has taken place in Singapore in terms of infrastructural development, political stability and sustained urban development. What were the forces that helped to shape the development of Singapore as an economic power house within Asia? One factor was certainly the government-directed planned development. As Senior Minister Lee Kuan Yew in reviewing Singapore's progress states, the Republic's success did not come "naturally", it was "contrived, designed" and "man-made" (*Straits Times* 23 March 1996,25). Like all man-made artifacts it can be "unscrambled". In the same light Singapore's nation building process is synthetic, a product of conscious Government social and cultural engineering programs.

Singapore's Vital Statistics

In 1990 Singapore's population was 3,128,966. This was 1.25 times the total population in 1980 or an increase of 2.3 percent per annum over the 1980-90 period. Of the 3.1 million, 90 percent (2.7 million) comprised citizens and permanent residents, while the other 300,000 were foreigners working in Singapore (Lau 1992). With a total area of 646 sq km, the city state had an average population density in 1993 of 4,400 persons per sq km. Singapore is an extremely small country even by Southeast Asian standards.

Despite its small size and population, Singapore is economically an important country. Its per capita income is the highest in the region. In 1994, the gross national product per capita as given by the World Bank was US$19,310 or S$28,960. Using per capita GNP, Singapore is the 18th highest ranking country in the world. However, in terms of purchasing power per person (US$20,470), Singapore is ranked 9th. In 1993, the city-state was the 14th largest exporting country (exports totalling US$74 billion) and is currently the 4th largest foreign exchange center in the world. Singapore is also a major investor in many countries in the Asia Pacific Area such as China, Vietnam, Myanmar, Malaysia, Indonesia, Thailand, Cambodia, the Philippines, India, Australia and New Zealand. Singapore is a good example of a 'global city' (Rajaratnam, 1972) or a transnational city.

Despite its impressive economic development and purchasing power, Singapore has also done well in terms of maintaining an environmentally friendly city. For example, the noted botanist and conservationist David Bellamy gave Singapore 7 out of 10 points for its environmental management. He noted, "Singapore comes out very very well in the assessment and as a role model for other cities in sustainable development" (*Straits Times*). Singapore's ability to maintain its high standard of economic development and sustained environmentally friendly policies often begs the question of how the city state has achieved this.

The success of Singapore's urban development, economic progress and successful environmental management has been and continues to be a product of pragmatic, perceptive, planned policies that have been implemented with political will and economic foresight. When Singapore established the Ministry of the Environment, there were only 10 countries in the world that had similar Government ministries. One of the central themes in the early years of the Ministry had to do with maintaining public health and hygiene, but over the years, the Ministry has expanded its role in the country's environmental management program.

Faith in Planning

Singapore inherited planning approaches from the British colonial system. In 1958, the British colonial government produced a colonial master plan and over the last several decades, the PAP government has reviewed and revised this master plan at roughly five year intervals (1965, 1971, 1975, 1980, 1985 and 1991) (Table 16-1). The master plans not only reflect changing goals and visions but to a large extent encapsulate the broader visions and goals of the government at various points in time. Specifically, Singapore's master plans have embodied three strands of thought: the vision of Singapore as a global city; the issue of Singapore's survival in the light of its physical limitations and economic and political viability; and state goals of achievement in terms of attaining excellence both individually and nationally (Teo 1992).

Despite the changes in master plans over the years, one of the critical governing factors in the planning process, is the small size, spatial limitations and limited natural resources within the country. Given its spatial limitations and small size, the government has felt compelled to ensure that there is an effective and efficient use of land in Singapore. This has clearly meant the city state has to be planned to meet all the various social, economic, cultural, recreational, and political demands of its citizenry. Unlike other countries that have huge land resources, Singapore has no alternatives. Its population has to be housed comfortably within its limited city-state political boundaries.

Two major concerns, population planning and housing, marked Singapore's early development. The first which also became a major issue at the population conference in Cairo (September 1994) is the issue of the relationship of population vis-a-vis natural resources/environment. In the 1950s and 1960s, Singapore's growth rate was exceedingly high by global standards. The crude

Table 16-1 Chronological Order of Official Plans in Singapore

Historical Period	Plan Name	Plan Period	Comments
Colonial (1819-1959)	Raffles Plan (1822)		
		Master Plan (1955) (adopted in 1958)	•General Improvement Plan of the Singapore Improvement Trust (SIT): road widening and construction/open space provision •Municipal Improvement Schemes: piecemeal House improvement schemes/ backlane clearance
Singapore in the Federation of Malaysia (1963-1965) Post-independence (post-1965)	Concept Plan, 1971	Master Plan (1965) Master Plan (1990) Master Plan (1985) Master Plan (1980) Master Plan (1975) Master Plan (1970)	
	Revised Concept Plan, 1971	New Master Plan (based on com-pleted Developement Guide Plans (DGPs)	•The completion of the 55 DGPs will lead to a new Master Plan

Source: Modified from Teo (1992:164)

birth rate in 1957 was 43 per 1,000 and Kandang Kerbau Hospital had the distinction of being listed in the *Guinness Book of Records* as the hospital with the highest number of new-born deliveries. This concern in the early years with population growth led to draconian measures on family planning and population control. The government initiated in 1966, its "stop at two" family planning program which led to a drastic drop in population growth. The crude birth rate dropped sharply to 17 per 1,000 in 1977 (Saw 1991,224), thus explaining the low population growth rate of 1.3 percent for the period 1970-80 (Lau 1992,2). Prime Minister Lee Kuan Yew in 1983 acknowledged that the government's population policy was a mistake. In fact, the government is now following the opposite policy, trying to increase population growth and at the same time trying to increase the quality of population. Given its limited land resources, the government's overall population plans for the future are targeted at maintaining an ultimate population of about 4 million people. The fundamental principle here is that there can be no environmental plans or effective environmental management if countries or cities do not decide on an upper population limit. In Singapore, the population dimension has always been kept in clear perspective in the government's national plans.

Planned Singapore: Ideology and National Identity

Singapore's planning process has been a product of three important considerations. First, the cardinal rule by which all planning in Singapore operates is the law of finite space. With competing claims for land and a growing population, the government has to continually maximize land utilization. Pragmatic and economic concerns often outweigh environmental issues (for example, nature conservation) when there are competing land claims. Second, the government has been governed by a no-nonsense approach in dealing with major problems such as housing, transport, and economic needs. All these issues have spatial ramifications in the planning process.

The third area that has greatly shaped political thinking on Singapore's planning program has to do with political ideology. Despite the government's adherence to meritocracy and pragmatism, the political leadership is governed by various ideological underpinnings in its state policies and plans. Five ideological statements by the leadership over the years that govern the planning process in Singapore can be included under the following headings: (a) Socialism, (b) Environmental Determinism, (c) Possibilism, (d) Multiracialism, and (e) Communitarianism.

Socialism

The early years of planning were partially influenced by the governing party's socialist ideology. Chua (*1995*) in *Communitarian Ideology and Democracy in Singapore* has argued that despite the Peoples Action Party's (PAP) self-proclaimed belief in pragmatism, they have followed a policy of ideological commitment, let alone ideological planning, over a sustained period of time. He goes on to argue that the PAP's ideological success may be interpreted within a "neo-marxist conception of the process of ideological formation in the development of a new social order". Riding on the democratic socialist ideological platform, the ruling PAP underscored their socialist ideals in two ways. One was the belief in creating a more equitable society that was not based on class distinctions. The public housing schemes were thus developed with the idea of mixing people of different economic abilities within one neighbourhood and estate. At a public housing neighbourhood/estate level mixing of economic classes upholds Government policy, but sociologists have noted that the distinction of 1-room, 2-room, 3-room and 4-room Housing and Development Board (HDB) public flats creates stratification of Singapore society (Hassan 1970).

The second outcome of the socialist leanings of the PAP was their endorsement of providing subsidies for basic public amenities—housing, health and education. The three services were the pillars of PAP's success. Their ability to translate effectively the public housing, health and education programs provided the fundamentals for Singapore's rapid development in the first 20 years (1959-1979) of the PAP rule.

Environmental Determinism

Modern urban planning is often said to be quasi-deterministic because urban planning is based on control functions, the regulation of urban form, location and character, conscious layouts and applied designs (Cherry 1988). At the root of urban planning is the belief not only in the creation of a new cultural landscape, an ideal city, but also that the ideal urban form can generate or elicit certain behavioral as well as social and political responses for a community. The search for an ideal community is also tacitly, if not explicitly, reflected in urban planning.

Overall planning in Singapore has been designed with some expected results clearly in mind. The Urban Redevelopment Authority, for example, was established in 1974 with the "overall objective of creating a new and gracious City with a better environment for residence, business and recreation" (Ministry of Culture 1975,224). The best exemplification of government planning must be with public housing. This is one area where extensive design and planning has been used for flats, housing estates and satellite towns all over the island, not only to provide housing but to create "a better living environment" for residents (Teh 1975,11).

The quasi-environmentalist thinking in public housing was, however, not only a question of solving the slum and squatter problem, it was also a program of planned national social change and development. The Housing and Development Board's housing estates were planned first according to the neighborhood concept and since 1979 with the precinct concept. The expected response for the neighborhood concept was evidently to ensure that residents uprooted from their former old and established environments would be able to retain ecological, social and Asian-type community orientated relationships in their new vertically-aligned housing estates (Yeung and Yeh 1975, 262-280). Much of the neighborhood concept was anchored on ensuring that residents had easy access to neighborhood shopping and community-related activity centres.

The precinct planning concept, however, with its emphasis on well-defined boundaries is meant to create a sense of place consciousness and an attachment to place. Each precinct revolves around a "community focal point" that should help residents "feel a sense of identity with the precincts" (*Straits Times* 25 February 1987,10). The promotion of the precinct concept is in line with the Government's attempts, through public housing to cultivate a greater "sense of rootedness" (*Straits Times* 9 February 1990, 20).

Not withstanding the quasi-determinist thinking involved in housing, urban or national planning, Singapore's small size, lack of natural resources, total dependence on imported energy supplies, and dependency on water and food resources have remained the most important perennial deterministic influences. Environmental problems are the crux of the leadership's survival obsession in political speeches. The bottom line is that given Singapore's environmental limitations, the country has very little margin for error. Any major mistake the country makes will have catastrophic results. Singaporeans need to find the right solutions to national problems for imperfect solutions will mean certain disaster. "And in the case of Singapore", warns Rajaratnam, "the disaster if it comes must be total, irreparable and final. There is no second chance for us" (Chan and Haq 1987, 210).

The belief in environmentalist influences, however, has given rise to technological changes, political adjustments and economic adaptations to overcome natural impediments. Obvious examples of official responses to Singapore's limited land area are the Government's ambitious land reclamation projects

(Wong 1985) land optimization planning policies and high rise high density living (Teh 1979,68-70). Externally, however, Singapore has been overcoming its spatial limitations by creating "multi-dimensional space" according to the Minister of State (Finance and Foreign Affairs) (*Straits Times* 9 February 1990,20). This was space created by peaceful diplomatic and economic means such as the twinning of Singapore and Hong Kong and the development triangle of Singapore, Johor and Batam:

> We need political space to live in peace and to do business. We need economic space for Singaporean capital to move overseas since we are so short of land and manpower here. We need military space to train the SAF. We need space for recreation, and so on (*Straits Times* 9 February 1990,20).

Possibilism: The Pragmatic Ideology

Singapore's development over the last 35 years is reflected in drastic changes in the landscape. During this period, 10 percent of the land area (about 6,000 ha) has been reclaimed, over 86 percent of the population has been housed in government public housing flats and estates, several new reservoirs have been created, rivers have been canalized and dammed, industrial estates established, over 5 million trees and shrubs planted, close to 1,800 hectares of manicured parks, gardens and recreational spaces have been developed, over 2,000 kilometres of roads, highways and expressways constructed and 2,145 km of central sewers network around the whole island (Figure 16-1). For many observers, the modification of landscape is best seen as an endorsement of possibilism. It also reflects what Gutkind (1956,11-12) calls "I-It" relationships where the earlier organic, symbolic and magical human-environment bonds are replaced by abstract, confident, aggressive, utilitarian, and exploitative attitudes towards nature. In the process, humans have imposed, through science, technology, huge investments of time, energy and money, culturally engineered signatures on the landscape.

If extensive and rapid landscape changes characterize Singapore, they emanate from the Singapore leadership's very "unsentimental, pragmatic" approach to problems and development. Singapore's vigorous development has been firmly anchored in "rational" public administration and "instrumental rationality" (Chua 1985,31), faith in science and technology, and the concern for an efficient, no-nonsense approach to issues. All these underlie the possibilist view of human-nature relations. Despite the natural resource and spatial limitations, the Singapore leadership sees its ability to overcome such inherent problems and make the country successful by moving Singapore into the technological age, investing in its people and placing importance in maintaining stability on the social, cultural and political fabric of society.

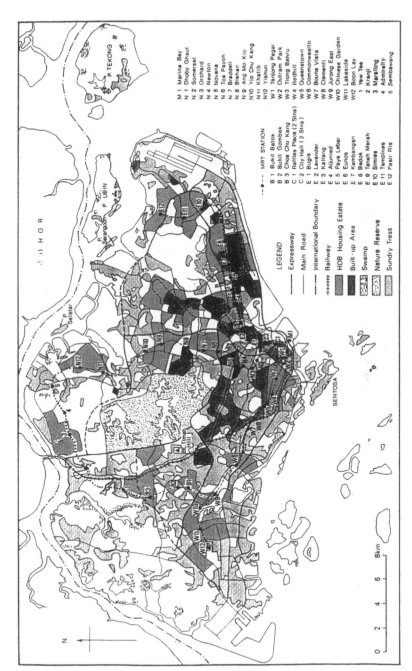

Figure 16-1 Singapore: Current Land Use

Faith in science and technology propelled Singapore into massive industrialization. The leadership fervently believes the only way of circumventing the country's environmental limitations is by making Singapore a technological and industrial country. One rationale is that the strongest, most prosperous and most dynamic societies are countries with the most advanced technologies (Chan and Haq 1987,255). The second reason is more central to Singapore. The pessimistic view of Singapore's future, based on physical limitations (e.g. smallness of area; no agricultural base; no raw materials), becomes irrelevant if Singapore becomes a technological and industrial country. "Size then becomes no handicap to progress and prosperity" observed a former Foreign Minister. Switzerland, Denmark, Holland, Belgium, and even Japan and the U.K., are not large countries, by world standards, but these countries enjoy higher standards of living than many bigger countries. The government emphasis on telecommunications, biotechnology, computer literacy, information technology, and hi-tech industrialization are all reflective of the Government's pro-possibilist thinking.

Multiracialism

Besides meritocracy and pragmatism, Singapore's nation-building since 1965 has revolved around the concept of multi-racialism which also encompasses multi-lingualism, cultural tolerance and the acceptance of religious diversity (Lau 1995,179). Multi-racialism is one of the founding tenants of Singapore and over the years has taken different directions. Initially the multi-racial policy was nurtured and pursued because of Singapore's stormy relations within Malaysia and the political sensitivities of being a predominantly Chinese island in a Muslim-Malay sea. Since the mid-1980s Singapore's renewed interest in ethnic identity has been pursued as a means of maintaining Asian identity and values in response to the leadership's perceived growing tide of Western materialism and decadence (Hill and Liau 1995, 36).

Like all other public issues, the state plays a critical role in ensuring that multi-ethnic issues and ethnic relations are harmoniously intertwined with nation-building. One of the ways of ensuring multi-racial intermingling has been through public housing. In the past, certain public housing estates had a predominant concentration of ethnic groups, but in the last 10 years the government is attempting to spread the different ethnic groups over various public housing estates. What it means is that the minorities (Malays, Indians, Eurasians and other groups) are given priority in areas where they are currently under-represented. The same policy is being used, especially for the Malays, with regard to schools.

Communitarianism

Since the mid-1980s, the political leadership has been promoting ethnic cultures. This policy partly endorses the government's interest in furthering an Asian paternalistic tradition. Government leaders have accepted the notion of a paternalistic democracy and the communitarian ideology prevalent in Asian cultures. Such an ideology is diametrically different from Western positions. In the USA, for example, political ideologies are based on individualism, competition and the barest minimum of state planning, while in Singapore the government played a more extensive role as vision setter, planner and consensus maker (*Straits Times* 8 January 1991, 40).

The enthusiasm for a communitarian ideology has received impetus in the recent White Paper proposal of shared values - an attempt to provide an identity for Singaporeans. The government has proposed five values: nation before community and society above self, family as the basic unit of society, community support for the individual, consensus instead of contention, and racial and religious harmony. It is these Asian values and communitarian ideology, now made politically explicit, that the government is encouraging to ensure Singapore's development as a viable urban ecosystem. It recognises that if the individualist, competitive, capitalist laissez-faire system of the USA takes over completely, Singapore will end up facing a tragedy-each individual for himself and God for us all (Savage and Kong 1993, 48-49).

The government's communitarian ideology, not only provides a counterweight to the Western beliefs in individualism and liberation, it also makes it "ideologically possible to rationalise the conflation of state/Government/society" (Chua 1995, 210). If the ideology of communitarianism justifies state interventions in social life as pre-emptive measures for ensuring the collective well-being, then state planning is thus legitimized. The uplifting of the collective well-being is often used by the planning authorities and political leaders as a defense for government policies when there is a head-on clash with pressure groups over the preservation of historical buildings or the conservation of nature reserves.

Deliberate Urbanization: Landscape Change

In his address at the OECD Conference in Melbourne in 1994, the Australian Prime Minister Paul Keating lamented that cities were "losing their soul" and were becoming "cold monuments slashed by expressways that depersonalise life and reduce one's self esteem" (*Straits Times* 23 November 1994, 6). Despite Singapore's rapid economic growth, on balance the political leadership's legendary faith in planning has ensured that the Singapore urban landscape remains efficient, clean, green, and relatively beautiful. In an attempt to eradicate slums and squatters, critics feel the government authorities have been over-enthusiastic in tearing down old buildings. Indeed, the shop-house which was once a ubiquitous landscape feature has now become an endangered landscape artifact. Even the lively traditional ethnic districts of Chinatown (Chinese), Little India (Indians) and Kampong Gelam (Malay) have been upgraded and changed to the extent that the spirit of these communities is lost even to tourism (Yeoh and Lau 1995).

Singapore is one of the most densely populated countries in the world, but considered simply as a city, its high density is justifiable. In 1967, its population density was 3,400 persons per square kilometer which rose to 4,800 per square kilometer in 1990, and it is likely to increase to 6,500 per square kilometer by around 2020 (Figure 16-2). This high density population is viewed by some as an impediment for Singapore's neo-utopian agenda, but despite its high population density, Singapore does not experience the urban problems found in many other cities. There are no slums and squatters, relatively little traffic congestion, and no over-crowding and no major urban infrastructural shortcomings.

About 50 percent of Singapore's total land area is built-up and over 10 percent is reclaimed from the sea. The human imprint on the landscape is evident everywhere. Even the government's goal to turn Singapore into a "model green city" (Foo and Rocha 1995, 222) is a product of human design. The leadership's early garden city goals helped to implement consciously its tree-planting campaign over the last 25 years. Up to 1994, the Park and Recreation Department maintained 877,964 trees and 4,773 hectares of parkland (Foo and Rocha 1995, 229). While there are numerous compliments of the government's efforts to make Singapore green, clean and pollution-free based on a built-up "garden" environment, the Singapore Nature Society has been critical of the way the Government is trading off its pristine nature assets for development. While the Government in its Green Plan has heeded some of the Nature Society's requests for conservation (Figure 16-3), there have been other areas where the government has clearly decided on a different agenda (Briffett 1990). Only time will tell whether the Singapore landscape and society will be poorer due to the dwindling of its bio-diversity. In the final analysis, Singapore scores well in creating an artificial, built-up environment, but can be faulted as to whether it is doing enough to conserve its natural environment. Singapore's Green City is not based on preserving nature areas per se, but rather on ensuring that high standards of public health and hygiene are maintained in its built environment and that its citizens enjoy a high quality environment for living, working and recreation.

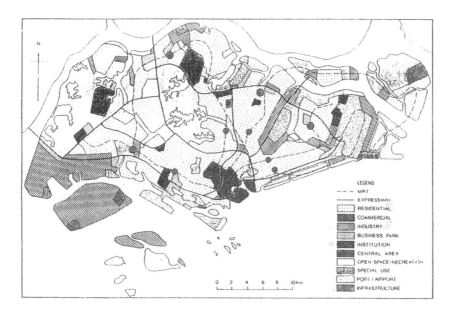

Figure 16-2 Singapore: Probable Land Use in the Year 2020

Despite its spatial handicap, the government wants to turn Singapore from its current status as an entrepot into a "hub city". This would mean that Singapore must offer first-class products and services based on an infrastructure that must rank with the best in the world (Government of Singapore 1991, 59). In spite of being dependent on imported energy, water, food and natural resources, Singapore's infrastructural development is impressive by world standards.

Singapore has created an urban landscape that is comfortable, clean, green and aesthetically pleasing. Singaporeans also reside in decent houses. While 13 percent of the population have private houses, the bulk (86 percent) of Singapore's population lives in public housing flats. This ensures that the population, no matter how poor, lives in minimum housing standards set by the Government — with good water, sanitation, sewerage and garbage disposal facilities.

Despite being reliant on imported energy, Singapore's power supply infrastructure is best among developing countries in meeting business needs in 1992 according to the World Competitiveness Report (*Straits Times* 19 March 1993, 2). The Report scored 95.7 percent for Singapore, as against 90.3 percent for Hong Kong and 78.8 percent for Korea.

Singapore's economic vitality is also predicated on its transportation systems. Singapore is the busiest port in the world in terms of shipping tonnage

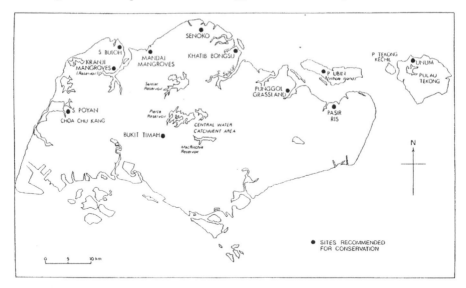

Figure 16-3 Nature Conservation Sites in the 1991 Concept Plan (after Briffett 1990)

and the world's top bunkering port with over 17 million tons of bunker supplied in 1994. Singapore's Changi airport is not only busy, but also a favorite with air passengers. It was voted the best in the world by several business traveler magazines, which voted Changi the "World's Favourite Airport".

But the critical test of any neo-utopian claim for an urban area is whether there is an efficient, comprehensive, comfortable and relatively cheap, land transportation system in place. In Singapore, land transportation receives a mixed review. The cost of owning a car in Singapore is much too costly. The government is seen as trying to rake in huge profits at the expense of car owners. Despite the legitimate reasons for reducing the vehicle population, some justification exists for the idea that the government is making good money out of its "pay for your car" policy. The registry of vehicles, for example made over S$4.2 billion in 1994, much of the revenue coming from the certificate of entitlement (S$1.9 billion) and the additional registration fee (S$1.1 billion) (*Straits Times* 23 September 1995,2).

Aside from its automobile policy, the government has put in place a comprehensive land transport system (Figure 16-4). The introduction of the Mass Rapid Transit (MRT) System, an integrated and complementary bus system together with a taxi service has made commuting relatively easy and efficient (Figure 16-5). The introduction of the Light Rail Transport (LRT) system is likely to further enhance mobility in key nodes in Singapore.

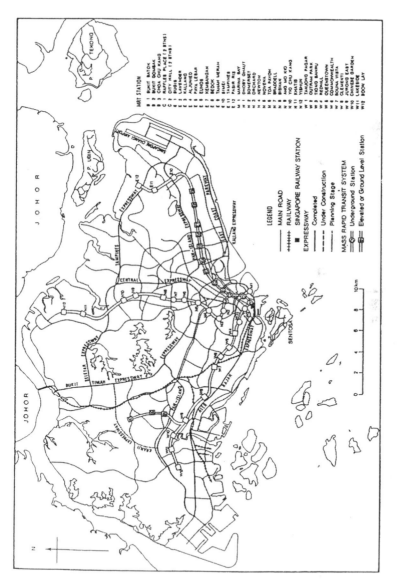

Figure 16-4 Singapore: Major Land Transportation Network

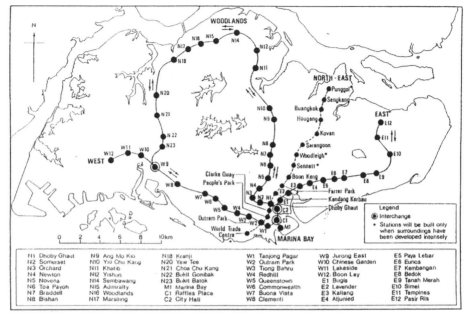

Figure 16-5 Singapore: Mass Rapid Transport (MRT) System

Public Housing: Designed Landscapes, Healthy Environments

If there is any single policy that has shaped the urban landscape of Singapore, it certainly has been public housing and the development of satellite towns throughout the island. The issue of public housing goes beyond the mere provision of homes for people. Public housing in Singapore struck at the very core of environment related issues that many developing cities like Singapore are facing. To understand the rationale for public housing, one needs to look at the scenario when it was initiated. Public housing had a significant impact in three areas that are environmentally related: (a) it provided an alternative to slum and squatter dwellings that were rampant in Singapore in the 1950s and 1960s; (b) it provided an alternative means of improving public health and hygiene; and (c) it helped to create healthy environments that provided the underpinnings for social and economic development.

The development of public housing was also a direct response to the long history of disease-prone, colonial Singapore. Mortality rates in colonial Singapore were indisputably high among the immigrants and indigenous communities. Death rates within the municipal area averaged over 40 per thousand in the late 19th and early 20th centuries, only falling below 40 per thousand in the mid-1920s (Savage and Yeoh 1993, 37). Right up to the 1950s,

Singapore was the hotbed of malaria, dysentery, tuberculosis and beri-beri and was ravaged from time to time by infectious diseases such as cholera, small-pox and bubonic plague (Tan 1991, 341-345). The reasons for the prevalence of such diseases and infectious diseases had very much to do with the insanitary environments, lack of public hygiene, sewerage disposal, sanitation and water purification in the crowded slum and squatter dwellings (Savage and Yeoh 1993, 37). Yet despite all the efforts and high costs to maintain public health, Singapore has not been able to eradicate environmentally-related diseases. In 1992 there were 2,878 cases of dengue fever and dengue haemorrhagic fever and 242 cases of cockle-associated indigenous hepatitis (*Annual Report 1992*, 25-27).

Public housing to provide clean, healthy and hygienic environments for citizens was to a large extent a reflection of the government's quasi-determinist planning notions of controlled functions and regulations arising from designed urban form (Cherry 1988). Public housing in Singapore, with its extensive plans and designs, was meant to create a better living environment for residents (Teh 1975, 11). The official thinking at this time was to produce low cost public housing which would offer Singaporeans a cleaner, hygienic and better environment which would in turn mean a healthier population requiring lower medical expenditure (Savage 1992, 193). The leadership also believed that the better residential environments would ensure a more productive labor force resulting in greater prosperity and a higher standard of living.

Public housing provided in the government's plans, the creation of a human designed and manipulated environment. Public housing estates were tailored to ensure that all the social, cultural, recreational and economic requirements of residents were adequately catered for within the new satellite towns. This exercise in infrastructural development, meant the creation of new environmentally friendly landscapes with provision for proper refuse disposal, modern sewerage systems and a clean purified water system. Without such public housing schemes, contemporary Singapore would never have achieved its image of a clean and green city. Public housing and satellite towns not only house a large percentage of Singapore's population in modern, environmentally friendly and sanitary environments, but they also have created artificial environments for the recreational and aesthetic needs of residents.

The Green Plan: Tropical City of Excellence

In the 1990s, Singapore's national plans have taken on more environmentally friendly and green perspectives. Having eradicated the filth and squalor of unhygenic slum and squatter landscapes and successfully implemented its Garden City concept, the government goals seemed to be oriented towards providing high quality life with environmental issues integrated in these plans. In 1991, the Urban Redevelopment Authority (URA) put forward their blueprint based on the Concept Plan of "Towards a Tropical City of Excellence". The 1991 Concept Plan was a revision of the 1971 Concept Plan. Unlike the 1971 Concept Plan that focused on infrastructural development and economic growth strategies, the 1991 Concept Plan covered more far reaching issues. Specifically, it was concerned with addressing the following questions: (a) How to sustain economic growth? (b) What is the best transport system? (c) How to accommodate comfortably 4 million people? (d) How to maintain a city of Asian character? and (e) How to improve the quality of life? This plan was intended to upgrade housing conditions and the quality of life (Figure 16-6).

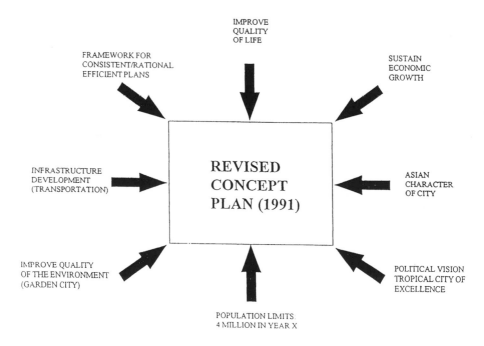

Figure 16-6 Major Themes in the Revised Concept Plan, 1991

Unlike the earlier Master Plans, the 1991 Concept Plan divides Singapore into 55 planning areas with specific localized proposals and plans called Development Guide Plans (DGPs) (Figure 16-7). When completed each new DGP will replace the 1985 Master Plan it covers. The government hopes to stimulate Singapore's economic growth in several areas: (i) maintain its pivotal role as a banking center in Asia; (ii) maintain hi-tech industries; (iii) create an environment for research and development; (iv) promote the city as a regional base and headquarters for MNCs; (v) develop its "hub city" status (Government of Singapore 1991,59); and continue to make Singapore an attractive tourist destination. To ensure the country's sustained economic development, the government has created regional centers, business parks and technology corridors in the 1991 revised Concept Plan (Figure 16-7).

The government's 1991 Concept Plan and 1992 Green Plan (Figure 16-8) reflect changing demands and interest amongst Singaporeans given their increasing affluence and rising standards of living. One outcome of this has been the government's plans to build more golf courses to meet the growing need for recreational pursuits and changing lifestyles. In addition to the 14 existing golf courses, the Urban Redevelopment Authority has identified 15 more sites for golf courses to be built gradually as Singapore's population stabilizes at 4 million. This issue once again has brought the government in direct conflict with the Singapore Nature Society. The proposal assumes that golf will be the future sport of choice of Singaporeans and that all Singaporeans will be wealthy enough to afford it. At current rates one needs at least S$100,000 to join a golf club.

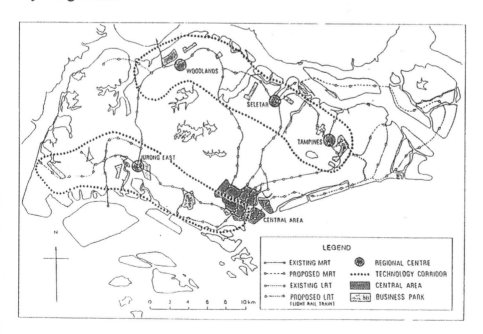

Figure 16-7 The Revised Concept Plan, 1991

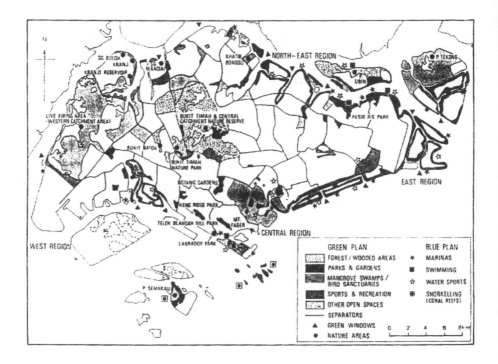

Figure 16-8 The Green Plan

Singapore's Neo-Utopian Goals

> Singapore must become a synonym for quality, reliability and excellence. We will become a business hub of the Asia Pacific (Government of Singapore 1991, 59).

The quest for excellence, better standards of living and high levels of achievement are all part of a neo-utopian ideology. In Singapore, the political leadership has adopted the neo-utopian ideals as part of political ideology as well as national goals. The best example of this is found when the Peoples Action Party leadership under Prime Minister Goh decided to chart a new political agenda. When the younger generation leaders took over in 1984, they unveiled the "Vision 1999" programme that set out to achieve the 1984 Swiss standard of living by 1999, thereby also underscoring in the public's eyes that the Swiss are living in some sort of neo-utopian environment. Specifically the younger generation leadership had envisioned their "City of Excellence" by 1999 as such:

The whole country will be clean and green, brightened by flowers and laden with fruits. There will be no unsewered households, slums or environmental pollution. We will build a city of expressways and free-flowing traffic. Of time-saving Mass Rapid Transit, cool and comfortable. Of electronics, telecommunications and computers (*Straits Times* 12 December 1984, 16).

Given this government endorsement, it is no wonder that in several surveys of school children, Switzerland is often seen as the ideal country, politically, economically and socially. In June 1995, the Deputy Prime Minister Lee Hsien Loong noted that Singapore had achieved the 1984 Swiss standard of living in housing, education, technology and economic development, four years in advance of the target date set by the Government (*Straits Times* 29 June 1995, 3). This achievement once again endorsed the people's faith in the government and the political leadership of the Peoples Action Party.

The longer term vision of Singapore's second generation leaders, however, is best enshrined in the book, *The Next Lap* (Government of Singapore 1991). In this government-produced book, all the hallmarks of the vision of a new society and a neo-utopian state are found in various key goals that the leadership hopes to create. The goal of the Singapore leaders is to make Singapore a city of excellence, quality and reliability; a people's city; a business city; an environment city; a city of culture and grace; in short a city for all seasons (Government of Singapore 1991). This is a leadership that hopes "to facilitate and regulate the physical development of Singapore into a tropical city of excellence" (Urban Redevelopment Authority 1995). The official concern with excellence has been implemented at two levels: maintenance of a world class infrastructure for quality living, and developing a quality of public service that is responsive to the demand of a globalizing economy and a better-educated population. In May 1995, the public service launched its PS 21 to ensure that public servants welcome change and ideas to improve the quality of the service and enhance their own work satisfaction (*Straits Times* 6 May 1991, 1).

Of all the concepts, it is the Garden City symbol that best captures the neo-utopian ideal because when taken as the sum of two key words (Garden and City), the concept demonstrates not only the power of human creativity, but also an achievable man-made landscape ideal (Savage 1991, 74). Singapore's Garden City experiment involves creation of a landscape of excellence based on human vision, creativity and management. Never mind that the garden landscapes are contrived, artificial and unnatural, or that order and neatness are the aesthetic rule. Specifically, the Garden City demonstrates in Singapore man's power over nature, the human abilities for landscape transformation, the cultural imprint of human creativity and technological prowess in the high-rise built landscape and massive land reclamation. Cities and gardens are not imaginary ideals or pristine landscapes, but pragmatic human designed and engineered realities. With order and neatness as its key aesthetic norms, Singapore's garden landscape embodies all the government's planning ideals of creating and maintaining a clean and green city, a non-polluted environment, an aesthetically pleasing landscape, a healthy living milieu, more open spaces and a variety of recreational outlets to enhance the quality of life. The word city, however, is synonymous with civilization. Singapore's vision is a city that underscores progress, modernity, good infrastructure and high technology. That view has been modified for the 1990s to include a city for quality living.

Conclusion

For many developing countries, national plans and programs have become a major vehicle of political propaganda of governments and political parties. To a large extent, national plans are a means of legitimizing political power. Plans that provide a vision of the future give citizens a sense of hope that their livelihoods will be improved if the political party and government is maintained in power. In the region, the existing governments in Singapore, Malaysia and the Philippines have all put forward plans as part of the future visions of their society. Singapore subscribes to achieving the city of excellence status by 1999, the Philippines has a vision 2000 Plan, and Malaysia has a vision 2020 Plan.

Singapore's success in long-term planning, and the execution and implementation of national plans have been made possible due to the PAP's continuation in political power (since 1959). Few countries in the world, save the communist states, exist where a government has remained in power for over 30 years. With a democratic system, the PAP's reign in power over the last 35 years is an exceptional feat. The maintenance of political stability has been a major factor in enhancing the success of Singapore's long-term national plans. While the top-down approach to national plans governed the first two decades of independence, the current national and green plans have increasingly tried to solicit grassroots participation and inputs. By publicizing plans and seeking public participation in them, the government has used the planning process as an endorsement of its political rule. State plans legitimize a political patron-client relationship in Singapore.

Unlike Western democratic countries, Singapore's top-down planning process as vision setter, planner and consensus maker has been successful because of its political culture. This top-down approach has been made possible because of the government's stress of a communitarian ideology which has received impetus in the recent White Paper proposal of shared values — an attempt to provide an identity for Singaporeans (Savage and Kong 1993, 48). Government leaders accept that the communitarian ideology is part of the Asian cultural tradition. It is thus critical of the Western position that stresses individualism, competition and the barest minimum of state planning. Singapore leaders see the state as above the individual and are thus wary of the dangers of individualist demands on state resources. This thinking underscores the maintenance of not only a harmonious human to human relationship as in the Confucian tradition, but also a harmonious human and environment relationship that is evident in the Taoist tradition.

Armed with a battery of professional architects, urban planners, bureaucrats and administrators, Singapore's planning process in the first 20 years was an unqualified success in creating a clean and green environment for healthy living. This top-down approach was made possible because the government was dealing with a society that was essentially poor, where unemployment was rampant and where expectations were low. Given these social conditions, the government's plans to develop infrastructure and upgrade the essential creature comforts for people was met with unanimous support. The landslide elections the PAP won in the 1970s and 1980s clearly demonstrated public approval of the government's plans and policies.

The 1991 Concept Plan and 1992 Green Plan show a change in Government planning strategies. Infrastructural development has less urgency, and the focus has shifted to a softer development. The focus is now on citizens, in shaping values (family values), enhancing communitarianism, developing a Singaporean identity and a sense of place, promoting courtesy, and developing a green consciousness and environmentally pro-active society. Environmental education is now viewed as a lifelong process to develop environmentally responsible behavior (Ministry of the Environment, Singapore 1993, 14).

The Model Green City is meant to be conducive to gracious living, with people who are concerned about and take a personal interest in the case of both the local and the global environment (Ministry of the Environment, Singapore 1993, 10). The new people-oriented planning agenda is likely to remain more contentious and more difficult to please. With three decades of good economic growth, Singaporeans as a nouveau riche society are still clamouring for consumer products and status symbols. We have a long way to go to reach the post-consumer Western-developed societies, where green issues are taken with greater personal involvement and seen in a more altruistic light.

References

Briffett, Clive, 1990, *Master Plan For The Conservation Of Nature In Singapore*, Singapore: Malayan Nature Society (Singapore Branch).

Chan Heng Chee and Obaid Ul Haq, (eds), 1987, *S. Rajaratnam: the Prophetic and the Political*, Singapore: Graham Brash and New York: St. Martins Press.

Cherry, Gordon E. 1988, *Cities and Plans*, London: Edward Arnold.

Chua, Beng Huat, 1985, "Pragmatism of the People's Action Party Government in Singapore: A Critical Assessment", *Southeast Asian Journal of Social Science*, 13:2:29-46.

Chua, Beng Huat, 1995, *Communitarian Ideology and Democracy in Singapore*, New York: Routledge.

Foo, Siang Luen and John Rocha, (eds), 1995, *Singapore 1995*, Singapore: Ministry of Information and the Arts.

Government of Singapore, 1991, *The Next Lap*, Singapore: Government of Singapore.

Gutkind, E.A. 1956, "Our World From the Air: Conflict and Adaptation", pp. 1-44 in Thomas, William L., (ed.), *Man's Role in Changing the Face of the Earth*, Chicago: The University of Chicago Press.

Hassan, Riaz, 1970, "Class, Ethnicity and Occupational Structure in Singapore," *Civilisations*, xx:496-515.

Hill, Michael and Liau Kwen Fee, 1995, *The Politics of Nation Building and Citizenship in Singapore*, London: Routledge.

Lau, Kak En. 1992, *Singapore Census of Population 1990*, Singapore: Department of Statistics.

Lau, Ah Eng. 1995, *Meanings of Multiethnicity*, Kuala Lumpur: Oxford University Press.

Ministry of Culture, 1975, *Singapore '75*, Singapore: Ministry of Culture.

Ministry of the Environment, 1992, *Annual Report 1992*, Singapore: Ministry of the Environment.

Ministry of the Environment, 1993, *The Singapore Green Plan - Action Programme*, Singapore: Times Edition.

Rajaratnam, S. 1972, *Singapore: Global City*, Text of Address by Mr S. Rajaratnam, Minister for Foreign Affairs, to the Singapore Press Club, 6 February.

Savage, V.R. 1991, "Singapore's Garden City: Reality, Symbol, Ideal," *Solidarity*, 131-132, (July-December) 67-75.

Savage, Victor R. 1992, "Landscape Change: From Kampung to Global City", pp. 5-31 in Avijit Gupta & John Pitts, (eds), *The Singapore Story Physical Adjustments in a Changing Landscape*, Singapore: Singapore University Press.

Savage, Victor R. and Lily Kong, 1993, "Urban Constraints, Political Imperatives: Environmental 'design,'" *Landscape and Urban Planning*, 25:37-52.

Savage, Victor R. and Brenda Yeoh, 1993, "Urban Development and Industrialisation in Singapore: An Historical Overview of Problems and Policy Responses", pp. 22-56 in Lee Boon Hiok and K S Susan Oorjitham, (eds), *Malaysia and Singapore: Experience in Industrialisation and Urban Development*, Kuala Lumpur: University of Malaya.

Saw, Swee Hock, 1991, "Population growth and control", pp. 21-241 in Ernest C. T. Chew and Edwin Lee, (eds), *A History of Singapore*, Singapore: Oxford University Press.

Straits Times, Singapore, Several issues.

Tan, Nalla, 1991, "Health and Welfare", pp, 339-356 in Chew, Ernest C.T. and Edwin Lee, (eds), *A History of Singapore*, Singapore: Oxford University Press.

Teh, Cheang Wan, 1975, "Public Housing in Singapore: An Overview", pp. 1-21 in Yeh, Stephen H. K., (ed.), *Public Housing in Singapore*, Singapore: Singapore University Press.

Teo, Siew Eng. 1992, "Planning Principles in Pre- and Post-Independence Singapore", *Town Planning Review*, 63:163-185.

Wong, P.P. 1985, "Artificial coastlines: the example of Singapore", *Zeitschrift fur Geomorphologie*, 57:175-92.

Yeoh, Brenda S.A. and Wei Peng Lau, 1995, "Historic District, Contemporary Meanings: Urban Conservation and the Creation and Consumption of Landscape Spectacle in Tanjong Pagar", pp. 46-67 in Yeoh, Brenda S.A. and Lily Kong. (eds), *Portraits of Places*, Singapore: Times Editions.

17 Seven Decades of New Land Planning: The IJsselmeer Polder Experience

COENRAD VAN DER WAL

Virtually all publications on spatial planning in the Netherlands refer to the high degree of national governmental involvement in the planning process (Faludi and Van der Valk 1994). Since the early 1960s, all spatial planning in Holland has been formalized in the *Wet op de Ruimteljke Ordening*, generally translated as the National Planning Act. In this law spatial planning procedures have been laid down with regard to the three levels of government: municipal, provincial and national. The hierarchical authority of state over province and province over municipality, the protection of the public through citizen participation in the planning and recourse on unjust planning decisions, are all part of this law and its later amendments.

The establishment of the law is, of course, the result of a very long development in planning, the idea of which can be said to have gained importance in the first two decades of the twentieth century. That was the time when it became abundantly clear that ever greater governmental involvement was necessary to regulate the erection of large new housing districts added to the large towns, which had been going on since the fourth quarter of the nineteenth century. The Netherlands Institute of Housing and Planning, founded in 1918, was instrumental in the advancement of national planning involvement (De Ruijter 1987). Its members were planning, building, and housing professionals and, probably more important, officials of municipal and other governmental organizations concerned with adequate housing and living circumstances for the masses. The 1924 international planning conference in Amsterdam, organized by the institute, had also emphasized the importance of urban centers in a regional context, or considering regional planning as an essential part of local problem solving. Out of this approach came forth the idea of a national plan, actively pursued since the formation of the State Authority for the National Plan in 1941.

The Reclamation of the Zuiderzee

In 1919 a new department of the Ministry of Transport and Water Management officially started planning the reclamation of large parts of the Zuiderzee, the large inland sea in open connection with the North Sea. Extensive floods in 1916 and the country's isolation during the First World War had shown the necessity of the closing off and the reclamation of this sea for flood control and to gain arable land. Eventually the Zuiderzee project would result in the reclamation of almost 170,000 hectares of land in which twenty new towns would be erected, that in 1986 would form the twelfth province of the Netherlands, Flevoland, and at the end of 1997 would have almost 300,000 inhabitants (Figure 17-1). So far four polders have been reclaimed. In 1930, the first polder, the Wieringermeer (20,000 ha.), fell dry, two years before the Barrier Dam closed off the Zuiderzee in 1932. In 1942, the Noordoostpolder (48,000 ha.) was drained, in 1956 Eastern Flevoland (54,000 ha.), and in 1968 Southern Flevoland (44,000 ha.). These three polders now form the province of Flevoland, whereas the Wieringermeer became (remained) part of the province of Noord-Holland (Van Duin and De Kaste 1990).

Planning and the IJsselmeer Polders

The development of the IJsselmeerpolders area parallels the development of the planning profession, planning methods, and governmental planning control in the Netherlands. However, interesting and unique in this whole development is the fact that the land was reclaimed by the national authority, and, therefore belonged to the State. Making the land habitable was not subject to local, municipal, or provincial interference. Two independent authorities: The Zuiderzee Project Department (ZPD) and the IJsselmeerpolders Development Authority (IJDA) acted as developing agents throughout the seventy years of the IJsselmeer polders' reclamation and development. Both departments were part of the Ministry of Transport and Water Management, and had working connections with the Ministry of Finance and the Ministry of the Interior. The ZPD was responsible for technical aspects of the reclamation: erecting the dikes, drainage, construction of water and road infrastructure, general plan layout of a reclaimed polder, and farm parcellation, while the IJDA took care of the human settlement aspects: planning and building of new towns, distribution of farmland and the selection of farmers, initiating business, social, educational, and even religious opportunities. IJsselmeerpolder development and

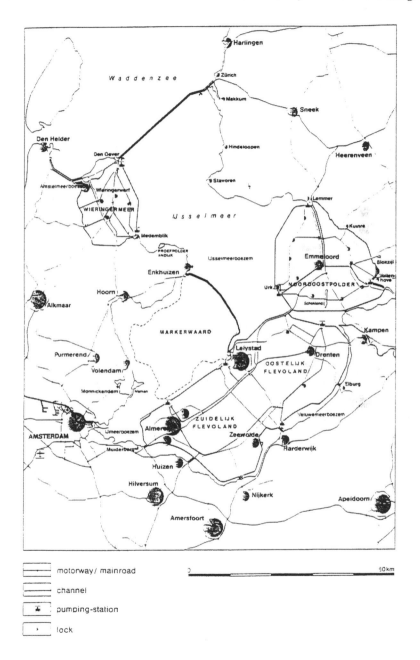

Figure 17-1   The Zuider Zee project, as it was about 1985

planning aspects have been dealt with by several scholars (Constandse, 1976; Dutt et al., 1985; Van der Wal, 1985).

Since the whole endeavor was financed by the State, it had control over the spatial expression of the physical environment, and to some degree over the social structure. This control extended to the selection of farmers, businessmen, shopkeepers, and by virtue of its control over the construction of housing: to the flow and social and economic composition of new immigrants into the IJsselmeer polders. Of course, as time advanced and nationally social acceptance of this paternalistic attitude diminished, strict IJDA control also diminished, and gradually these controls were abolished and citizens participated in the planning decisions. However, that is part of the interesting changes described in the following account.

When, in the course of the first ten years of its existence, it became evident that the technically oriented ZPD was not adequately adressing the human aspects of the polder reclamation, another department, which eventually became the IJDA, was formed. It was part of the same ministry and had a status equal to the ZPD. Both departments did their utmost to bring the work with which they were charged, to a successful conclusion. In their zeal to make the venture successful the IJDA not only built (public) housing and public spaces, but also schools, community buildings and other facilities. They were also drawn into procedures for the settlement and selection of shopkeepers, physicians, and other personnel to get a normal community going in the villages. Important herein is that the national government not only condoned this expansion of the IJDA's duties, but was also willing to finance it. It was a sign that Dutch society in general had advanced to the point where this was deemed important. Obviously, the liberal laissez-faire attitude of the nineteenth century had been left behind. In the previous large reclamation, where the State had drained the Haarlemmermeer (1852, 18,000 ha.), the residents had been left to their own devices. In that polder many settlers had gone bankrupt and had suffered ill health before the municipality finally intervened.

The Wieringermeer Polder

After the drainage of the Wieringermeer, first the arable land was prepared. Optimal drainage was provided by means of drain tiles and ditches, and the desalination process was carefully observed. Roads and canals were constructed and villages were erected, providing the basic amenities to the new population (Figure 17-2). All this took place ahead of the arrival of the carefully selected farmers and the first residents of the villages. The IJDA were

Wieringermeer

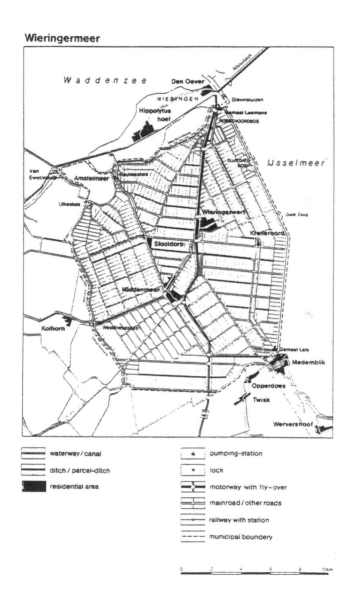

waterway / canal
ditch / parcel-ditch
residential area

pumping-station
lock
motorway with fly-over
mainroad / other roads
railway with station
municipal boundery

Figure 17-2 The Wieringermeer Polder

set to create as few obstacles as possible for the instant birth of a new polder society. They may be forgiven for acting in an authoritarian way, the more so since in that period the State's authority was naturally accepted. Idealism and confidence in the future were certainly part of the IJDA's modus operandi. As an example, the IJDA, contrary to usual practice on the old land, provided elementary schools on a non-sectarian basis, hoping to break down religious barriers in the new polder. However, within a few years, Catholics and Protestants set up their own schools, and by and by society developed along the traditional old-land line.

Local representation was not established until a municipality was formed in 1941. From the first settlers until 1937, the polder was divided between the bordering municipalities. This was very inconvenient as events, such as births and deaths, had to be recorded in neighboring towns. In 1937, therefore, the Wieringermeer was designated a "public authority", in effect a municipality without local representation. The IJDA director served as mayor under the title of *landdrost* or bailiff. This lasted up to 1941.

The Noordoost Polder

When, at the end of the nineteen-thirties, the ZPD and the IJDA turned their attention to the next polder, the Noordoost polder. The Wieringermeer had been completely developed, three new towns had been established, and the population was, economically and socially speaking, in good condition. Although initially the Noordoost polder was being planned along the same lines as the Wieringermeer, there are marked differences which can be ascribed, on the one hand, to experience gained in the Wieringermeer, but on the other, must be attributed to changing times or a shift in social perception (Figure 17-3). A better distribution of the towns over the polder must be seen as a reaction to the unfortunate central placing of the villages in the Wieringermeer, where families located at the polder's edge were too far from shops, schools and churches.

The attention to social issues, such as the relationship of farmer-farmhand, the question as to whether farm-laborers' housing should be provided in town or near the farm, were considerations new to polder planning. In the Wieringermeer only a physical planner had been consulted, in the Noordoost polder sociologists were added to the planning staff. This was done under pressure from the newly formed (1941) State Authority for the National Plan. In the discussion on the social status of farm laborers, forming the bulk of the future population the sociologists (or social geographers as they were called)

Noordoostpolder

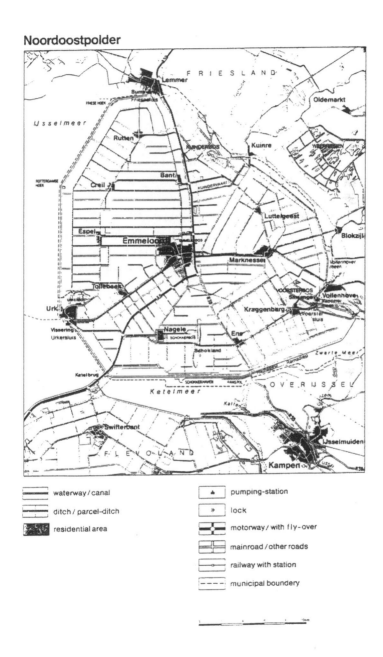

waterway / canal

ditch / parcel-ditch

residential area

pumping-station

lock

motorway / with fly-over

mainroad / other roads

railway with station

municipal boundery

Figure 17-3 The Noordoost Polder

evidently took the progressive position against the IJDA directorate and the physical planners. This discussion could be so thorough, because it took place during World War II, when construction was slowed down considerably anyway. After the war, construction of the towns was resumed in large part thanks to Marshall Plan aid.

The central town of Emmeloord had a regional function, the ring of ten villages around it took care of the daily needs of the population, who were no more than about a fifteen minute bike ride from the nearest village. Having learned their lesson in the Wieringermeer, the IJDA based the minimum village size on the presence of three elementary schools: Catholic, Protestant, and public (non-religious). This would amount to about 600 residents per village feeder area. The paternalistic and all encompassing procedures of the IJDA in the previous polder continued during the Noordoost polder's development.

Civil government was handled in the "public-authority" manner. The IJDA's director was again the bailiff, but the day to day management was laid in the hands of an executive secretary, who often defended the polder's interest in an independent manner. Full representation did not occur until the establishment of the municipality in 1962, although from 1951 onwards some popular representation was established by the formation of an elected polder Commission and an Executive Advisory Board (town council). The polder became part of the adjacent province of Overijssel when the municipality was established (Verkaik and Van Royen 1993).

Eastern Flevoland

Except for the discussions on the annual budget in the Second Chamber (House of Representatives) the Wieringermeer and the Noordoost polder were being developed, relatively speaking, in complete isolation as regards the public eye. However, as Eastern Flevoland was coming on the drawing board and Southern Flevoland and the Markerwaard were drawn in place, the rest of the country started to take some notice of the possibilities the new polders would have to alleviate some of the problems of the overcrowded Randstad.

The Randstad is the densely populated and economically most important area in the west of the Netherlands. It encloses the large towns of Rotterdam, The Hague, Amsterdam and Utrecht and the area (and smaller towns) in between. Where the previous two polders had had an agricultural function only, in the next polders the option of the settlement of commuters became feasible. When Eastern Flevoland had been drained in 1956 this possibility was not yet considered, but it became an issue as the plans were being developed over the years. Initially, having been set up on a town pattern similar to the Noordoost polder, as time progressed and society and the economy changed, the number of planned villages was gradually reduced. This was caused by the reduction in required farm labor due to a drastic increase in farm mechanization. The proximity of a town became less urgent as private motorized transportation increased. In addition, villages of 600 souls could not support a modern local retail structure. For that, a minimum feeder area of 3,500 population per village would be required. From the originally projected ten villages only two were eventually executed.

A regional support central town, Dronten, was also realized, and on the western shore the capital of Flevoland, Lelystad, was projected as a middle sized city (50,000 to 100,000). Since the regional labor force could not even support these towns and villages, commuters were encouraged to come and live in the villages.

While this process of change took place, another phenomenon could be observed, namely a change in how the rural area was being viewed and therefore planned. The Wieringermeer and the Noordoost polder had been planned (or designed) purely for their agricultural function. Any embellishments in the landscape were subservient to this function. The central town and the villages were constructed purely for the benefit of the polder's population. This was not to say that woods and treed lanes were not realized, they were. However, the planted woods had a function either as small recreation areas for the villages, or they were planted on land less suitable for other purposes. The polders were planned for themselves!

Eastern Flevoland was planned/designed in a somewhat different way (Figure 17-4). Much consideration went into catering to the visual experience of a passing motorist/cyclist. When the plan was first presented this was precisely the criticism the (ZPD) designer received from the IJDA who looked at the plan from an agricultural point of view. The IJDA were not against an aesthetic treatment per se, they were against an aesthetic from the urbanite point of view. Where in the other polders the towns were there for the benefit of the rural area, in Eastern Flevoland (with some exaggeration) the rural land was there for the benefit of the towns. This situation, although in a more subdued way, was later continued in Southern Flevoland.

The IJDA continued to exercise tight control over all construction, and landscaping, and it kept on selecting farmers, retail-business persons, and so on, as it had been doing since the 1930s. Even in Lelystad, a town envisioned for 100,000 residents, tight control over all aspects of the physical environment was maintained. Of course, control over the social and economic development of the city was impossible, although in providing the physical means the IJDA tried to influence those aspects also.

In 1955 a Public Authority was installed for the Southern IJsselmeer polders. Again the bailiff was the IJDA's director. However, an executive advisory board was installed at an early date in Dronten's development. In 1962, Dronten, occupying a large part of Eastern Flevoland (and the town of Dronten itself and two villages) became a municipality. In 1976 the offices of IJDA director and bailiff were split. The new bailiff, naturally taking his task seriously, immediately began to try and exert influence on the (society) building efforts of the IJDA, in accordance with democratic rules of local representation and participation in the establishment of one's own fate and future. This was particularly applicable to the town of Lelystad, in full process of development at the time. A person's place in society had long since changed from an obedient, authority abiding, citizen to a self-assertive, but willing to cooperate, participant in his/her own destiny. The IJDA was loath to let go of its own creation, afraid the inexperienced new residents would commit critical errors. Lelystad became a full-fledged municipality on January 1, 1980. Neither Lelystad nor Dronten were part of a province, therefore not subject to provincial, or even some State regulations, until the new province of Flevoland was established on January 1, 1986.

Southern Flevoland

A natural consequence of the reclamation of Eastern Flevoland was the reclamation of Southern Flevoland rather than the Markerwaard. Southern Flevoland would form the connection between the Randstad and the northern part of the country on a national level, and between Lelystad and the Randstad on a regional level. Earlier reports had made mention of it, but the 1961 Structure Plan for the southern IJsselmeer polders by the ZPD featured the relationship of the polders with the surrounding old land, particularly the Randstad. The plan emphasized the new polders as expansion areas for the penned-up Randstad. Housing, industry, and recreational areas would become available.

The outward urban thrust of the Randstad ring could be guided comfortably into the polders along a broad swath between Amsterdam and Lelystad. Near Amsterdam a large new town could be realized. That new town, later called Almere, together with Lelystad, could become an urban overspill center, similar to certain especially dedicated existing small towns along the Randstad's periphery. These towns, known as "urban growth centers" (*groeisteden*) re-

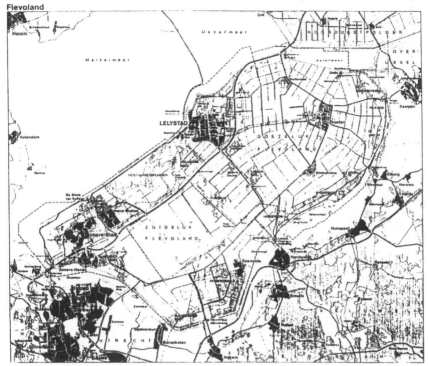

Figure 17-4 The Flevoland Polders, showing the location of Lelystad and Almere

ceived extensive subsidies, special housing consideration and other State financed facilities to grow to self-sufficient medium-sized cities.

In between Lelystad and Almere large scale industrial estates could be located, and water- and motorways would provide adequate connections with the centers in the west, and on to the east. A railroad could connect the Randstad with the northern provinces and Germany and Scandinavia beyond. Of course, the Structure Plan fully anticipated the Markerwaard to be reclaimed on time for Lelystad to develop as an important industrial and transportation hub in the middle of the country. The Structure Plan and subsequent IJDA and ZPD reports breathed a different atmosphere than those before, the new polders would be for the benefit and use of an urbanized Netherlands rather than be limited to agricultural use. The IJsselmeer polder planners were in action before it could become a political issue. In the well-known 1966 Second Report, a government report on the planning of the Netherlands for the next twenty-five years, this vision of the IJsselmeer polders was made into government policy, and ever since the polders have been an integral part of the country.

Planning Southern Flevoland in the 1960s had taken on a different form and planning approach. Even the way in which plans were presented had changed. During the development of the Noordoost polder, polder- and town plans had been designed on the basis of a definitive number of inhabitants, and on the assumption that change would be slow. Town plans had been drawn as architectural design problems. Although it must be said, that the physical planner of the Wieringermeer towns considered growth as inherent in town planning. He, therefore, designed the towns so they could expand into the surrounding country. However, considerations of this sort had a low status in the planning of the Noordoost polder towns.

The first town in Eastern Flevoland, Dronten, first took on the form of a structure plan, i.e., the plan was drawn as a presentation of urban functions and their relationships, and exact physical features were not filled in. That was something to be done at a later stage. This way of planning indicated the acceptance of uncertainty about the future, and the willingness to leave planning design decisions to later, when they needed to be solved. That not all planners agreed to such an approach is demonstrated by the first Lelystad plans, drawn in the spirit of a final architectural image. This image would be the framework on which the urban structure eventually would be completed. This planning approach caused the plan to be rejected, in spite of the planner's status in the profession. Times had changed, the future could not be predicted. In the planning of Southern Flevoland and its towns this fact was taken into account.

The ZPD made the polder parcellation plan, which was commented on and adjusted by the IJDA. Only the central portion was apportioned to large scale agriculture. The fact that this would be the last possibility for a large open landscape in the Netherlands figured heavily in the design considerations. The western quarter of the polder was reserved for a large new town. Large areas were reserved for recreational and natural purposes: deciduous woods and recreational natural areas in the southeast, nature reserves and wetlands in the northwest (the industrial estates never materialized).

As already mentioned, the 1961 Structure Plan indicated two areas for new towns: a western city and a rural support center for the agricultural middle area. The national government, however, placed great priority on the realization of the western new town to help alleviate the immense housing shortage in the Randstad area. The IJDA was charged with making the plans. For that purpose a "Project Bureau" was formed. The bureau, consisting of experts on a variety of planning aspects, such as physical planners, traffic planners, quantity surveyors, landscape planners, and social geographers, had a separate status within the IJDA organization, directly under the staff of directors. Most members were freshly hired as promising planners and other experts.

It was the IJDA's solution to present a more or less independent planning force that would have the confidence of the IJDA, and at the same time withstand the critical review of the many that would be following the whole procedure from the outside. The formation of a more or less independent planning force was necessary because (a) the development of the physical environment in the polders had gained attention from those whose spatial problems were promised to be solved in the new polder, (b) society had changed, the work of all governmental agencies was being followed with a critical eye, as people wished to exercise influence on what previously had been held beyond their scope, and (c) the IJDA had lost some of its credibility with the way in which it had previously operated. Its last big project, Lelystad, had not been well received.

The physical result, the manner and organization of the planning process, and the authoritarian manner in which it had operated, had been criticized. The IJDA realized that it could no longer work on the same basis as it had done for the previous thirty years. All the same it did not wish to relinquish its hold on the planning process, as had happened in the case of Lelystad's first planning, when a private consultant had been an independent planner. In that case the IJDA had experienced the planning process, as the director put it: "as a portrait of one' wife, like it or not, one had to take it"(Nawijn 1989, 12).

The project bureau form turned out to be a viable compromise. The bureau was able to work independently, to develop a creative structural scheme for Almere, and from there on to more detail on the various townships, districts, neighborhoods, and eventually to provide the details of streets and squares and control over construction. Meanwhile, the bureau had to constantly explain its activities. A large number of reports, studies, accounts, and justifications were published at every step of the procedure. Some of these reports were meant for formal approval from government agencies, national or from surrounding municipalities. Others served to inform the planning profession and academicians for their comment and discussion (Van der Wal 1997, 195-200).

Once the first residents came to the town the regular participation procedures were followed, and plans were made public for review and comment, all according to the rules of the national planning act. Oddly, this law did not apply in an area which had not yet become part of the legal framework of a municipality or province. Nonetheless, the procedures were followed as soon as people settled in the new town. From their side the residents, in the spirit of the 1970s, assertively took the opportunity to let their voice be heard, particularly when it came to the realization of their own neighborhood environment.

Almere

Almere is the largest of all planned new towns. Its plan reflected the concern for the issues of the time. The environment was such an issue. Careful planning of optimal public transport, efforts to reduce automobile use (even auto ownership) and travel/commuting distance were considered. Idealistic miscalculations occurred: the premise that with the residents, employment would come to Almere also, proved false, resulting in large traffic jams every day caused by those who lived in the polder and continued to hold their jobs in the Randstad. Nonetheless, Almere has an exemplary public transport system with a higher per capita usage than any other municipality in Holland.

Another contemporary issue was the question of how to deal with a high general affluence and much leisure time. In a way it is this issue that caused the multi-nuclear scheme of Almere. Almere was conceived not as one large urban area, but as a number of nuclei (or townships, not unlike Howard's idea of the garden city), separated by broad green areas for parks, some agriculture, and recreational uses. The idea for several nuclei came from the concept of creating Almere as a series of villages, of which each would have a large recreational center (something like the British leisure centers) in the middle. This scheme was adjusted at an early stage into a hierarchy of nuclei, with a large central nucleus that would contain all those facilities of a large downtown area. The green areas in between the nuclei, the so-called "internuclear areas", would be easily accessible from the neighborhoods. This was another naive attempt to reduce the need for travel to recreational facilities. The internuclear spaces and all the land within the municipal border, agricultural or otherwise, were viewed as the playground of the residents: an urban view.

Interestingly, this drew the ire of the farming lobby and the Ministry of Agriculture, so that the planners were forced to make a clear distinction between agricultural and non-agricultural (woods, recreation) destinations, so a set percentage would remain reserved for farming. The importance of the environmental issue can also be seen in the treatment of the large wetlands area along the northwestern dike. As has been mentioned, this area was originally intended for other use. However, it being the lowest area, when the polder was being drained, it remained wet for a long time. This caused wetland vegetation and fauna to grow abundantly, soon birds diverted their migratory routes to use it as a resting area, and the IJDA of its own volition decided to maintain it as a nature reserve. This would have been considered an inconceivable waste of good agricultural land in the Wieringermeer or Noordoost polder. Some years later, when the railroad was planned along the "Oostvaardersplassen", as the area was called, environmental activists succeeded in altering the track right-of-way away from the wetlands, so that resting and foraging areas could be kept together.

Almere thrived, to an extent at the expense of Lelystad, which was farther away from the Randstad. Lelystad also suffered from the fact that the railroad and freeway reached that town two years later than Almere. The first residents came to Almere in 1976. Until January 1, 1984, when it became a municipality, Almere was being developed by the IJDA in much the same authoritarian manner as had been the case in Lelystad. The project bureau had a firm grip on all construction not only by making the detailed neighborhood plans, but also by controlling architects and developers in everything from a building's form and function to the color of its exterior paints.

In an environmental spirit, the free movement of cars was arrogantly hampered, to the detriment of efficiency. Granted, more use was being made of public transit, but it did not keep the residents from owning more cars than elsewhere. The municipal government, having to listen to its electorate, was more inclined to make concessions to irreversible changes in society. Gradually bottlenecks were removed and traffic flows were corrected to more natural channels.

Similar developments can be noticed in the area of housing construction. In the beginning, more than 75 percent of all housing starts were corporation (rental) housing. This was in large part due to the national policy of subsidizing public housing. However, it was also a result of the Almere Planning Bureau's idealistic attitude that Almere would provide an attractive area for the laboring class. As time passed, the Netherlands became more affluent, and labor class solidarity was replaced by Dutch "Reaganomics" in the 1980s. This had its consequence in the housing market: privately owned homes increased drastically, duplex and freestanding home sites also increased. What's more: housing in suburban surroundings, far from shops, schools, and service facilities, no public transport, all deemed socialistically unacceptable in the early 1970s, became regular features in Almere of the late 1980s and early 1990s. In the 1990s, public housing construction comprises only about a quarter of all housing starts in Almere.

Local Municipality Planning

Although the Netherlands have a reputation for strong centralized government, traditionally the cities have had a large degree of autonomy and influence in state matters. Municipalities, particularly the larger and older ones, have maintained much of their independence even in modern times. In the

Dutch three-tiered planning system the national planning agency prescribes the general planning policies, the provincial planning agencies make the regional plans, but the responsibility for detailed spatial planning has been placed with the municipalities. These plans must be approved by the provincial governments, nonetheless in practice municipalities have the autonomy and the initiative for spatial developments on their side.

This situation often creates curious juxtapositions. As an example: the Randstad to an outsider presented as a coherent spatial planning phenomenon, consists of municipalities which can be at odds with each other about basic planning procedures, and compete for the establishment of industries and business. Similarly, national planning policies, as laid down in the national reports on national planning are not always followed gladly by municipalities. A case in point is the Fourth Report's emphasis on reducing the use of the private car by encouraging the construction of work areas, such as business and industrial estates close to public transport stations/nodes. Businesses, service industries and the like generally wish to locate their buildings along the freeway for easy access and for exposure. Municipalities have a tendency to go along with this, if, by doing so, they can attract business and industry to their town. In the same vein, municipalities are willing to provide housing at places where they can attract the affluent and business opportunity. Almere is no exception, and in that respect has become a place like any other.

Zeewolde

The last new town in the IJsselmeer polders, Zeewolde, was meant as a service center for the central rural area of Flevoland. Originally located in the middle of its feeder area, it was moved to the edge of the water. The argument used to do so was the greater attraction of a water front location. Just as the small new towns in Eastern Flevoland, Zeewolde could not attain a reasonable size or economic solvency if it depended purely on the rural population. The site at the water and the presence of large wooded areas in the south of Southern Flevoland would contribute to making the town attractive to commuters. Farm labor had become virtually non-existent.

Zeewolde proved a very desirable town for typical suburbanites. Although, strictly speaking, against government policy, the planners had no choice but to design the town as a suburban center. In fact, the difference between rural and urban culture, still deemed recognizable in the 1970s by IJDA social geographers, had virtually disappeared in the middle 1980s. The NIMBY (not in my backyard) effect could also soon be observed, as the autonomous growth of the town was looked at with suspicion by the newly arrived residents. Small town ambiance, peace and quiet, were deemed much more important than local conveniences, such as a high school (for which the children had to travel at least ten kilometers), or extensive medical facilities. Adequate bus service was also found not necessary; most Zeewolde homes have more than one car.

The old ways of the IJDA had long gone by the wayside, people could determine their own future and environment. In the case of Zeewolde this was all the more evident, because it was the first town in IJsselmeer polder history that was responsible for its own development. Before a single home had been built in the town, Zeewolde had become a municipality in 1984, with a mayor and a city council elected by the farming families who already resided in the area. Since a planning department did not exist (and all of the land was owned by the State), the IJDA acted as the developing agent for the town, but under the responsibility of the municipality. Ten years later all IJDA involvement ceased. In 1996, the small department (in a completely reorganized-since-1989-IJDA-ZPD combination) still busy on the finalization of its development task, was disbanded. Thus ended seventy years of State involvement in the development of the IJsselmeer polders.

Conclusion

For most of the seventy years of national involvement in the development of the IJsselmeer polders, the ZPD and IJDA had almost complete control, at least over the creation of the physical environment. The State Treasury guaranteed an independent and steady flow of cash, and the newly created land belonged to the State. Especially when the Wieringermeer and the Noordoost polders were reclaimed there was very little attention from the public, or indeed from the professionals, for this unequalled project. Therefore, one might expect that the IJsselmeer polders, having been developed independently, would not only look different, but also *be* different in a socio-economic sense. However, looking back and retracing the historic developments one can only conclude that the polders have followed the social, cultural, and economic changes that also occurred in the rest of the Netherlands and changing planning concepts are also reflected in the new towns. In spite of the fact that a large part of the polder was not reclaimed until thirty years ago, the IJsselmeer polders have undergone a complete integration with the old land on the other side of the former Zuiderzee shores.

References

Constandse, A.K. 1976, *Planning the Creation of an Environment*, Lelystad: IJsselmeer Polder Authority.
De Ruijter, P. 1987, *Voor Volkshuisvesting en Stedebouw*, Utrecht: Matrijs.
Dienst der Zuiderzeewerken, 1961, *Een Structuurplan voor de Zuidelijke IJsselmeerpolders*, Gravenhage.
Dutt, Ashok K., Frank J. Costa, Coenrad van der Wal and William Lutz, 1985, "Evolution of Land Uses and Settlement Policies in Zuiderzee Project Planning", pp. 203-230 in Ashok K. Dutt and Frank J. Costa, (eds), *Public Planning in the Netherlands*, Oxford: Oxford University Press.
Faludi, A., A.J. Van der Valk, 1994, *Rule and Order, Dutch Planning Doctrine in the Twentieth Century*, Dordrecht: Kluwer Academic Publishers.
Nawijn, K.E. 1989, *Almere, hoe het begon*, Lelystad: Rijksdienst voor de IJsselmeerpolders.
Van der Wal, Coenrad, 1997, *In Praise of Common Sense, Planning the Ordinary*, Rotterdam: 010 Publishers.
Van der Wal, Coenrad, 1985, "New Towns of the IJsselmeer Polders", pp. 231-50 in Ashok K. Dutt and Frank J. Costa, (eds), *Public Planning in the Netherlands*, Oxford: Oxford University Press.
Van Duin, R.H.A., G. De Kaste, 1990, *The Pocket Guide to the Zuyder Zee Project*, Lelystad: Rijksdienst voor de IJsselmeerpolders.
Verkaik, J.P., P.C. Van Royen, 1993, *50 Jaar Bestuur in Flevoland Noordoostpolder en Wieringermeer, Het Openbaar Lichaam in de Zuiderzee- en IJsselmeerpolders 1937-1987*, Zutphen: Walburg Pers.

18 Decentralization, Popular Participation, and Changing Patterns of Urban and Regional Development in Bolivia

ROBERT B. KENT, EDGAR GUARDIA, OLAV K. SIBILLE

Like much of Latin America, Bolivia has experienced considerable social, economic, and political change during the last twenty years. In Bolivia's case, these changes have occurred as a result of a combination of forces, including but not limited to, a steep decline in the importance of traditional exports, sweeping neoliberal economic reforms, and a return to democratic government in the early 1980s, which has continued uninterrupted to the present (Mayorga 1997).

The process of regional and urban planning and development in Bolivia has been significantly transformed as a result of these events. On the one hand, the decline of mining as the principal source of export income, the dissolution of the state mining company, COMIBOL, and the precipitous reduction in mining jobs all contributed to a reduction in the traditional power and dominance of the national capital, La Paz, and other highland centers (Nickson 1995, 109). Political and economic power shifted downward on the urban hierarchy toward competing urban centers and out of the highlands, particularly to Santa Cruz and to a lesser extent Cochabamba, as commercial agricultural development, the illegal drug trade, and internal population migration favored their development.

Neo-liberal economic reforms have also impacted urban and regional development and regional planning. These reforms have sought to streamline the functioning of the state and the economy. The privatization, downsizing, or outright closure of a range of inefficient state-owned businesses and industries have reduced government expenditures drastically in some areas, stemming the flow of red ink and freeing resources for other critical public sector needs. Modernization of the tax system and an emphasis on the fair and consistent collection of taxes has played an important role in these reforms as well (Rowat 1996, 403). These kinds of reforms have also contributed to an increase in public revenues, and the ability of the state to respond more effectively to urban and regional development needs.

The installation of a democratically-elected government in the early 1980s and the institutionalization of the democratic process in Bolivia since that time have also had important impacts on the process of urban and regional planning and development. The attitudes of citizens and public officials began to change. Concepts emphasizing local control, decentralization of political power, citizen participation, revenue sharing, and a more equitable distribution of central government income began to enjoy wide currency. In the mid-1990s the government of President Gonzalo Sánchez de Lozada (1993-1997) succeeded in passing two key critical pieces of legislation, notably the *Ley de Participación Popular* and the *Ley de Descentralización Administrativa* which are intended to bring Bolivia a more democratic and participatory system of government. This chapter examines how these two laws have transformed the process of urban and regional planning and development in Bolivia, focusing particularly on the changing relationships between the central government and local governments.

Prior to the Reforms of the Mid-1990s

Prior to the initiation of the principal structural reforms in Bolivia in 1994, regional development was characterized by an excessive concentration of resources in three principal departments (La Paz, Santa Cruz, Cochabamba) and especially in their metropolitan centers of the same name, limited participation of the local population in decision making, and a process of highly centralized planning directed from above. The principal institutions involved in this process were the Regional Development Corporations (*Corporaciones Regionales de Desarrollo*, CORDES), municipal governments in the department capitals and largest provincial cities, and the central government through the Prefectures and other public institutions, ministries, and agencies.

Following a model utilized in several other South American countries (Colombia and Peru) during the 1960s and 1970s, the Regional Development Corporations were decentralized institutions of the central government with the responsibility to plan regional development and to program and execute public investment at the department level. The *"Ley de Corporaciones"* promulgated in 1974, provided the legal basis for this structure as well as transferring the ownership and management of some publicly-owned companies of the defunct *Corporación Boliviana de Fomento* to the corporations. A Regional Development Corporation functioned in each of Bolivia's nine departments, with its seat of operations in the department capital. A board of directors, led by a President named by the central government, and including other central government officials as well as civic, business, and labor representatives from each department, managed the regional development corporations. Despite the appearance of local participation in the governance of the corporations, they were basically agencies of the central government, dependent almost entirely on the executive branch. Almost all funding came directly from the central government, except in a few departments where royalties on local mineral and petroleum production supplemented Corporation budgets (notably in the Department of Santa Cruz). The Corporations produced multisectoral regional plans (i.e. CORDECO, 1983) and either constructed or contracted the construction of a wide range of public infrastructure projects including health posts, schools, roads, and irrigation works.

The municipal governments, as established in the *Ley Orgánica de Municipalidades* (the Law of Municipalities) in 1985, are autonomous governmental units. Municipalities are governed by a town council whose members are elected by voters. The council in turn elects the mayor from among its members. The law assigns municipal governments a wide range of responsibilities which include the planning and promotion of urban development, the construction and maintenance of local public infrastructure (schools, markets, health centers, parks, and streets), the regulation of public markets and marketing, and the provision of water, sewerage, and public lighting.

Revenue sources for municipal governments were scarce. The central government assigned 10 percent of its revenue sharing funds (known as *coparticipación* in Bolivia) to them, refunded a portion of the property tax municipal governments collected but paid to the central government, and permitted local governments to levy a range of modest taxes, registration fees, and user fees which generated limited revenues. These included business license fees, fees charged to periodic marketers for street vending, charges for the slaughter of animals, a tax on *chicha* (an indigenous alcoholic beverage), and in a few cases tolls on local access roads.

However, prior to the reforms of the mid-1990s, most municipal governments did not have the financial resources or the technical capacity to accomplish the responsibilities with which they were charged. Generally, municipal governments only functioned adequately in the department capitals and some larger intermediate-sized cities. Municipal governments in peripheral regions muddled through, often only operating small ineffectual municipal offices, supervising local markets, and using their scant financial resources to maintain and repair streets and public parks. The Regional Development Corporations and central government ministries sometimes provided or usurped those services which local governments could not provide, but as often as not, such services might not be provided at all.

The territorial jurisdiction of the municipalities was often not clearly defined. In many other Latin American nations, municipalities (known variously as *municipios, provincias, cantones,* and *distritos*) are defined as territorial units incorporating both urban and rural areas. Typically, in these countries, municipalities will be comprised of an urban center and its rural hinterland. In Bolivia, often only the urban areas were defined as municipalities. The rural hinterlands frequently had no political status whatsoever, or they were defined as belonging to indigenous or peasant communities which operated outside of the formal local government structure.

Prefectures formed another level of government's administrative structure. Before the reforms of the mid-1990s, the Prefectures exercised a limited role in regional development. The Prefects, named by the central government, were the representatives of the executive branch in each department. As such, their principal functions included political activities, the maintenance of public order, and the administration of limited financial resources in the name of the central government at the department level — notably the Departmental Treasury (*Sub-Tesoro Departamental*), responsible for paying the public employees, the Civil Registry (*Registro Civil*), the State Notary (*Notaria de Hacienda*), and several other entities. Within each department, the Prefects named the Sub-Prefects of each province, who in turn named the *Corregidores* or deputies for the smallest administrative units, the cantons. The role of the Sub-Prefects and the *Corregidores* within their respective territories was even more restricted, and was generally limited to conflict mediation and resolution. In addition, it is important to note that these officials and their offices had no budgetary support from the Prefect or central government.

Despite being appointed by the chief executive of the central government, Prefects and the Presidents of the Regional Development Corporations often found themselves in disagreement and conflict. These power conflicts were the cause of many difficulties, especially in times of crisis when it was desirable that a single authority speak for the executive branch of government within a department.

The central government exercised its influence in regional planning and development principally through ministries and other state agencies with narrow sectoral responsibilities like health, education, agriculture, irrigation, and highways. Nevertheless, since 1985 when a "structural adjustment" was applied to the Bolivian government and economy, many functions of the central government have been reduced.

To further complicate the institutional panorama, the central government implemented a series of narrowly focused economic development funds. The first of these was the Fund for Social Emergency (*Fondo Social de Emergencia*), created in 1985 as a temporary effort to mitigate the effects of the nation's structural adjustment measures. This fund was later converted into the permanent Fund for Social Investment (*Fondo de Inversión Social*). Using it as a model, the government also created the Fund for Peasant Development (*Fondo de Desarrollo Campesino*), the National Fund for Regional Development (*Fondo Nacional de Desarrollo Regional*), the National Fund for Alternative Development (*Fondo Nacional de Desarrollo Alternativo*), the National Fund for the Environment (*Fondo Nacional de Medio Ambiente*), and several others. These funds, each with attributes, functions, and specific goals, distorted the assignment of financial resources destined for public investment and caused inequities between regions. Because the management of the funds was directly dependent upon the central government, and their individual operations were not regionally or locally coordinated, their operation interfered with a rational process of planning at both the regional (department) and local level (provinces and municipalities).

In terms of taxation and finance, the Tax Reform Law (*Ley de Reforma Tributaria*, #843), approved in 1986, defined a program of revenue sharing (*co-participación*), distributing central government tax revenues in the following manner: 10 percent for the Regional Development Corporations, 10 percent for municipal governments, 5 percent for universities, and 75 percent for the central government.

In conclusion, the process of planning and regional development prior to the reforms of the mid-1990s was a true institutional tangle, involving the participation of various public institutions, the central government, and municipalities whose responsibilities were often overlapping or poorly defined. The result was a patchwork effort at regional planning and development, in which the players frequently found themselves pursuing divergent agendas leading to significant conflict. The population, especially in peripheral and rural areas, had no capacity to influence the decision making process nor the assignment of financial resources, which were habitually distributed from the central government as political favors.

The Reforms of the Mid-1990s

The reforms promulgated into law in 1994 and 1995, the Law of Popular Participation (*Ley de Participación Popular, Ley #1551*) (Bolivia, 1994a) and the Law of Administrative Decentralization (*Ley de Descentralización Administrativa, Ley #1654*) (Bolivia, 1995a), respectively, have had a significant impact on the social and economic structure of Bolivia — as well as on the focus given to regional development strategies — in terms of both the process of planning and the assignment of resources.

Significantly, the *Ley de Participación Popular* has defined all parts of the national territory as belonging to a specific municipality, and municipalities are now defined in a manner consistent with most other Latin American countries. Urban centers, towns, and villages still form the focus of the municipal unit, but rural hinterlands are also incorporated within the municipal boundaries (Bolivia 1994a, Art. 2). Traditional indigenous communities and peasant communities now form part of municipalities. Bolivia has been divided into 311 autonomous municipal governments according to the previous territorial divisions within each province (the sub units into which departments were and are divided), called *secciones* (Bolivia 1994a, Art. 2). A three-tiered system of territorial division now exists in the country — departments, provinces, and municipalities. The department of Cochabamba, for example, has 16 provinces with a total of 44 municipalities. The number of municipalities in each province can vary greatly. In Cochabamba, for instance, the provinces of Quillacollo, Punata, and Carrasco each have five municipalities, while the provinces of Tapacarí, Bolívar, and Tiraque each are co-incident with a single municipality of the same name. Similarly, the province of Cercado, in which the city of Cochabamba is located, is also a single municipality.

Despite the official designation of a three-tiered system of territorial division in Bolivia, at only one level, that of the municipality, is there a democratic local government structure. Governance at the department and provincial level is simply an extension of the executive branch of the national government, with Prefects and Sub-Prefects serving as the representatives of the nation's chief executive. Governance at the municipal level however is dependent on the direct election of municipal council members who choose the municipal executive, the mayor, from their own ranks. Municipalities are empowered to collect tax revenues, control their own budgets, plan and execute public works, and legislate on local matters.

The process of the delimitation of municipal boundaries has caused many problems, resulting primarily from old territorial conflicts which were never resolved, or because provincial and sectional boundaries were never accurately delimited. The neighboring municipalities of Cochabamba and Sacaba have been engaged in a legal dispute for several years over the demarcation of their territorial limits, as is the case with several other municipalities throughout the country.

The *Ley de Participación Popular* led to the transf from the central government to municipal governments of the responsibilities for the construction and maintenance of the public sector infrastructure in the areas of health, education, sports, culture, and local roads (*caminos vecinales*) (Bolivia 1994a, Art. 14). However, teachers and health professionals who staff schools, hospitals, and health posts are still employees of the central government and not municipal governments. Nevertheless, the reforms have meant that municipal governments are now responsible for some services and activities previously undertaken by the central government. This transfer of responsibilities has not been without difficulties and the promise of local control and autonomy still falls short of reality (Darras 1997).

In terms of public finance, the *Ley de Participación Popular* specifically identifies the central government revenues subject to revenue sharing (*coparticipación*) and establishes a new structure for allocating these resources. The key national revenues subject to revenue sharing include the value added tax, a tax on business profits, transactions' taxes, import taxes, and several other lesser categories (Bolivia 1994a, Art. 19). As part of the new system of revenue sharing, these funds are now distributed as follows: 75 percent for the central government treasury, 20 percent for the municipalities, and 5 percent for the universities (Bolivia 1994a, Art. 20). The percentage allocated to the municipalities is distributed according to the population of each municipality as reported in the national census enumerated in 1992 (Bolivia 1994a, Art. 21 and 24). In addition, the law assigns municipal governments the right to levy and keep all taxes on urban and rural real property. Prior to these changes, municipalities collected the property tax, remitted it to the central government, and then the central government returned only a portion to them. Municipal governments also now levy and keep the property taxes on vehicles, airplanes, and boats.

This distribution of resources is much more equitable than it had been. Before the approval of the law, 91 percent of the central government revenue sharing transfers were destined for the municipalities of the three principal departments, La Paz, Cochabamba, and Santa Cruz and the rest of the country's municipalities received only 9 percent. Under the new system of fiscal transfers the percentage allocated to the municipalities of the three principal departments has been reduced to 68 percent, with the balance, 32 percent, distributed to the remainder of the countries municipalities (Bolivia 1995b, 49). Furthermore, the overall percentage of the government's revenue sharing monies going to municipalities has been increased from 10 percent to 20 percent.

The department of Cochabamba provides a useful example of how this change plays out. Before the passage of the law in 1994, the department received Bs 34.2 million (in 1996 $1 = Bolivianos (Bs) 5.19) as its share of central government revenue sharing. Over 87 percent of these monies remained in the departmental capital, the city of Cochabamba, and slightly less than 13 percent was divided among the remaining municipal governments in the department. Subsequent to the passage of the *Ley de Participación Popular* in 1994, total revenues increased to Bs 70.4 million, with 49.8 percent going to the city of Cochabamba and the balance, 50.2 percent, being assigned to the remaining departmental municipalities. In 1995 and 1996, revenue transfers from the central government grew to Bs 117.4 million and Bs. 137.7, respectively. The relative proportion assigned to the city of Cochabamba declined to just 37.3 percent, while the remaining municipalities received the balance (Bolivia 1995, 49 and Bolivia 1996).

The relative percentage of funds allocated to the departments and municipalities in 1994 represented a transition between the old system and the methodology mandated in the new law. The percentage assigned in 1995 and in subsequent years is a function of the percentage of population in each municipality, based on the results of the population census in 1992. These percentages will be revised when new population statistics are available from the next scheduled national census in 2000. Subsequently, revisions of those percentages will be calculated halfway between each decennial census based on official population estimates provided by the Instituto Nacional de Estadística (Bolivia 1994a, Art. 24).

In addition, the *Ley de Participación Popular* has increased the participation of civil society in the process of regional planning and development by recognizing the legal status (*personería jurídica*) of territorial based organizations (TBOs). These territorial based organizations include peasant communities defined as the basic unit of the social organization in the rural areas and their equivalents in the urban areas, neighborhood organizations, the *juntas vecinales* (Bolivia 1994b, Art. 1). In addition, the law includes indigenous communities (i.e. *ayllus*) and their traditional authorities (i.e. *curacas, mallcus, and jilacatas*) (Bolivia 1994a, Art. 2 and 3).

These territorial based organizations have the legal right to oversee and to intervene in municipal planning and administration through the establishment of vigilance committees (*Comités de Vigilancia*) which can be created in each municipality (Bolivia 1994a, Art 10). The *Comités de Vigilancia* can force the suspension of the utilization of government revenue sharing funds if they can demonstrate that these monies are being mismanaged or not allocated according to the law's guidelines. For instance, the *Ley de Participación Popular* requires that municipal funds be spent equitably among urban and rural populations and that no more than 10 percent are expended for administrative expenses (Bolivia 1994a, Art. 10 and 11).

The territorial based organizations must register with the central government to legitimize their participation in the municipal governance process (Bolivia 1994a, Art 5). Over 10,500 organizations have registered with the government, but the vigilance committees have only been established in less than half, 140, of the nation's municipalities. Some traditional organizations, particularly those representing indigenous peoples, have been suspicious of the government's motives, suspecting that the government is attempting to co-opt them and destroy their traditional forms of social organization through this mechanism (Albó 1996, 19).

To date, experiences with the new legal structure have been a bit cumbersome. The implementation of the new law has been slow, in part because some municipal governments are unaware of the full extent of its ramifications or have misunderstood its mandate. In other instances, participation by the public has been extremely limited by disinterest or by local authorities who have worked to block it. The law has meant increased financial resources for all municipalities, and many have used these resources responsibly to undertake long deferred local development projects. Others, however, have been overwhelmed by the increased level of fiscal transfers and have invested these windfalls in projects which have not corresponded to real local needs (IDB 1997, 11-12). For example, one can note a geographic concentration in the municipal capitals where revenue sharing monies have been utilized to build impressive municipal office buildings or improve the central plaza. Meanwhile, small towns and rural areas in many municipalities have received almost none of the increased revenues.

But without a doubt, this new context has stimulated the process of local participation in planning and has permitted the involvement of a wide range of public and private organizations working in a particular municipality. This has helped to avoid the duplication of effort which often characterized the situation before the law's passage. In this sense, the *Ley de Participación Popular* has created profound changes in the way non-governmental organizations (NGOs), bilateral and multilateral organizations, and other institutions implement projects. The changes have generally signified an improvement in local planning processes. For instance, the wide range of funds established by the central government prior to the reforms still operate, but their operations have been decentralized from the national capital to regional centers. In operational terms they are required to coordinate their investment plans and decisions directly with municipal governments.

One of the outcomes of this coordination and local participatory planning is a planning document, the municipal development plan (*Plan de Desarrollo Municipal* — *PDM*). Each municipality is expected to prepare a *PDM* which is intended to be a five-year strategic plan. Using these medium-term plans as a framework, each year municipal governments are to produce annual operating plans (*Planes Operativos Anuales—POA*). The annual operating plans are detailed documents which specify the projects to be executed with specific budgetary detail.

The central government has actively encouraged the elaboration of *PDM*s at the municipal level and has provided revenue sharing monies to accomplish their completion. Nevertheless, many municipalities still have not successfully produced these plans. Government regulations, spelled out in the planning norms for municipal participatory planning (*Norma de Planificación Participativa Municipal*) require that all municipalities complete a *PDM* before the end of 1998. Fifty percent of the costs of the implementation of these are covered using revenue sharing funds from the central government. However, the balance must be generated from each municipal government's own revenue sources, credits, or donations.

In an initial phase of the application of the *Ley de Participación Popular*, the *PDM*s prioritize infrastructure projects which fill basic social needs, principally in education, health, and public sanitation, sectors often ignored by the central government during the last 30 years. In addition to these basic types of public infrastructure, the public is demanding greater investment as well, in productive sectors and infrastructure — local roads and bridges, marketing facilities, irrigation works, and support for small scale private sector businesses.

The reforms implemented in the national constitution of 1994 and the *Ley de Descentralización Administrativa* (the Law of Administrative Decentralization), approved in 1995, have transformed the Prefecture into almost a department government, with a Prefect, still named by the central government's executive, but also with a departmental council, whose members are elected from within each province of the department (Bolivia 1995a, Art. 3 and 4).

The new legal structure assigns specific revenues to the Prefectures from the central government insuring a more dependable financial base than previously existed. The principal revenue sources for the Prefectures include 25 percent of the national tax on hydrocarbons, the Departmental Compensation Fund, salary monies for health, education and social welfare agencies, and special natural resource extraction royalties assigned to some departments (Bolivia 1995a, Art. 20).

The *Ley de Descentralización Administrativa* expands the previous responsibilities of the departmental Prefecture to include the management of resources to formulate and execute regional development plans focusing on social and economic development and to execute major public investment projects, especially highways and secondary roads, rural electrification, and irrigation projects (Bolivia 1995a, Art. 5). In addition, the Prefectures have the responsibility to promote popular participation (Bolivia 1994a, Art. 25), support municipal governments and their development plans, and to undertake other activities which previously corresponded with the responsibilities of the Regional Development Corporations, institutions eliminated as a consequence of the *Ley de Descentralización Administrativa* (Bolivia 1995a, Art. 26 and 29). The public agencies dependent on the corporations, their projects, and the better part of their employees were transferred to the Prefectures. The businesses operated by them, such as the milk processing plant in Cochabamba department known as PIL (*Plant Industrializadora de Leche*), were privatized.

Furthermore, all regional planning work now is realized in accordance with the regulations established by the *Sistema de Planificación Nacional* (*SISPLAN*) in coordination with the Prefecture, municipal governments, and the respective sectoral ministries. With the aim of implementing the basic norms of *SISPLAN*, procedures, operational mechanisms, and other technical and administrative structures for municipal planning and development, have been codified in norms for participatory municipal planning (*Norma de la Planificación Participativa Municipal*). The objective of this type of regulation has been to institutionalize the process of participatory planning at the municipal level and to guarantee that the *PDMs* represent true instruments of local planning.

In this context and as a result of the reforms, the Prefectures have become important players within the process of regional planning and development. Nevertheless, planning at the local level is now the complete responsibility of local governments—the municipalities.

Conclusions

Bolivia's efforts to increase the participation of its citizens in local government affairs and to decentralize its governmental structure through the passage of the *Ley de Participación Popular* and the *Ley de Descentralización Administrativa* have provided the means for dramatic changes in the process of regional planning and development. Perhaps most significant has been the increased autonomy that this legislation has granted to the regional (department) and local (municipal) governments. Fiscal autonomy for local government units is a critical issue and these reforms insure the assignment of specific central government revenues to departmental prefectures and municipal governments through the revenue sharing process (*co-participación*).   In addition, these fiscal reforms have assigned municipalities all revenues generated from local property taxes.   Property taxes have a tremendous potential to generate local resources if municipalities establish effective systems of property mapping, assessment, and collection.

Legislative reform has also meant increased responsibility and control at the local level as the central government has transferred some important governmental functions to local government.   Municipalities are now responsible for maintaining the physical infrastructure of primary and secondary schools and most public health clinics and hospitals.   At the same time, departmental prefectures are now responsible for the personnel who staff these facilities.   In addition, the departmental prefectures have now absorbed most of the regional planning and development functions formally held by the Regional Development Corporations.

Not only have these legislative reforms transferred resources and responsibilities to the local governments, they have provided the mechanisms for more effective participation of the public in the process of local government.   While Prefects are still named by the nation's chief executive, the Prefect's independence is tempered by a department council, whose members are elected from each of the department's provinces.   Municipal governance has also been transformed in important ways.   While municipal council members are still elected, the inclusion of all the national territory into municipal units ensures effective local suffrage of the country's rural inhabitants and indigenous groups.   Furthermore, the legal recognition of local organizations, and their ability to form part of vigilance committees with a legal right to oversee municipal governance, contributes to more responsive local government.

Finally, the reforms have opened the way for more consistent and coordinated forms of regional planning and development at the department and municipal levels.   A clearer, more transparent system of planning has been established, with prefectures undertaking the responsibility for planning at the department level and coordinating the planning activities of municipal government as they relate to the department.   Municipalities are required to plan by developing strategic five-year plans and annual operating plans.   Development responsibilities are more clearly identified than in previous years.   For instance, prefectures plan and construct the primary and secondary road networks, while local roads are within the purview of municipal governments.

These legislative reforms hold out the promise for dramatically changing the relationships between different levels of the urban hierarchy—decentralizing power and moving tremendous resources to small urban places and peripheral regions and ensuring greater local participation in planning and development. Only a few years have passed since these reforms have been promulgated into law and it is far too early to render definitive statements about their impact. However, it is clear that until the passage of this legislation the vast majority of Bolivian local governments had never experienced the possibility of deciding how to invest public monies. If these new responsibilities are exercised wisely by local governments and their citizens, Bolivia will have made a critical step toward a brighter future for its people.

References

Albó, Javier, 1996, "Making the leap from local mobilization to national politics", *NACLA Report on the Americas*, 29:5:15-20.

Bolivia, Congreso de la República, 1994a, *Ley de Participación Popular*, No. 1551 del 20 de abril de 1994.

Bolivia, Presidencia de la República, 1994b, *Reglamento de las Organizaciones Territoriales de Base*, Decreto Supremo No. 23858 del 9 de septiembre de 1994.

Bolivia, Congreso de la República, 1995a, *Ley de Descentralización Administrativa*, No. 1654 del 28 de julio de 1995.

Bolivia, Secretaría Nacional de Participación Popular, 1995b, *Realidad numérica de Cochabamba*, Cochabamba: Centro de Investigación Multidisciplinario.

Bolivia, Instituto Nacional de Estadística, 1996, Sistema Nacional de Inversión Pública, La Paz: INE. (http:\\www.ine.gov.bo).

CORDECO, 1983, *Estrategia para el desarrollo regional de Cochabamba: La macroestrategia, tomo I*, Cochabamba: Ediciones CORDECO.

Darras, Christian, 1997, "Local health services: Some lessons from their evolution in Bolivia", *Tropical Medicine and International Health*, 2:4:356-362.

Inter-America Development Bank, 1997, "Local officials taking charge", *The IDB*, 24:9/10:10-11.

Nickson, R. Andrew, 1995, "Bolivia", pp. 107-116 in Nickson, R. Andrew, (ed.), *Local Government in Latin America*, Boulder, CO: Lynne Riener.

Mayorga, René Antonio, 1997, "Bolivia's silent revolution", *Journal of Democracy*, 8:1:142-156.

Rowat, Malcom D. 1996, "Public sector reform in the Latin American and Caribbean Region—issues and contrasts", *Public Administration and Development*, 16:4:397-411.

19 Future Trends: Globalism and Regionalism

CHRISTOPHER CUSACK

Even a cursory review of current literature reveals a plethora of terminology involving the word "global", yet the catalyst for globalization took place nearly 30 years ago. Indeed, the early 1970s dissolution of the Bretton Woods system of national economic control is generally considered to have launched the global era. Over the past several decades, the global economy has developed such that now it connects internationally all facets of society, including those that are political and geographical.

Economically, globalization is highlighted by growth of transnational corporations (TNCs), mobility of capital and labor, influence of high technology industries, and the role of innovation and knowledge. Politically, the global society is evidenced by devolution of authority, declining influence of the nation-state, and an increase in number of non-governmental organizations (NGOs) and international governmental organizations (IGOs). These political trends have impacted the global geography leading to "a devolution of regulatory practices to local governments at regional and urban levels on one hand and a transfer to supranational institutions, such as the EU or NAFTA, on the other" (Sykora 1994, 1152).

Geographically, then, globalization has given rise to two trends of regionalism. One trend is characterized by the joining of nations into large, multi- or supra-national regions. Complementing the supra-national regions, however, is the "micro-region" which, anchored by a city, is based on economic rather than political borders. Currently, these mutually reinforcing economic, political, and geographical processes involved in the globalization of society (Figure 19-1) are altering the bases of power.

A discussion of the current global society necessarily includes analysis of these political, economic, and geographic change agents. An understanding of these considerations may then give rise to plausible scenarios regarding the future trends of globalization.

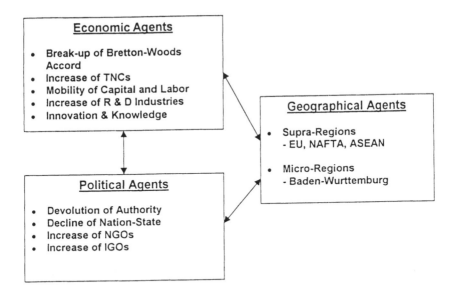

Figure 19-1 Economic, Political and Geographical Agents Involved in the Globalization of Society

Geographic Regionalism

Globalization of the economy results in two distinct, though complementary, trends towards regionalism. At the supra-national scale, countries are aligning themselves into alliances primarily based on their geographic situation. Indeed, increases in global trade during the latter half of the twentieth century have led to concentrations of economic development in three regions: North America, Europe, and East Asia (Spybey 1996, 86).

Multinational economic alliances, which represent these regions, include among others the European Union (EU), the North American Free Trade Agreement (NAFTA), the Association of Southeast Asian Nations (ASEAN), as well as the newly formed Latin American alliance (Mercosur). Each of these regional alliances is comprised of groups of countries which:

> essentially give up elements of their economic sovereignty in exchange for wider access to resources and markets under strong contractual and institutional guarantees of cooperation (Scott 1996, 396).

These economic alliances constitute a form of international governmental organization (IGO), the objective of which is to coordinate relations and attain common objectives among states (Knight 1989, 33). Notably, the processes of globalization have corresponded with the growth of IGOs. Though negligible in number in the 1960s, subsequently a veritable explosion has occurred in the number of IGOs. As a result, a total of 5,401 IGOs were recorded by 1994 (Scott 1994, 396).

Thus, the internationalization of the economy has precipitated a restructuring of the global order from the nation-state to regional alliances. The nation-state has recognized the limitations of "going it alone", and mergers of nation-states, based in part on geographic proximity, have become commonplace. Though the EU, NAFTA, and ASEAN constitute the "Big Three" of regional alliances, unification of the newly industrializing countries (NICs) in Asia could likely result in yet another powerful regional competitor.

However, the growth and prosperity enjoyed by some regions remains as yet unattainable for others. Though the economy may be global, it is not all-inclusive and the forces driving the global economic engine may leave behind many parts of the world:

> Between 1960 and 1991, the share of the world income for the richest 20 percent of the global population rose from 70 percent to nearly 85 percent, while the share of the poorest 20 percent declined from 2.3 percent to 1.4 percent...The fracture between poverty and affluence has proved to be one of the most enduring and alarming features of the emerging global order (Sagasti 1995, 596).

Not only is the gap between the "haves" and "have nots" widening, it is also being recognized at a regional scale within the global society. Regional alliances and regional disparity are characteristics of globalization at the supra-national scale.

The forces of globalization are also related to the promotion of regionalism at a sub-national scale. Indeed, there is a complementary effect of the movement towards regional trading blocs and the development of micro-regions within a single country. While global processes have fostered development of regional alliances, these same stimuli have necessitated a restructuring of individual cities and their hinterlands. The promotion of regional trade agreements and multi-national trading blocs has led cities into the new economic space of an international market.

As a result, cities have become centers of micro-regions, which offer better potential for success in the international economy. Rather than exist as rivals, neighboring cities and their surrounding areas have come to understand that they are not competing with each other, but with other cities and sub-regions across the globe.

The principal components of such city-based micro-regions are readily identifiable:

- Regions compete in the global economy; a nation is best understood economically as an assemblage of regions;
- Economically strong regions are characterized by vital central cities;
- Successful regions compete best by strategically combining policies regarding development of human resources, land use, social issues, university research, and philanthropy, among other elements, in ways that allow effective competition in niches of the global economy (Kirlin 1993, 371).

An approach based on sub-regions within the nation-state is now being recognized and advocated as offering the primary potential for successful competition on a global scale.

Examples of regional economic focal points abound across the globe. One regional approach is evident in the German State of Baden-Wurttemberg, which has confirmed hundreds of agreements with other regions and entities in making its own foreign and trade policies (Newhouse 1997, 68-69). Baltimore provides a second example of strategic regional networking by promoting its reputation as a "Renaissance city" in order to attract investment and vault into global competitiveness (Levine 1989, 149).

While both supra-national and micro-regions are growing in vitality, their increasing prominence is diminishing the traditional authority of the nation-state. The combination of these regional approaches to globalization calls into question the precise status of the nation-state (Spybey 1996, 64). This waning role of the nation-state in the global society warrants review.

Decline of the Nation-State

The arrival of regionalism as a potentially powerful competitor in the global economy highlights the declining role of the traditional nation-state. The mutually reinforcing processes of the decline of the nation-state and rise of regionalism are characterized by a shift in political and socio-economic perspective from a national to regional scale. The nation-state, which reached its apex during the Cold War, is already descending from its position of dominance. Indeed, nation-states emerged primarily for military purposes, therefore with the end of the Cold War a major justification for the necessity of the nation-state has been eroded (Kirlin 1993, 373).

A replacement, or at least rival, for the nation-state has emerged in the form of the region. At the supra-national level, the forfeiture of some national power so as to join a limited number of like countries reflects the declining influence of the nation-state. Correspondingly, the micro-region offers the potential for proactive, strategic development on a local, more intimate level. Furthermore, reactive planning may be designed more quickly and efficiently when focused on the micro-region than if dependent on the nation-state for response. What has become evident is that in this age of global trade, supranational and local alliances are replacing centralized governments as the geographic units of control.

The future of the nation-state remains in question. Sentimentality and illusion is posited by some to be the sole remaining foundation of the nation-state (Miyoshi 1993, 744). Others argue that because of its grip on taxation and defense, "the nation-state is not going anywhere, anytime soon" (Newhouse 1997, 84). Thus, although the future of the nation-state remains under debate, global changes will unquestionably continue to induce changes in its form and function.

Government to Governance

The rise of the global society and the decline of the nation-state has, in turn, impacted the form and function of government throughout the world. Currently, nations around the world are moving towards less involvement from the national government in favor of more control and initiative at the local level. With this devolution of federal authority, local governments are valuing the need for a more holistic approach to local rule, and are therefore making the transition from government to governance. Indeed,

> The scope and scale of the problems being dealt with at the local level now require all three sectors—government, business and community-based organizations— to jointly take responsibility for meeting these challenges. Government has become a convener and catalyzer in the broad process of community governance (Gates 1996, 3).

Increased involvement from all segments of society is recognized as the best, and perhaps only, way for local success in the global society. Such understanding has given rise to an abundance of new structures of governance, the foremost of which are nongovernmental organizations (NGOs).

The increasing prominence of such organizations and other community-based groups is a notable result of the devolution of national authority. Nongovernmental organizations found in both developed and developing countries are expected to number over 10,000 by the year 2,000 (Figure 19-2).

Currently, the role of NGOs has been broadened to include functions formerly associated with government. One such example is evidenced by NGO efforts in Karachi, Pakistan to develop a water and sanitation system in a low-income area. After three years, the system was in operation and at costs nearly ten times lower than the commercial or government planned rates. Other successful NGO projects include an anti-littering effort in Bangkok and the development of basic housing projects in Metro Manila. The role and success of NGOs is, however, not limited to developing countries. In North America, police efforts to combat crime have been substantially aided by "neighborhood watch" efforts and other anti-crime NGOs (Laquian 1994, 208-209).

Nongovernmental Organizations

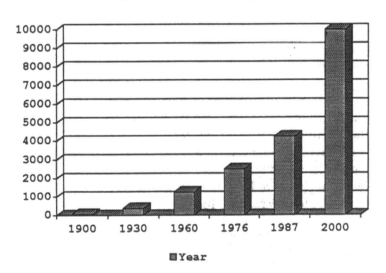

Figure 19-2 Growth of Nongovernmental Organizations, 1900-2000
Source: Knight, 1989.

Nongovernmental organizations exist in close communication with the populace, and have traditionally served dual roles as "enablers" of the community, as well as mediators with the government (Elliot 1994, 103). However, as NGOs continue to grow in number and influence, they have become more formal structures of governance. Indeed, while efforts of NGOs have traditionally been limited to neighborhood concerns, they are now likely to be involved with issues which are citywide in nature. Nongovernmental organizations represent the movement from government to governance in the transition of global order.

Transnational Corporations

A critical factor in the trend towards globalization is the transnational corporation (TNC), which may be defined as centrally controlled, multi-locational, multinational enterprises that are special purpose and profit-making. They have become perhaps the most powerful change agents in the current global economy (Knight 1989, 35). The power of TNCs is evidenced in their number and economic influence. Transnational corporations have increased from approximately 7,000 in the 1960s to 37,000 by the late 1980s. Furthermore, the sales of TNCs outside their home country reached $5.5 trillion in 1992, greater than the total global trade of goods and services for that year (Fry 1995, 22).

The TNC should be viewed not as "unpatriotic" or "anti-nationalistic", but rather as a natural product of this global era. It is now survival of the fittest on a global scale for TNC's, which have had to adapt to compete. TNC's have had no choice but to "go global" for three primary reasons: new methods of production have been brought about by accelerated technological change; greater transnational mobility of capital has made investing abroad easier, quicker and cheaper; major changes in the ease of transport and communication have been developed (Amin and Thrift 1995, 94).

In union with the trend towards regionalism, the increase in number and power of the TNC has further reduced the role and image of the nation-state. The assumption that the nation-state has the ability to control its private business sector has been largely invalidated with the rise of the transnational corporation. Loyalty of the TNC does not lie with any one nation, but rather is devoted to a fraternity of stockholders in whose interests the TNC acts. In effect, the mobility of the TNC enables it to "veto" whatever governments order (Tyler 1993, 30).

A growing overlap between international and domestic issues has become so consequential that "intermestic policies" dealing with the local-global fusion are now commonplace (Fry 1995, 25). Such legislation increases potential for global competitiveness at the local level. However, by usurping activity traditionally directed by the national government, intermestic policies result in a diminished role of the nation-state. Therefore, regions are necessarily adapting to the transnational corporation in order to achieve greater economic and political vitality, and consequently have become the strategic centers of the global society.

Technology

Globalization of the economy and the rise of the transnational corporation may in turn be attributed to progress in technological development. Indeed, the "enabling" technologies of production, transaction, and circulation have resulted in a transformation of the relationship between corporate capital and labor, the emergence of new roles for the state and the public sector, and the development of new international and inter-metropolitan divisions of labor (Knox 1994, 57).

Furthermore, technology may be related to regionalism as it represents a growing component of the economic base of an increasing number of micro-regions. As regions seek their niche in the global economy they are "organizing, strategizing, mobilizing, and, increasingly, staking their local economic future on high technology" (Muniak 1994, 803).

Thus, in addition to changing regional form, technology is also associated with changing its functional basis as well. Research and design centers and areas of technological innovation have been developed in the continuing transition to a post-industrial society. Examples of such centers include the complex of high-tech firms in the Docklands of East London, near Route 128 in Greater Boston, Silicon Valley in California, and around Grenoble, France (Marshall 1991, 98).

Regions based on technology must be aware that dangers of technological specialization exist. The primary challenge to such 'technoburbs' is to control, rather than be controlled by, the technological revolution. As the processes of globalization widen the gap between rich and poor members of society, care must be taken to ensure that technological innovation does not become an additional factor of exclusion. The optimal strategy to provide equal access to and benefits from technology therefore is to take coordinated action and develop communication both within and between micro-regions (Mercadal 1991, 314).

Technology, however, need not be considered a liability, as the development of an integrated technological base offers opportunity for micro-regions to derive significant benefits. Indeed, numerous areas are already enjoying substantial economic profit based on high technology. Synthesizing technology with the linkages of its knowledge base provides the city-region with the optimum ability to secure and augment its economic advantage.

Knowledge-Based Development

An integral component of the theory focusing on the new global regionalism addresses the knowledge base of the region. Indeed, the old, industrial base has been proven limited, and regions must search for appropriate alternatives offering a more diversified and therefore more vigorous foundation. Areas striving to position themselves in the global arena must not await a paradigm shift, but must be proactive in their development policies and work to strengthen and diversify their economic base.

A new framework—one based on knowledge—has been promoted as a viable successor to production as the economic base of the region. As the aegis of national barriers is reduced, traditional advantages based on location and position are declining in relevance. In their stead, human, technological, and cultural resources are becoming critical to successful regional development. Three basic strategies have been identified as methods for the potential development of the knowledge base: to deepen knowledge in specific fields; to spread technology and extend existing networks; and, to anchor knowledge resources based in the city and region (Knight and Leitner 1993, 43).

The region is the geographical unit that has become the focal point for knowledge-based development. In this age of global capitalism, it is the region that possesses the infrastructure, including the necessary physical, communication, manufacturing, and human elements, to facilitate innovation, production, and learning. The infrastructure of the knowledge-base differs from that of the earlier industrial age, as current focus is placed on developing knowledge-oriented facilities, such as universities, medical complexes, and cultural facilities (Knight 1991, 11). Regions developing their knowledge-base may therefore be understood as learning regions.

> Learning regions, as their name implies, function as collectors and repositories of knowledge and ideas, and provide an underlying environment or infrastructure which facilitates the flow of knowledge, ideas and learning. Learning regions are increasingly important sources of innovation and economic growth, and are vehicles for globalization (Florida 1995, 528).

The development of the knowledge-based learning region is founded upon open dialogue and communication. Such a framework offers the optimum opportunity to understand and address the needs of the community. Successful communication requires the formation of horizontal linkages among all segments of the city (Figure 19-3). Linkages provide a network of information and communication from which decisions may be made and actions may be taken.

A number of micro-regions have already recognized the importance of the knowledge base as the economic foundation of the future. These areas have worked to establish their niche in the global economy by specializing in some aspect of the knowledge base. Examples of this from the United States may be found in the concentration of polymer research around Akron, Ohio, the well known Silicon Valley in California and the clustering of high-technology firms primarily based on the aircraft industry in the metroplex of Dallas-Ft. Worth.

Knowledge may therefore serve as the foundation of the city. As a result, the post-industrial city vying for global significance is altering its structure in order to afford itself an optimal opportunity for knowledge accumulation.

Future Trends

The future is not only one of change, it is also one of potential. By retarding the dominion of traditional economic and political structures, the processes of globalization are opening new opportunities for cities and regions. Declining authority of the nation-state, increasing influence of transnational corporations, technological innovation, and a shift from government to governance are indicative of the global change currently underway. As these trends open new markets and create new linkages, they also provide locales with the ability to carve out their niche and become competitive in the global market. Indeed, the benefits to heretofore non-viable areas are worthy of note, as:

> Globalization has enabled small communities and individuals in remote, undeveloped areas to have more opportunities and choices due to the free flow of information, better communications, lower costs of transportation and expansion of communications and transportation networks (Samudavanija 1995, 9-10).

Therefore, potential exists for those regions that traditionally have been economically discounted to attain some measure of competitiveness. Through careful cultivation, global processes may be harvested through proactive regional response.

Yet, as noted, the prospect for increased marginalization and polarization of regions remains considerable. Developing regions generally lack the necessary stability and regional collaboration to compete globally. Furthermore, efforts at global economic competitiveness do not ensure prosperity. Areas suc-

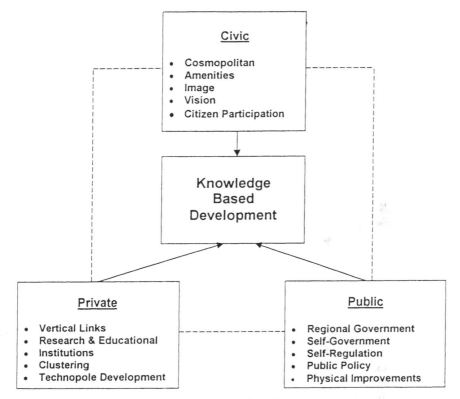

Figure 19-3 Linkages Among Segments of the City
Source: Author.

cessful in attracting transnational corporations must also subsequently oversee a distribution of benefits to all segments of society. Indeed, arrangements between local elites and TNC capitalists to "split the spoils" will only harm the local populace (Miller 1995, 130).

The problems and potential presented by globalization prevail over the economic and political efforts of both the supra-region as well as the micro-region. At the supra-national scale, the principal trading blocs of NAFTA, the EU, and the Pacific Rim, dominate world trade in a triad of global power. As a result, regions unaligned with these blocs are becoming increasingly isolated (Michalak and Gibb 1997, 265).

Such isolation is exemplified by Sub-Saharan Africa which, highlighted by domestic political instability, continues economic regress (Chege 1995, 323). Until the region secures its domestic situation it will be generally disregarded by international investment and fall further behind its more economically viable counterparts.

Regions at the sub-national scale are encountering similar circumstances as they strive to improve their global position. One blueprint for economic competitiveness advocates that the micro-region "begin to function in much the same ways as competitive firms" (Kresl 1992, 197). Yet while many local communities have the ability to deal on a global scale, care must be taken to ensure that the needs of the indigenous population is not subjugated by the desires of external investors. Moreover, the number of regions as yet unable to enter the global market highlights growing global inequality.

As evidenced by extreme regional variation, local responses to global processes vary significantly. Although there is growing substitution of governance for government, there is disparity in the role and influence of nongovernmental agencies. Additionally, the increase in NGOs may enable local governments to transfer duties, yet refrain from relinquishing power. Also, NGOs may pose problems of accountability by obscuring the connection between agency and policy. Overall, however, NGOs serve an important and ever increasing role of assisting the local populace.

Variation is also highlighted by differences in the technological base and institutional innovation of regions. These factors enable areas to integrate into the global society by providing a foundation for transnational corporations. Although benefits provided by TNCs are not evenly distributed, transnational corporations often supply additional economic income through the service sector.

Furthermore, a mutually beneficial synergy exists between TNCs and the knowledge base of the city or region. The knowledge base requires input from all sectors of society, and therefore relies on strong horizontal linkages. Regions that have the necessary base for these sectors to connect will further enhance their global opportunity.

These trends represent not only the current situation, but are also indicative of the future. The rise of regionalism may be related to the global economy, which in turn is heavily dependent on transnational corporations. Subsequently, the importance of TNCs may be correlated with an increase in nongovernmental agencies and the partial substitution of governance for government. Finally, the role of high technology and an extensive knowledge base, which assist regions in securing dominance, will continue to grow. Yet as these trends are taking place at different rates and to a different degree around the world, they are resulting in growing global disparity.

The quintessential challenge of the future will be to economically incorporate regions marginalized by these processes. This objective may be attained if profit seeking is tempered and humanitarian concerns promoted. The globalization of capitalism, however, is likely to thwart the potential for change of economic philosophy. As a result, the polarization of incomes and the difference between "have" and "have not" regions will continue to grow on a global scale.

References

Amin, Ash and Nigel Thrift, 1995, "Globalisation, Institutional 'Thickness' and the Local Economy", pp 91-108 in Patsy Healy, et al., (eds), *Managing Cities: The New Urban Context*, Chichester, U.K.: John Wiley.

Chege, Michael, 1995, "Sub-Saharan Africa: underdevelopment's last stand", pp. 309-345 in Barbara Stallings, (ed.), *Global Change, Regional Response*, New York: Cambridge University Press.

Elliot, Jennifer A. 1994, *An Introduction to Sustainable Development: The Developing World*, London: Routledge.

Florida, Richard, 1995, "Toward the Learning Region", *Futures*, 27:5:527-536.

Fry, Earl H. 1995, "North American Municipalities and Their Involvement in the Global Economy", pp. 21-43 in Kresl, Peter Karl, and Gary Gappert, (eds), *North American Cities and the Global Economy*, Thousand Oaks, CA: Sage Publications.

Gates, Christopher, 1996, "Introduction", *National Civic Review*, 85:3:1-4.

Kirlin, John J. 1993, "Citistates and Regional Governance", *National Civic Review*, 82:4:371-379.

Knight, Richard V. 1991, *Amsterdam Knowledge City: Knowledge-Based Development with Recommendations and Actions*, Amsterdam: University of Amsterdam.

Knight, Richard V. 1989, "City Development and Urbanization: Building the Knowledge-Based City", pp. 233-242 in Knight, Richard V. and Gary Gappert, (eds), *Cities in a Global Society*, Newbury Park, CA: Sage Publications.

Knight, Richard V. and Kurt Leitner, 1993, *Werstattberichte: The Potentials of Vienna's Knowledge Base for City Development*, Vienna: Institute for Higher Studies.

Knox, Paul L. 1994, *Urbanization: An Introduction to Urban Geography*, Englewood Cliffs, NJ: Prentice-Hall.

Kresl, Peter Karl, 1992, *The Urban Economy and Regional Trade Liberalization*, New York: Praeger Publishers.

Laquian, Aprodicio, 1994, "Social and Welfare Impacts of Mega-City Development", pp. 192-214 in Fuchs, Ronald J., et al., (eds), *Mega-City Growth and the Future*, Tokyo: United Nations University Press.

Levine, Marc, 1989, "Urban Redevelopment in a Global Economy: The Cases of Montreal and Baltimore", pp. 141-152 in Knight, Richard V. and Gary Gappert, (eds), *Cities in a Global Society*, Newbury Park, CA: Sage Publications.

Marshall, Bruce, (ed.), 1991, *The Real World: Understanding the Modern World Through the New Geography*, Boston: Houghton Mifflin.

Mercadal, Georges, 1991, "Evolution of Cities in the Post Industrial Society Era", *Ekistics*, 58:350/351:308-314.

Michalak, Wieslaw, and Richard Gibb, 1996, "Trading Blocs and Multilateralism in the World Economy", *Annals of the Association of American Geographers*, 87:2:264-279.

Miller, Morris, 1995, "Where is Globalization Taking Us?", *Futures*, 27:2:125-144.

Miyoshi, Masao, 1993, "A Borderless World? From Colonialism to Transnationalism and the Decline of the Nation-State", *Critical Inquiry*, 19:4:726-751.

Muniak, Dennis C. 1994, "Economic Development, National High Technology Policy and America's Cities", *Regional Studies*, 28:8:803-809.

Newhouse, John, 1997, "Europe's Rising Regionalism", *Foreign Affairs*, 76:1:67-84.

Sagasti, Francisco R. 1995, "Knowledge and Development in a Fractured Global Order", *Futures*, 27:6:591-610.

Samudavanija, Chai-Anan, 1995, "Bypassing the State of Asia", *New Perspectives Quarterly*, 12:1:9-12.

Scott, Allen J. 1996, "Regional Motors of the Global Economy", *Futures*, 28:5:391-411.

Spybey, Tony, 1996, *Globalization and World Society*, Cambridge, MA: Polity Press.

Sykora, Ludek, 1994, "Local Restructuring as a Mirror of Globalisation Processes: Prague in the 1990's", *Urban Studies*, 31:7:1149-1166.

Tyler, Gus, 1993, "The Nation-State vs. the Global Economy", *Challenge*, 36:2:26-32.

Index

Printed and bound by CPI Group (UK) Ltd, Croydon, CR0 4YY

21/10/2024

01777085-0003